Raman Spectroscopy

Raman Spectroscopy

D. A. Long

Professor of Structural Chemistry
University of Bradford, England.

McGraw-Hill International Book Company

New York · St. Louis · San Francisco · Auckland · Bogatá · Düsseldorf · Johannesburg
London · Madrid · Mexico · Montreal · New Delhi · Panama · Paris · São Paulo
Singapore · Sydney · Tokyo · Toronto

Library of Congress Cataloging in Publication Data

Long, Derek Albert.
Raman spectroscopy.

Bibliography: p.
Includes index.
1. Raman spectroscopy. I. Title.
QC454.R36L66 535'.846 77-22488
ISBN 0-07-038675-7

5 4 3 2 1 JWA 7 9 8 7 6

Printed and bound in Great Britain

Contents

Preface

"Comme quelqu'un pourrait dire de moi que j'ai seulement fait ici un amas de fleurs étrangères, n'y ayant fourni du mien que le filet à les lier."

M. E. Montaigne

As a result of the development of lasers as powerful sources of monochromatic radiation and substantial improvements in detection techniques, Raman spectra may now be obtained at least as readily as infrared spectra from a very wide range of materials. In consequence, Raman spectroscopy has become a major area of spectroscopic research. The number of papers published on this subject has increased considerably, and many specialist reviews have recently appeared dealing in depth with particular aspects of the Raman effect. However, I have come to feel that there would be a welcome for a different kind of review: an up-to-date survey of the whole subject, setting Raman spectroscopy in perspective, unifying the basic theory, illustrating the applications and potential of the technique, and guiding the reader towards the specialist literature.

Such a survey I have endeavoured to provide with this book. Its structure and organization are explained in chapter 1, so only a few general matters need be mentioned here. SI units are used throughout except that the Å and cm^{-1} have been retained. To increase its value as a working reference book, many key formulae have been collected together in a central section; and the appendices include symmetry tables and guided reading lists. The special effort directed to making the book visually attractive should also have increased its clarity.

It is hoped that this book will prove useful to a wide variety of people. The gentle pace of the early stages of the theoretical development and the many examples should commend it to undergraduates; those beginning research in this field should find that the book provides a convenient collection of the necessary basic tools and a guided introduction to more specialized skills; and even those experienced in the subject will hopefully be interested in my selection and organization of familiar material.

Nobody writing on Raman scattering could fail to owe an incalculable debt to the perceptive account of the theory published by G. Placzek in 1934. It is a measure of his insight that he predicted almost all of the new phenomena that have been observed in recent years using laser sources.

I am greatly indebted to many people for help in the preparation of this book. My colleagues at the University of Bradford, Dr B. M. Chadwick, Dr H. G. M. Edwards, Dr V. Fawcett, and Dr M. J. French, have assisted in many ways and have made substantial contributions to chapters 6, 7, and 8. Professor A. D. Buckingham, Dr L. D. Barron, Dr J. L. Duncan, Dr N. Krauzman, Professor J. P. Mathieu, Dr L. Stanton, Dr L. A. K. Staveley, and the late Dr L. A. Woodward very kindly read part or all of the typescript and made valuable suggestions for improvement. The typing (and retyping!) of the manuscript was undertaken by Mrs S. Galloway and Mrs J. Stevenson, with patience, skill, and cheerfulness. Mr D. Farwell, Mr J. Goldberg, Mr S. Teal, Miss S. J. Alexander, and Mr J. Merrick have prepared many of the diagrams and plates. Mr Ronald Lowe, the artist, and the University of Bradford, the owners, have generously allowed me to reproduce on the dust cover the painting, *Misty Flight*, which captures so happily the atmosphere of Pembrokeshire where much of this book was written. The staff of McGraw-Hill have been most cooperative in meeting my wishes regarding style and format and have shown commendable efficiency at all times. Richard and Andrew have helped with the proof-reading; and my wife Moira not only sustained me throughout the rigours of authorship but also assisted materially in many aspects of the preparation of this book.

D. A. LONG

School of Chemistry
University of Bradford
December 1976

Acknowledgements

Acknowledgment is made to the editors, authors and publishers of the following works and journals for permission to use figures. *Nature* (Macmillan Journals Ltd, London) for Fig. 1.1(a): *Die Naturwissenschaften* for Fig. 1.1(b); The Royal Society for Figs. 1.2a and b; Mr A. J. Mitteldorf and Spex Industries Inc. for Figs. 6.6 and 6.7; Professor B. P. Stoicheff and John Wiley & Sons Inc. for Fig. 7.9 (from *Advances in Spectroscopy*, Volume 1, H. W. Thompson, Editor): Dr W. J. Jones and *Journal of the Chemical Society, Faraday Transactions II* for Fig. 7.12; Professor M. Delhaye and *Journal of Raman Spectroscopy* (D. Reidel, Dordrecht, Holland) for Fig. 7.48. Tables 7.1. and 7.2. and Fig. 7.18 and a number of examples in section 7.3. are taken from the article by Professor B. Schrader in *Angewandte Chemie*, volume 12, pages 882–908, 1973, with the kind permission of the author and the publisher, Verlag Chemie GMBH, Weinheim. Mr R. Carless, Spectra Physics Ltd., collaborated in the production of Plate 1 and Dr P. D. Maker of the Ford Research Laboratories, Dearborn, kindly provided Plate 2. Professor T. G. Spiro kindly provided the spectra on which Figs. 6.10 and 6.11 are based.

Most of the illustrations and spectra have been specially drawn for this book by the staff of Oxford Illustrators Ltd. Where spectra are based on information in the scientific literature, the source is indicated in the references.

1 Introduction

"One of the spectre order."

Lewis Carroll

1.1 The nature of Rayleigh and Raman scattering

When monochromatic radiation of wavenumber $\tilde{\nu}_0$ is incident on systems like dust-free, transparent gases and liquids, or optically perfect, transparent solids, most of it is transmitted without change, but, in addition, some scattering of the radiation occurs. If the frequency content of the scattered radiation is analysed, there will be observed to be present not only the wavenumber $\tilde{\nu}_0$ associated with the incident radiation but also, in general, pairs of new wavenumbers of the type $\tilde{\nu}' = \tilde{\nu}_0 \pm \tilde{\nu}_M$. In molecular systems, the wavenumbers $\tilde{\nu}_M$ are found to lie principally in the ranges associated with transitions between rotational, vibrational, and electronic levels. The scattered radiation usually has polarization characteristics different from those of the incident radiation, and both the intensity and the polarization of the scattered radiation depend on the direction of observation.

Such scattering of radiation with change of frequency (or wavenumber) is called Raman scattering, after the Indian scientist C. V. Raman[1] who, with K. S. Krishnan, first observed this phenomenon in liquids in 1928. The effect had been predicted on theoretical grounds in 1923 by A. Smekal.[2] Very shortly after the paper of Raman and Krishnan was published, Landsberg and Mandelstam[3] in Russia reported the observation of light scattering with change of frequency in quartz; and Cabannes[4] and Rocard[5] in France confirmed Raman and Krishnan's observations. By the end of 1928 some 60 papers had been published on the Raman effect! Facsimiles of the classic papers of Raman and Krishnan, and Landsberg and Mandelstam are shown in Fig. 1.1(a) and (b).

In the spectrum of the scattered radiation, the new wavenumbers are termed Raman lines, or bands, and collectively are said to constitute a Raman spectrum. Raman bands at wavenumbers less than the incident wavenumber (i.e., of the type $\tilde{\nu}_0 - \tilde{\nu}_M$) are referred to as Stokes bands, and those at wavenumbers greater than the incident wavenumber (i.e., of the type $\tilde{\nu}_0 + \tilde{\nu}_M$) as anti-Stokes bands.

Scattering with change of frequency (or wavenumber) originating, in essence, from a Doppler effect can also be observed with gases, liquids, and solids. Scattering of this kind was predicted in 1922 by Brillouin,[6] but was not observed until 1930 by Gross.[7] In Brillouin scattering, the change of wavenumber is very small and of the order of $0{\cdot}1\ \mathrm{cm}^{-1}$. Brillouin scattering is therefore not separated from the scattered radiation at $\tilde{\nu}_0$ under the experimental conditions used for most studies of Raman scattering. Brillouin scattering will not be treated in this book.

A New Type of Secondary Radiation.

If we assume that the X-ray scattering of the 'unmodified' type observed by Prof. Compton corresponds to the normal or average state of the atoms and molecules, while the 'modified' scattering of altered wave-length corresponds to their fluctuations from that state, it would follow that we should expect also in the case of ordinary light two types of scattering, one determined by the normal optical properties of the atoms or molecules, and another representing the effect of their fluctuations from their normal state. It accordingly becomes necessary to test whether this is actually the case. The experiments we have made have confirmed this anticipation, and shown that in every case in which light is scattered by the molecules in dust-free liquids or gases, the diffuse radiation of the ordinary kind, having the same wave-length as the incident beam, is accompanied by a modified scattered radiation of degraded frequency.

The new type of light scattering discovered by us naturally requires very powerful illumination for its observation. In our experiments, a beam of sunlight was converged successively by a telescope objective of 18 cm. aperture and 230 cm. focal length, and by a second lens of 5 cm. focal length. At the focus of the second lens was placed the scattering material, which is either a liquid (carefully purified by repeated distillation *in vacuo*) or its dust-free vapour. To detect the presence of a modified scattered radiation, the method of complementary light-filters was used. A blue-violet filter, when coupled with a yellow-green filter and placed in the incident light, completely extinguished the track of the light through the liquid or vapour. The reappearance of the track when the yellow filter is transferred to a place between it and the observer's eye is proof of the existence of a modified scattered radiation. Spectroscopic confirmation is also available.

Some sixty different common liquids have been examined in this way, and every one of them showed the effect in greater or less degree. That the effect is a true scattering and not a fluorescence is indicated in the first place by its feebleness in comparison with the ordinary scattering, and secondly by its polarisation, which is in many cases quite strong and comparable with the polarisation of the ordinary scattering. The investigation is naturally much more difficult in the case of gases and vapours, owing to the excessive feebleness of the effect. Nevertheless, when the vapour is of sufficient density, for example with ether or amylene, the modified scattering is readily demonstrable.

C. V. Raman.
K. S. Krishnan.

210 Bowbazar Street,
Calcutta, India,
Feb. 16.

Fig. 1.1 (a) Facsimile of paper by Raman and Krishnan (published 31 March 1928 in *Nature*, vol. 121, page 501)

Eine neue Erscheinung bei der Lichtzerstreuung in Krystallen.

Bei dem Studium der molekularen Lichtzerstreuung in festen Körpern, welches zur Klärung der Frage vorgenommen wurde, ob dabei eine Wellenlängenänderung stattfindet, was man nach der DEBYEschen Theorie der spezifischen Wärme vermuten kann, haben wir eine neue Erscheinung gefunden, die, wie es uns scheint, ein bedeutendes Interesse beansprucht.

Diese Erscheinung besteht in der Wellenlängenänderung, welche aber von anderer Größenordnung ist, als die von uns erwartete und welche einen ganz anderen Ursprung hat.

Ein intensives Lichtbündel von einer Quecksilberquarzlampe wurde durch ein Quarzkrystall gesandt, und das senkrecht zu dem primären Bündel zerstreute Licht wurde mittels eines Quarzspektrographen aufgenommen. Die gewöhnlichen Maßregeln gegen fremdes Licht wurden getroffen[1]. Als Vergleichsspektrum diente eine Aufnahme des von schwarzem Samt reflektierten Lichtes. Expositionszeit von 2—14 Stunden. Die Versuche wurden mit zwei verschiedenen Quarzstücken ausgeführt. Es erwies sich, daß auf allen Spektrogrammen alle Quecksilberlinien von je einem deutlich ausgeprägten Trabanten von etwas größerer Wellenlänge begleitet waren, und außerdem noch bei jeder Linie zwei oder drei weniger ausgeprägte andeutungsweise zum Vorschein kamen. In dem Vergleichsspektrum war keine Spur dieser Trabanten zu sehen. Fig. 1 gibt eines der Spektrogramme wieder. Die angenäherten Ausmessungen der Spektrogramme zeigen für den stärkeren Trabanten folgende Wellenlängenänderung:

Tabelle 1.

λ in Å	$\Delta\lambda$ beobachtet	$\Delta\lambda$ berechnet
2536	ca. 30	30,8
3126	„ 47	47,0
3650	„ 63	64,0

Wir haben verschiedene Kontrollversuche angestellt, um festzustellen, daß die beobachteten Linien nicht von einem zufälligen falschen Licht herrühren. Der folgende Versuch scheint uns entscheidend zu sein. Zwischen dem zerstreuenden Quarzkrystall und dem Spektrographspalt wurde ein Quarzgefäß mit Quecksilberdampf eingeschaltet, welcher die Linie 2536 Ångström vollständig absorbiert. Auf dem Spektrogramm haben wir diese Linie nicht mehr erhalten, wohl aber ihre Trabanten. Das beweist sicher, daß diese Trabanten wirklich eine andere Wellenlänge haben als die Grundlinie.

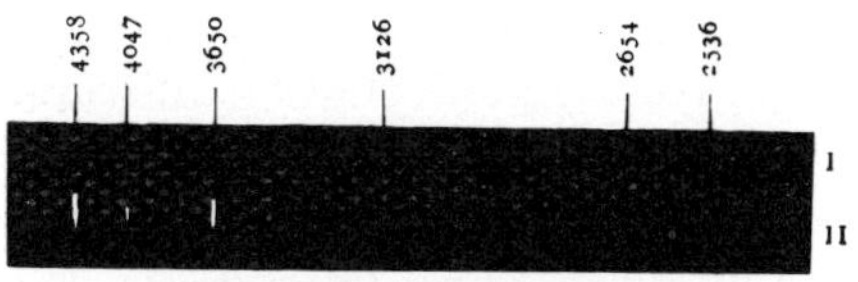

I. Das Spektrum des zerstreuten Lichtes.
II. Das Vergleichsspektrum.

Wir halten es für verfrüht, schon jetzt eine definitive Deutung der geschilderten Erscheinung zu geben. Eine der möglichen theoretischen Deutungen besteht vielleicht in folgendem: Bei der Zerstreuung des Lichtes können einige eigene ultrarote Frequenzen des Quarzes auf Kosten der Energie des zerstreuten Lichtes angeregt werden und dadurch würde die Energie der zerstreuten Quanten und folglich ihre Frequenz um die Größe der entsprechenden infraroten Quanten abnehmen. Geht man dabei von der Frequenz, welche der Wellenlänge $\lambda = 20{,}7\,\mu^1$ entspricht, aus, so erhält man eine gute Übereinstimmung zwischen den berechneten und tatsächlich beobachteten Werten (s. Tab. 1).

Ob und wieweit die von uns beobachtete Erscheinung mit der von RAMAN[2] erst kürzlich beschriebenen im Zusammenhang steht, können wir zur Zeit noch nicht beurteilen, weil seine Schilderung zu summarisch ist.

Moskau, Institut für theoretische Physik der I. Universität, den 6. Mai 1928.

G. LANDSBERG. L. MANDELSTAM.

Anmerkung bei der Korrektur. Wir haben inzwischen das im Kalkspat zerstreute Licht untersucht und dieselbe Erscheinung beobachtet. Die Wellenlängeänderung ist hier größer als im Quarz. Sie würde einer Infrarotenfrequenz von $\lambda = 9.1\,\mu$ entsprechen.

[1] G. LANDSBERG, Zeitschr. f. Physik **43**, 773. 1927; **45**, 442. 1927.

Fig. 1.1 (b) Facsimile of paper by Landsberg and Mandelstam (published 13 July 1928 in *Naturwissenschaften*, vol. 16, page 557)

The scattering of radiation without change of frequency (or wavenumber) had been known for some time prior to the discovery of the Raman effect. Where such scattering arises from scattering centres, like molecules, which are very much smaller than the wavelength of the incident radiation, it is called Rayleigh scattering, after Lord Rayleigh,[8] who explained the essential features of this phenomenon in terms of classical radiation theory in 1871. Rayleigh scattering *always* accompanies Raman scattering and so will always be observed with Raman scattering, unless selective filters are employed. The natural presence within a Raman spectrum of the Rayleigh band at wavenumber $\tilde{\nu}_0$ has some advantages since it can serve as a reference band for the determination of $\tilde{\nu}_M$ from $\tilde{\nu}'$.

Scattering of radiation without change of frequency (or wavenumber) can also arise from larger scattering centres like dust particles, and is generally referred to as Mie scattering.[9] Since such scattering can be very intense, it is very rarely totally absent unless extraordinary efforts are made to remove dust and other Mie scatterers from the system. Consequently, what is usually referred to, loosely, as Rayleigh scattering at $\tilde{\nu}_0$ normally consists in practice of true Rayleigh scattering, together with some Mie scattering at $\tilde{\nu}_0$ and unresolved Brillouin scattering. However, in the ensuing discussions of the theory, the term Rayleigh scattering will always refer to true Rayleigh scattering.

It is a little difficult to make generalizations about the relative intensities of Rayleigh scattering and Raman scattering since these depend on many factors like the physical state, the chemical composition, and the direction of observation relative to the direction of illumination. However, as an indication of the intensity levels involved, the intensity of Rayleigh scattering is generally about 10^{-3} of the intensity of the incident exciting radiation; and the intensity of strong Raman bands is generally about 10^{-3} of the intensity of Rayleigh scattering.

The phenomena of Rayleigh and Raman scattering are illustrated in Fig. 1.2, which presents facsimiles of two of the first spectra published by Raman and Krishnan.[10] Figure 1.2(a) shows the photographically recorded spectrum of the essentially monochromatic radiation from a mercury arc used to produce the scattering; the spectrum has one intense band at wavenumber $\tilde{\nu}_0 = 22\,938\ \mathrm{cm}^{-1}$ (or wavelength[11] 4358·3 Å, 435·83 nm). (The mercury spectrum contains other lines in this region, but they are much weaker and need not be considered here.) Figure 1.2(b) shows the photographically recorded spectrum of the radiation after scattering by carbon tetrachloride (liquid). This spectrum contains a strong band at $\tilde{\nu}_0 = 22\,938\ \mathrm{cm}^{-1}$ due to Rayleigh scattering of the incident radiation, and a number of weaker bands whose wavenumbers are given by $\tilde{\nu}_o \pm 218$, $\tilde{\nu}_0 \pm 314$, $\tilde{\nu}_0 \pm 459$, $\tilde{\nu}_0 - 762$, and $\tilde{\nu}_0 - 790\ \mathrm{cm}^{-1}$. The first three pairs of lines arise from Stokes and anti-Stokes Raman scattering associated with $\tilde{\nu}_M$ values of 218, 314, and 459 cm^{-1}. The remaining two lines arise from Stokes Raman scattering associated with $\tilde{\nu}_M$ values of 762 and 790 cm^{-1}; the corresponding anti-Stokes lines are not observed. These $\tilde{\nu}_M$ values all correspond to transitions in the carbon tetrachloride molecule associated with vibrational energy levels.

Rayleigh and Raman scattering are now more usually excited by monochromatic radiation from a suitable laser and detected photoelectrically. For a comparison with Fig. 1.2(b), the Rayleigh and Raman spectra of carbon tetrachloride (liquid) excited with 20 487 cm^{-1} radiation (4879·9 Å, 487·99 nm) from an argon ion laser and

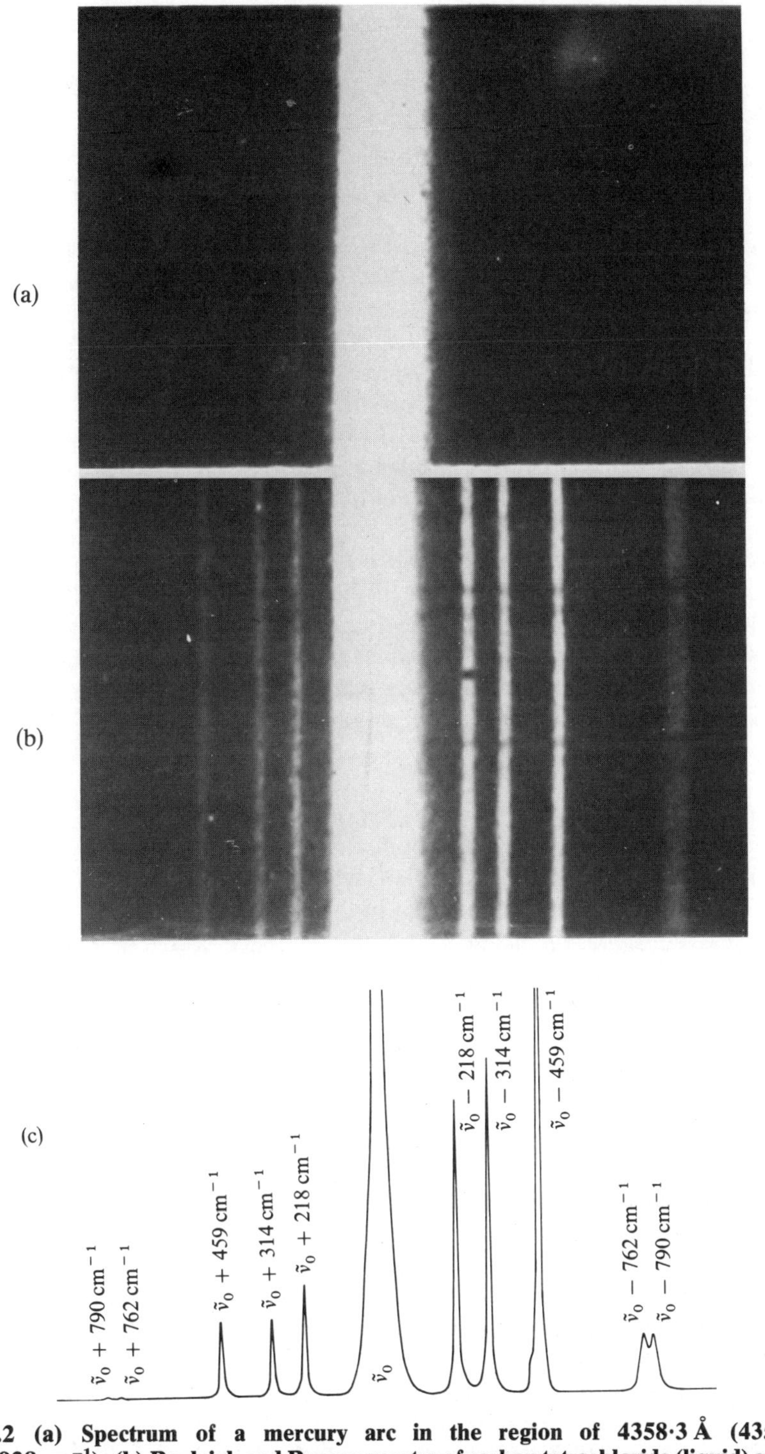

Fig. 1.2 (a) Spectrum of a mercury arc in the region of 4358·3 Å (435·83 nm, $\tilde{\nu}_0$=22 938 cm^{-1}); (b) Rayleigh and Raman spectra of carbon tetrachloride (liquid) excited by mercury arc radiation, $\tilde{\nu}_0$=22 938 cm^{-1}; (c) Rayleigh and Raman spectra of carbon tetrachloride (liquid) excited by an argon ion laser, ν_0 = 20 487 cm^{-1} (4879·9 Å, 487·99 nm). The spectra in Fig. 1.2(a) and (b) are facsimiles of spectra recorded by Raman and Krishnan (*Proc. Roy. Soc.*, vol. 122, p. 23. 1929) and were photographically recorded.

directly recorded are shown in Fig. 1.2(c). In addition to all the bands observed previously, this spectrum contains the anti-Stokes bands at $\tilde{\nu}_0+762$ and $\tilde{\nu}_0+790\,\mathrm{cm}^{-1}$, which were missing from the spectrum of Fig. 1.2(b). Their very low intensity explains why they were not observed in the early work.

1.2 An energy transfer model for Raman and Rayleigh scattering

The origin of the modified frequencies (or wavenumbers) found in Raman scattering may be explained in terms of energy transfer between the scattering system and the incident radiation. When a system interacts with radiation of wavenumber $\tilde{\nu}_0$, it may make an upward transition from a lower energy level E_1 to an upper energy level E_2. It must then acquire the necessary energy, $\Delta E = E_2 - E_1$, from the incident radiation. The energy ΔE may be expressed in terms of a wavenumber $\tilde{\nu}_M$ associated with the two levels involved, where $\Delta E = hc\tilde{\nu}_M$. This energy requirement may be regarded as being provided by the annihilation of one photon of the incident radiation of energy $hc\tilde{\nu}_0$ and the simultaneous creation of a photon of smaller energy $hc(\tilde{\nu}_0 - \tilde{\nu}_M)$, so that scattering of radiation of lower wavenumber, $\tilde{\nu}_0 - \tilde{\nu}_M$, occurs (see Fig. 1.3(a)). Alternatively, the interaction of the radiation with the system may cause a downward transition from a higher energy level E_2, if the system happens already to be in that excited level, to a lower energy level E_1, in which case it makes available energy $E_2 - E_1 = hc\tilde{\nu}_M$. Again a photon of the incident radiation of energy $hc\tilde{\nu}_0$ is annihilated, but in this instance there is simultaneously created a photon of higher energy $hc(\tilde{\nu}_0 + \tilde{\nu}_M)$, so that scattering of radiation of higher wavenumber, $\tilde{\nu}_0 + \tilde{\nu}_M$, occurs (see Fig. 1.3(b)).

In the case of Rayleigh scattering, although there is no resultant change in the energy state of the system, the system still participates directly in the scattering act, causing one photon of incident radiation $hc\tilde{\nu}_0$ to be annihilated and a photon of the same energy to be created simultaneously, so that scattering of radiation of unchanged wavenumber, $\tilde{\nu}_0$, occurs (see Fig. 1.3(c)). Although the involvement of the system is not apparent in the wavenumber of the scattered radiation, other properties of the scattered radiation are characteristic of the scattering system and confirm its participation in the scattering act.

It is clear that, as far as wavenumber is concerned, a Raman band is to be characterized not by its absolute wavenumber, $\tilde{\nu}' = \tilde{\nu}_0 \pm \tilde{\nu}_M$, but by the *magnitude* of its wavenumber *shift* $|\Delta\tilde{\nu}|$ from the incident wavenumber, where $|\Delta\tilde{\nu}| = |\tilde{\nu}_0 - \tilde{\nu}'| = \tilde{\nu}_M$. Such wavenumber shifts are often referred to, somewhat loosely, as Raman wavenumbers. Where it is necessary to distinguish Stokes and anti-Stokes Raman scattering we shall define $\Delta\tilde{\nu}$ to be positive for Stokes scattering and negative for anti-Stokes scattering, that is $\Delta\tilde{\nu} = (\tilde{\nu}_0 - \tilde{\nu}')$. Unless small values of the wavenumber shift are involved, it is Stokes Raman scattering which is mainly studied, since as Fig. 1.2(c) illustrates, the intensity of anti-Stokes relative to Stokes Raman scattering decreases rapidly with increase in the wavenumber shift. This is because anti-Stokes Raman scattering involves transitions to a lower energy state from a populated higher energy state. The thermal population of such higher states decreases exponentially as their energy, $hc\tilde{\nu}_M$, above the lower state increases. For example, at 300 K the population of a non-degenerate upper state lying $1000\,\mathrm{cm}^{-1}$ or more above the ground state is less than one per cent of the ground state population.

		Initial state	Process	Final state
(a) Stokes Raman scattering	Radiation	$hc\tilde{\nu}_0$	Annihilate $hc\tilde{\nu}_0$ Create $hc(\tilde{\nu}_0-\tilde{\nu}_M)$	$hc(\tilde{\nu}_0-\tilde{\nu}_M)$
	Scattering system	E_2 ——— E_1 ——●—	E_2 ——— E_1 —↑—	E_2 ——●— E_1 ———
	Total energy	$hc\tilde{\nu}_0+E_1$		$hc(\tilde{\nu}_0-\tilde{\nu}_M)+E_2$
(b) Anti-Stokes Raman scattering	Radiation	$hc\tilde{\nu}_0$	Annihilate $hc\tilde{\nu}_0$ Create $hc(\tilde{\nu}+\tilde{\nu}_M)$	$hc(\tilde{\nu}_0+\tilde{\nu}_M)$
	Scattering system	E_2 ——●— E_1 ———	E_2 ——— E_1 —↓—	E_2 ——— E_1 ——●—
	Total energy	$hc\tilde{\nu}_0+E_2$		$hc(\tilde{\nu}_0+\tilde{\nu}_M)+E_1$
(c) Rayleigh scattering	Radiation	$hc\tilde{\nu}_0$	Annihilate $hc\tilde{\nu}_0$ Create $hc\tilde{\nu}_0$	$hc\tilde{\nu}_0$
	Scattering system	E_2 ——— E_1 ——●—	E_2 ——— E_1 ——●—	E_2 ——— E_1 ——●—
	Total energy	$hc\tilde{\nu}_0+E_1$		$hc\tilde{\nu}_0+E_1$

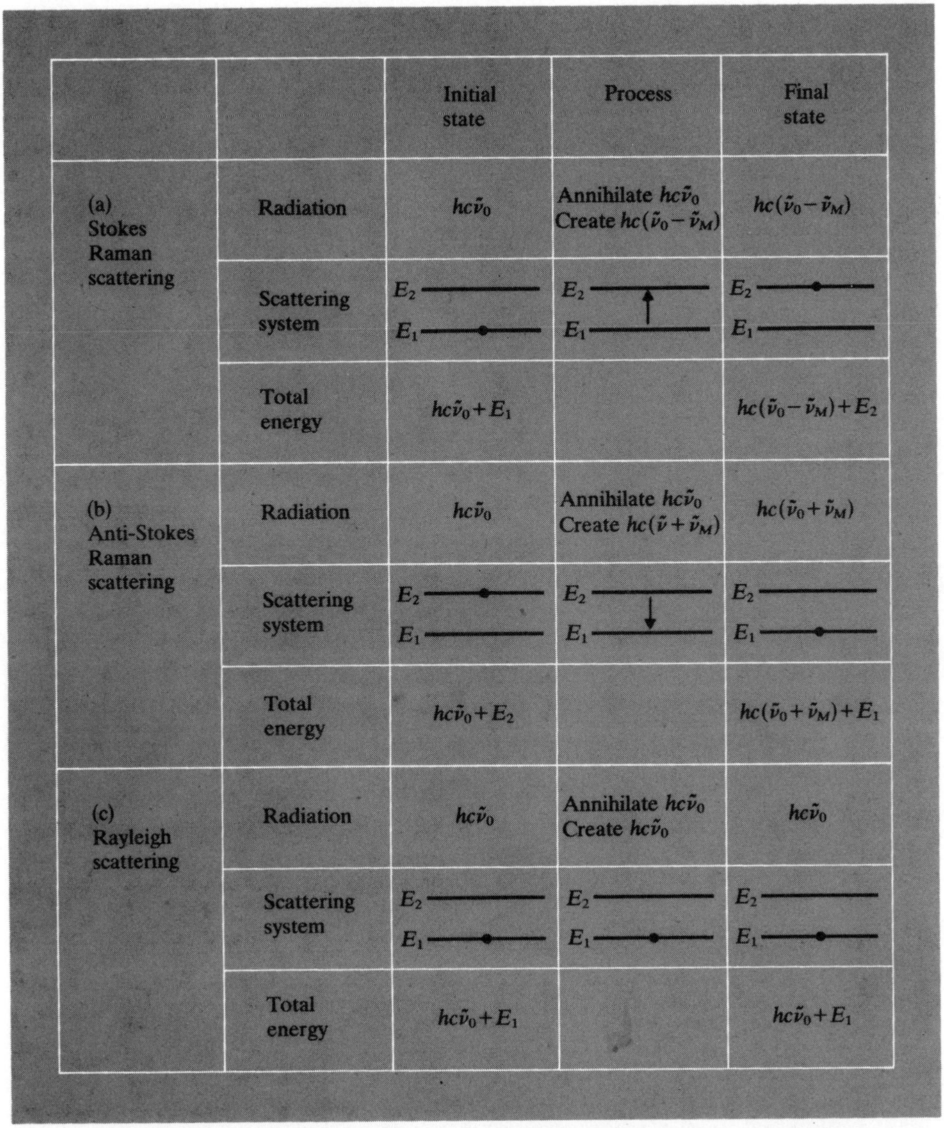

Fig. 1.3 Diagrammatic representation of an energy transfer model of (a) Stokes Raman scattering, (b) anti-Stokes Raman scattering, and (c) Rayleigh scattering

1.3 Some generalizations about the patterns of Raman spectra

The form of an observed Raman spectrum will depend on the energy levels in the scattering system, the transitions permitted between them, and the conditions of observation, particularly the resolving power of the instrument used to analyse the scattered radiation. Since the energy levels of molecular systems fall into particular patterns, it is possible to make some generalizations about the form of the Raman spectrum to be expected for such systems, under given conditions of observation.

Thus, for example, for an assembly of free molecules, if electronic transitions are neglected and if the resolving power is great enough, the Raman spectrum will consist of a series of rather finely spaced lines of relatively small wavenumber shift, arising from transitions between the closely spaced rotational levels of the molecules in the ground vibrational state and a number of other lines of larger wavenumber shift, arising from transitions which involve either a change of vibrational state or a simultaneous change of vibrational and rotational states (see Fig. 1.4(a)). The Stokes and anti-Stokes Raman lines arising from transitions between a given pair of

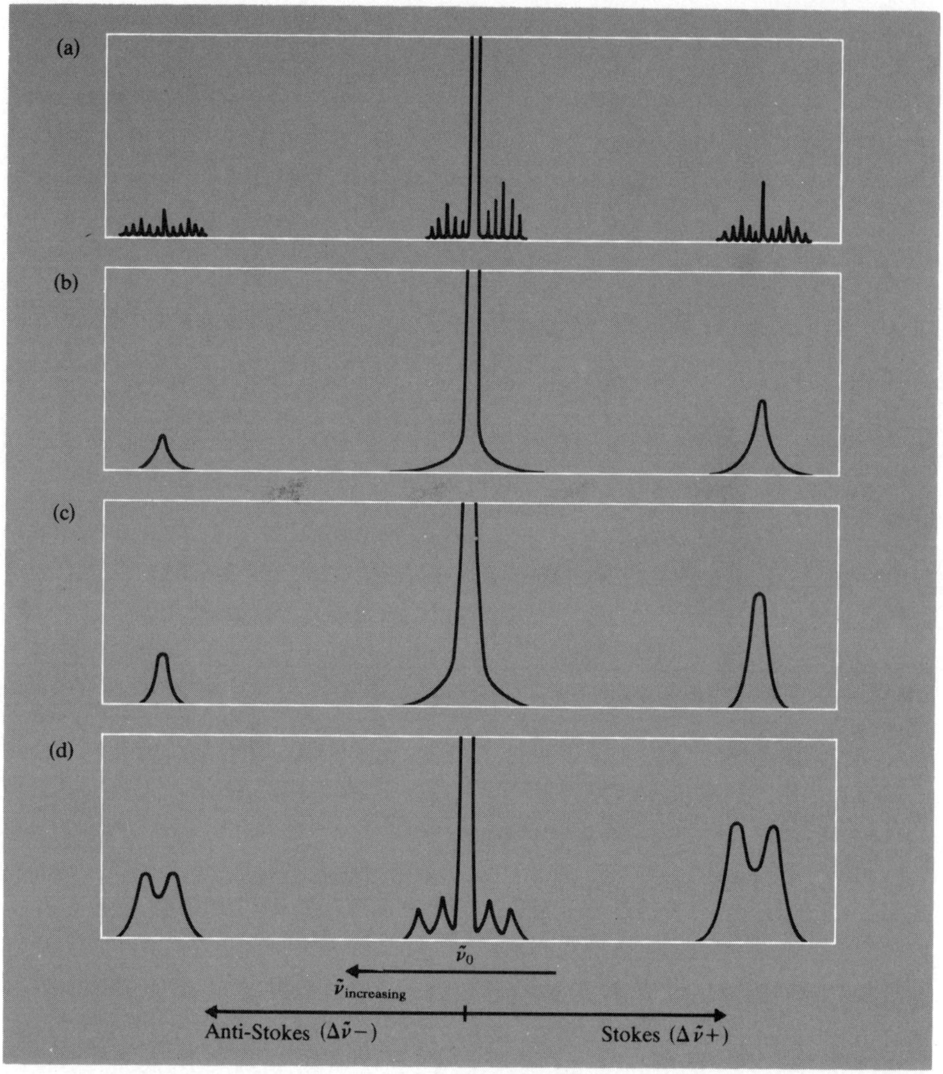

Fig. 1.4 Diagrammatic representation of Raman spectra for different states of matter: (a) gas under high resolution showing resolved Stokes and anti-Stokes rotation and vibration–rotation lines; (b) gas under low resolution showing unresolved Stokes and anti-Stokes rotation and vibration–rotation bands; (c) liquid, showing Stokes and anti-Stokes vibration bands; and (d) crystal, showing splitting of Stokes and anti-Stokes vibration bands and Stokes and anti-Stokes libration bands

rotational levels will have similar intensities since the separation of the rotational energy levels is small, and, in consequence, the Stokes and anti-Stokes rotational Raman spectra will be comparable in intensity. In contrast, the anti-Stokes Raman lines arising from transitions involving a change of vibrational quantum number will be very much weaker than the corresponding Stokes Raman lines, except for relatively small values of the Raman shift $\tilde{\nu}_M$. Thus, the Stokes Raman spectrum originating from vibrational transitions will generally be more extensive and more intense than the corresponding anti-Stokes spectrum. With moderate resolving power, the Raman spectrum of a gas will usually show only Raman lines arising from vibrational transitions (see Fig. 1.4(b)). The more finely spaced lines associated with changes of rotational quantum number will not be resolved, and will appear as unresolved wings, associated with the Rayleigh line for pure rotational transitions and with the Raman lines of vibrational origin for rotational transitions which also involve a change of vibrational quantum number.

For liquids, the Raman spectrum will be expected to show only bands of essentially vibrational origin, irrespective of the resolving power, since free rotation is inhibited (see Fig. 1.4(c)). The wavenumbers will usually show slight differences from those observed for the molecule in the gaseous state, because of environmental perturbation of the energy levels.

In the case of molecular crystals, the Raman spectrum might be expected to consist not only of the larger wavenumber shifts associated with the intramolecular vibrations but also additional smaller wavenumber shifts associated with the intermolecular motions, like the translation and rotation of one molecule relative to another. However, the symmetry of the environment in which the molecule finds itself, and interactions between molecules in a given unit cell and between inter- and intramolecular vibrations can result in the observation of still further bands. The Raman spectrum is thus strongly dependent on the crystal symmetry and the strength of the intermolecular interactions (see Fig. 1.4(d)). Further, in single crystals the Raman scattering can vary with the direction of observation in a manner related to the crystal symmetry. Nevertheless, the complexity of the Raman spectra of crystals, as compared with liquids, is balanced by the greater amount of information available.

Examples of the different patterns of Raman spectra of molecular origin shown in Fig. 1.4(a) to (d) will be discussed in chapter 7. The observation of Raman spectra is not, of course, confined to the molecular systems just discussed. There are many other systems whose energy levels can participate in energy exchange with the exciting radiation and so give a Raman spectrum. However, it is not profitable to attempt generalizations about the forms of the resulting spectra, since the possible energy patterns are so diverse. Specific examples will be considered in chapter 7.

1.4 Selection rules: complementary nature of Raman and infrared spectroscopy

Although the simple model discussed above accounts for the observed wavenumber shifts in terms of the exchange of energy between a scattering system and the incident radiation, it affords no explanation of the mechanism of the interaction. In consequence, it gives no information about other important characteristics of Raman scattering such as intensity, band contours, directional properties, and state of polarization, or about the selection rules that determine which transitions will participate in Raman scattering. It transpires that the selection rules for Raman scattering are different from those for infrared and microwave absorption spectroscopy; some transitions may be observed only through Raman scattering, some only in direct absorption, and some both in direct absorption and through Raman scattering; still others cannot be observed either in absorption or by Raman scattering. Even when the wavenumber associated with a transition can be observed, in principle, both in absorption and by Raman scattering, intensity factors may make it easier to observe in one effect than in the other. It is clear, therefore, that a complete knowledge of the energy levels of a system is likely to require, *at least*, a study of both Raman scattering and absorption spectroscopy. Thus, infrared and Raman spectroscopy are to be regarded as *complementary*, rather than alternative, methods of investigating transitions between energy levels of molecules.

1.5 Infrared and Raman spectroscopy in historical perspective

In principle, Raman scattering has important practical advantages as a method of studying rotational and vibrational transitions. For all rotational, and nearly all vibrational levels, $\tilde{\nu}_M$ will lie in the range 0–3500 cm^{-1}. A simple calculation will show that the complete Stokes Raman spectrum, covering shifts in the range 0–3500 cm^{-1}, lies in the visible region of the spectrum, for any exciting radiation in the range 4000–6000 Å (400–600 nm, 25 000–16 670 cm^{-1}). Consequently, in Raman spectroscopy, the study of the whole range of molecular rotational and vibrational wavenumbers need involve only one dispersing system and one detector, and the sample cells can be constructed of glass. This is in marked contrast to the observation of rotational and vibrational wavenumbers in direct absorption, where a variety of different techniques and appropriate cell materials must be used to cover the microwave and infrared regions of the spectrum in which these absorptions occur.

In 1928, at the time of the discovery of the Raman effect, the observation of infrared absorption involved specialized equipment, was difficult and tedious, and was restricted to wavenumbers greater than about 650 cm^{-1}. The far infrared and microwave regions covering the lower wavenumbers were not then accessible. By contrast, the new Raman scattering yielded information over the whole range of molecular wavenumbers and required only readily available equipment. A mercury arc lamp served as a source of monochromatic radiation and a routine laboratory spectrograph was adequate as a dispersing system. Not surprisingly, there was an immediate upsurge of interest in this new form of spectroscopy. Important structural information was obtained from studies of rotational spectra and characteristic vibrational wavenumbers were measured for a large number of liquids, mainly organic in nature.

However, experimental techniques in infrared spectroscopy underwent progressive development. By the early fifties, the direct recording of infrared spectra above 650 cm^{-1} was well established. Since that time, the wavenumber region below 650 cm^{-1} has become more and more readily accessible, although often more than one spectrometer and certainly a variety of cell materials and detectors are still needed to cover the whole vibrational wavenumber range. The great usefulness of infrared spectroscopy, particularly for analytical and diagnostic purposes, generated such a demand for instruments that it became economic to produce, in quantity, modestly priced instruments, albeit of modest performance, for routine work. By the late fifties, microwave spectroscopy had also been highly developed, although this still remained a specialized and relatively expensive experimental technique.

The development of experimental techniques in Raman spectroscopy did not keep pace with those in infrared spectroscopy. It is true that the availability of photomultipliers made possible the direct recording of Raman spectra and that commercial, direct recording Raman spectrometers began to be produced in the fifties, but there were still very considerable restrictions on the widespread application of Raman spectroscopy. These restrictions stemmed, in the main, from the shortcomings of the mercury arc which was almost the only practicable source of monochromatic radiation for the excitation of Raman spectra. The mercury arc provides only one strong excitation line in the visible region which lies in the blue region at 4358·3 Å (435·83 nm, 22 938 cm^{-1}); two other lines in the violet (4046·6 Å, 404·66 nm, 24 705 cm^{-1}) and green (5460·7 Å, 546·07 nm, 18 307 cm^{-1}) regions, while usable, are much weaker. There are two very much weaker lines in close proximity at 4347·5 Å (434·75 nm, 22 995 cm^{-1}) and 4339·2 Å (433·92 nm, 23 039 cm^{-1}). In the ultraviolet there is a strong resonance line at 2536·5 Å (253·65 nm, 39 412 cm^{-1}). This restricts Raman spectroscopy to samples which do not absorb in these regions, i.e., colourless or pale yellow materials. Although the mercury arc consumes several kilowatts of power, since it is a spatially extensive source, the power per unit area achievable at the sample is relatively low, usually of the order of 1.5×10^4 W m^{-2}. Because of the relative weakness of Raman scattering, it was therefore necessary to use relatively large samples (for liquids a few cm^3 and for gases several litres at s.t.p.). and to collect as much as possible of the Raman scattering, so involving the use of complicated collection optics. The relationship between the direction of incidence and direction of observation was therefore usually of a complicated and near-indefinable nature, and it was often not possible to study satisfactorily the directional and polarization properties of Raman scattering which theory shows can yield useful information. These restrictions were particularly limiting in the study of single crystals. Even in the most refined form of the mercury arc, the width of the 4358·3 Å (435·83 nm, 22 938 cm^{-1}) line is 0.24 cm^{-1}, which restricts the study of rotational spectra to relatively light molecules for which the rotational line spacing exceeds the excitation line width. Added to all these restrictions, the sample had to be of high quality, since scattering from dust particles or crystal imperfections, and fluorescence from impurities could easily swamp the relatively weak Raman scattering.

It is not surprising that in the fifties and early sixties Raman spectroscopy was far less widely used than infrared spectroscopy. In recent years, the situation has been dramatically transformed as a result of the availability of gas lasers. At the focus of a gas laser the power per unit area can be many orders of magnitude greater than that achievable with a mercury arc. Very small samples can therefore be used (10^{-6} cm^3 for liquids, 10^{-12} cm^3 for solids). The output from a laser forms a self-collimated beam with a defined state of polarization, so that the study of intensities and polarizations is greatly facilitated. In particular, the investigation of single crystals is made relatively straightforward, and the subtle polarization phenomena associated, for example, with chiral systems can now be explored. Further, a number of highly monochromatic excitation lines ranging over the whole visible region is available from appropriate lasers, and the frequency output of some lasers can be continuously varied over restricted ranges of the spectrum. Thus, not only may a suitable excitation wavenumber, well removed from an absorption band, be found for the observation of ordinary Raman scattering from all but the most intractable materials, but resonance Raman scattering produced where the excitation wavenumber is close to an absorption band can be effectively studied. The widths of lines from conventional laser sources are narrower than for the mercury arc, and with appropriate techniques can be made so narrow that no line width restriction exists in the study of rotational spectra. Thus, the gas laser is, without doubt, the long soughtafter source for the production of intense Raman spectra from minute amounts of sample of almost any nature.

In addition, there have been substantial developments in the dispersing and detection systems used for Raman spectroscopy. The use of double and even triple monochromators provides such good discrimination against stray light that Raman spectra can be readily obtained from samples of poor quality. Improvements in detectors and associated electronics have been such that very weak scattering can be effectively recorded. With the experimental difficulties now largely solved, Raman spectroscopy can fulfil its role as the complement of infrared spectroscopy. In addition, the special advantages of Raman spectroscopy over infrared spectroscopy in certain experimental situations can be fully exploited.

1.6 The plan for the rest of the book

To understand how Raman spectroscopy may be applied to a wide variety of chemical problems, it is essential to know not only how the selection rules operate but also how the intensity, directional properties, and state of polarization of Raman scattering are related to the nature of the scattering system. A substantial part of this book is therefore devoted to the theory of the Raman effect. In chapter 2, relevant parts of classical radiation theory are treated and basic formulae used in subsequent chapters are presented and explained. Chapters 3 and 4 deal, respectively, with classical approaches and a semi-quantum mechanical approach to the theory of Rayleigh and Raman scattering, with particular reference to rotational and vibrational transitions in transparent, non-chiral, molecular systems. A more general quantum mechanical treatment is developed in chapter 5, and aspects of scattering from absorbing systems and chiral systems are also considered. The emphasis in chapters 3, 4, and 5 is on scattering from isotropic gases and liquids. Current experimental techniques are briefly reviewed in chapter 6. A number of examples of the application of Raman spectroscopy are discussed in chapter 7. In chapter 8, a brief account is given of some related forms of light scattering, like the stimulated Raman effect, the inverse Raman effect, the hyper Raman effect, and coherent anti-Stokes Raman scattering. These novel effects, all of which have been discovered since 1962, are produced as the result of the non-linear response of the scattering system when it is subjected to the extremely intense radiation that is associated with certain kinds of lasers. The appendices include an extensive guide to the literature of Raman spectroscopy and symmetry Tables. A special central reference section contains Tables summarizing formulae for the intensity and polarization characteristics of Rayleigh and Raman scattering.

References

1. C. V. Raman and K. S. Krishnan, *Nature*, **121**, 501, 1928.
2. A. Smekal, *Naturwiss.*, **11**, 873, 1923.
3. G. Landsberg and L. Mandelstam, *Naturwiss.*, **16**, 557, 772, 1928.
4. J. Cabannes, *Compt. rend.*, **186**, 1201, 1928.
5. Y. Rocard, *Compt. rend.*, **186**, 1107, 1928.
6. L. Brillouin, *Ann. Phys.* (*Paris*), **88**, 17, 1922.
7. E. Gross, *Z. Physik*, **63**, 685, 1930.
8. Lord Rayleigh, *Phil. Mag.*, **XLI**, 274, 447, 1871.
9. G. Mie, *Ann. Physik*, **25**, 377, 1908.
10. C. V. Raman and K. S. Krishnan, *Proc. Roy. Soc. Lond.*, **122**, 23, 1929.
11. See Chapter 2 for a discussion of the relation between wavenumber, frequency and wavelength in free space (page 14) and in air (page 40).

2 The nature of electromagnetic radiation

"We all know *what light is, but it is not easy to tell what it is."*

Samuel Johnson

2.1 Plane harmonic electromagnetic waves: propagation equations

We now consider the laws governing the propagation of electromagnetic radiation, i.e., the propagation of the electric and magnetic fields which constitute such radiation. Maxwell's equations, which apply to all electromagnetic phenomena, reveal that the electric and magnetic fields associated with radiation propagate as waves in free space, and enable the equations for the wave propagation to be deduced. We shall summarize here only the results for the case of a plane harmonic wave in free space. Such a wave has special properties: the quantity being propagated varies only in the direction of propagation, and the variation is harmonic in time and displacement along the propagation direction.

2.1.1 *The trigonometric representation*

Consider such a plane electromagnetic wave travelling in free space along the positive direction of the z-axis. It is a consequence of Maxwell's equations that such a wave has no longitudinal components, only transverse components. If the system of axes is chosen so that the x-axis is parallel to the electric vector, then the electric field intensity has only the component E_x, and this is given by

$$E_x = E_{x_0} \cos \left\{ \omega\left(t - \frac{z}{c} \right) + \theta_x \right\} \tag{2.1}$$

where E_{x_0} is the maximum value or amplitude of E_x, ω is the circular frequency, t is the time, z is the displacement along the z-axis, c is the velocity of propagation, and θ_x the phase angle. In mathematical treatments of radiation or, indeed, any periodic phenomenon, it is convenient to use the circular (or angular) frequency ω. We note that $\omega = 2\pi\nu$, where ν is the frequency; that $\nu = c\lambda^{-1}$ where λ is the wavelength; that $\tilde{\nu} = \lambda^{-1}$ where $\tilde{\nu}$ is the wavenumber; and thus $\nu = c\tilde{\nu}$ and $\omega = 2\pi c\tilde{\nu}$.

The associated magnetic field vector is always perpendicular to the electric field vector and so, in this case, will be parallel to the y-axis. Thus, the magnetic field intensity has only the component H_y, and since it is in phase with E_x,

$$H_y = H_{yo} \cos \left\{ \omega \left(t - \frac{z}{c} \right) + \theta_x \right\} \tag{2.2}$$

where H_{yo} is the amplitude of H_y and $\theta_x = \theta_y$ in this case.

It can be seen that E_x and H_y have the same relative magnitudes at all points at all times. Thus, the ratio of the magnitudes is a constant which is called the impedance of free space and is given by

$$\frac{E_x}{H_y} = \mu_0 c = \mu_0^{\frac{1}{2}} \varepsilon_0^{-\frac{1}{2}} = 377 \text{ ohm} \tag{2.3}$$

where μ_0 is the permeability of free space and ε_0 the permittivity of free space. Figure 2.1 shows the variation of E_x and H_y with z at a particular instant of time.

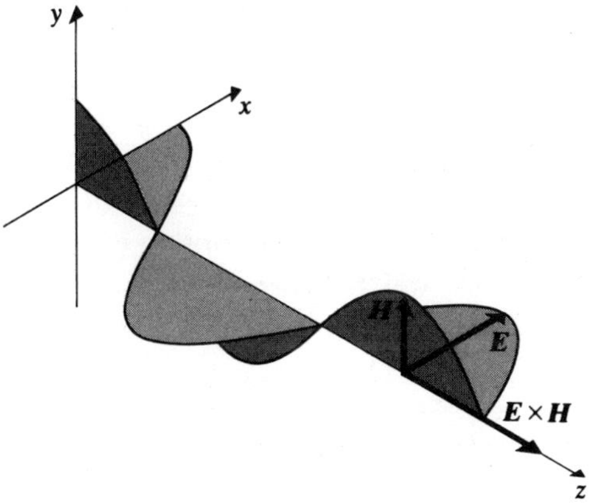

Fig. 2.1 The variation of *E* and *H* with *z* at a particular instant for a plane electromagnetic wave travelling in the positive direction along the *z*-axis. The vector *E*×*H* gives the direction of propagation

In many applications, we are concerned only with the variation with time of E (or H) at a chosen value of z; simplified forms of eqs. (2.1) and (2.2) can then be used. For example, if z is put equal to zero, eq. (2.1) reduces to

$$E_x = E_{xo} \cos (\omega t + \theta_x) \tag{2.4}$$

If phase considerations are not important, this equation can be further simplified by putting the phase angle θ_x equal to zero.

2.1.2 *The exponential representation*

It is often convenient to use an exponential rather than a trigonometric form as a representation of a plane wave. For example, since

$$\exp \{-i(\omega t + \theta_x)\} = \cos (\omega t + \theta_x) - i \sin (\omega t + \theta_x) \tag{2.5}$$

eq. (2.4) may be represented as

$$E_x = \text{Re } E_{xo} \exp \{-i(\omega t + \theta_x)\} \tag{2.6}$$

where Re is an operator which means 'take the real part of what follows'. Since

$$\exp\{i(\omega t+\theta_x)\}=\cos(\omega t+\theta_x)+i\sin(\omega t+\theta_x) \tag{2.7}$$

eq. (2.4) could equally well have been represented by

$$E_x=\mathrm{Re}\,E_{x_0}\exp\{+i(\omega t+\theta_x)\} \tag{2.8}$$

The choice of representation is arbitrary and we shall elect to use eq. (2.6) throughout. The operator Re is often omitted where it is clear from the context that the exponential representation is being used.

Two convenient properties of the exponential representation should be noted. A phase shift of $\pi/2$ is equivalent to multiplying by $-i$ and a phase shift of $-\pi/2$ is equivalent to multiplying by $+i$. For example,

$$\begin{aligned}\exp\{-i(\omega t+\theta_x+\pi/2)\}&=\exp\{-i\pi/2)\exp\{-i(\omega t+\theta_x)\}\\&=-i\exp\{-i(\omega t+\theta_x)\}\end{aligned} \tag{2.9}$$

since

$$\exp\{-i\pi/2\}=-i \tag{2.10}$$

It should be noted that the sign of i associated with a phase shift depends on the choice of sign in the exponential of the original representation.

Differentiation with respect to time can also be expressed very conveniently using this notation. Since

$$\frac{\mathrm{d}}{\mathrm{d}t}\{\exp -i(\omega t+\theta_x)\}=-i\omega\{\exp -i(\omega t+\theta_x)\} \tag{2.11}$$

we see that, in general,

$$\frac{\mathrm{d}^n}{\mathrm{d}t^n}\{\exp -i(\omega t+\theta_x)\}=(-i\omega)^n\{\exp -i(\omega t+\theta_x)\} \tag{2.12}$$

so that the operator $\mathrm{d}/\mathrm{d}t$ can be replaced by the factor $-i\omega$.

It should be noted particularly that the exponential representation is valid only for mathematical operations like addition, subtraction, differentiation, and integration, which do not mix the real and imaginary parts. Whenever multiplication is involved, one must revert to the trigonometric representation (that is take the real part) *before* multiplying.

2.2 Energy considerations

The presence of an electromagnetic field implies the presence of energy in the space occupied by the radiation. The electric energy density ρ_{elec} is given by $\frac{1}{2}\varepsilon_0 E_x^2$ and the magnetic energy density ρ_{mag} by $\frac{1}{2}\mu_0 H_y^2$. These energy densities are in phase and are equal since

$$\frac{\frac{1}{2}\varepsilon_0 E_x^2}{\frac{1}{2}\mu_0 H_y^2}=\varepsilon_0\mu_0 c^2=1 \tag{2.13}$$

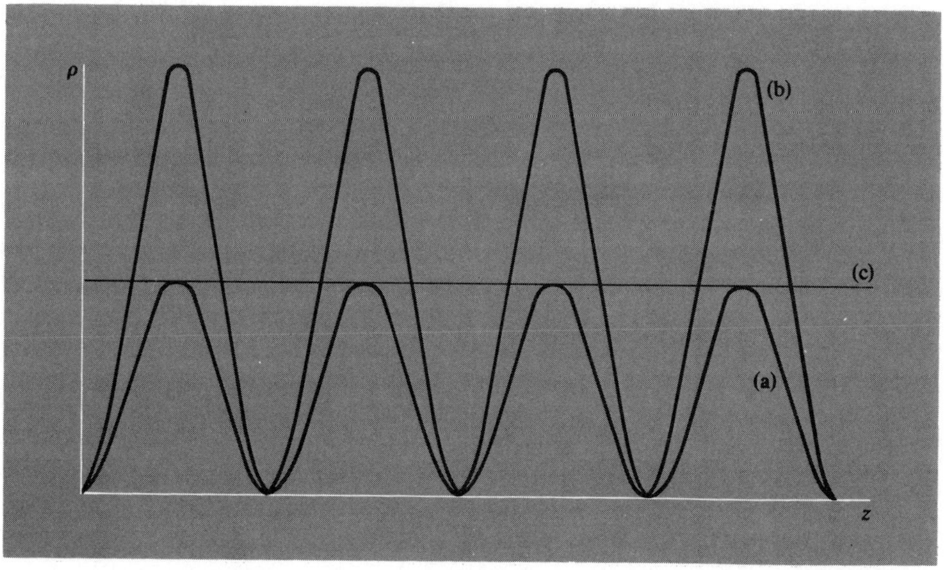

Fig. 2.2 Energy densities ρ as a function of z for a plane electromagnetic wave travelling along the z-axis: (a) $\rho_{electric}=\rho_{magnetic}$; (b) $\rho_{total}=\rho_{electric}+\rho_{magnetic}$; and (c) the time-averaged total energy density $\bar{\rho}_{total}$

The total energy density ρ_{total} at a given position along the propagation direction and at a given instant is thus given by

$$\rho_{total}=\tfrac{1}{2}\varepsilon_0 E_x^2+\tfrac{1}{2}\mu_0 H_y^2=\varepsilon_0 E_x^2=\mu_0 H_y^2 \qquad (2.14)$$

The variations with z of the electric, the magnetic, and the total energy densities at a given instant are shown in Fig. 2.2.

For radiation of the wavelength range considered in this book, the periodic time is many orders of magnitude less than the response time of any detector. For example, for a wavelength of 5000 Å (500 nm) the frequency is $6{\cdot}0\times10^{14}$ Hz and the periodic time is $1{\cdot}7\times10^{-15}$ s, whereas the response time of a photomultiplier may be of the order of 10^{-8} s. Consequently, only time-averaged properties of radiation can be measured. For the plane wave of eq. (2.1), $\overline{E_x^2}$ and $\overline{H_y^2}$, the time averages over one cycle of the squares of the electric and magnetic field intensities are given by

$$\overline{E_x^2}=\tfrac{1}{2}E_{xo}^2 \text{ and } \overline{H_y^2}=\tfrac{1}{2}H_{yo}^2 \qquad (2.15)$$

Thus, the time averages over one cycle of the electric, magnetic, and total energy densities, namely $\tfrac{1}{2}\varepsilon_0\overline{E_x^2}$, $\tfrac{1}{2}\mu_0\overline{H_y^2}$ and $\varepsilon_0\overline{E_x^2}$ (or $\mu_0\overline{H_y^2}$), respectively, are proportional to the squares of the amplitudes of the electric and magnetic field intensities. The average total energy density $\bar{\rho}_{total}$, for example, is given by

$$\bar{\rho}_{total}=\tfrac{1}{2}\varepsilon_0 E_{xo}^2=\tfrac{1}{2}\mu_0 H_{yo}^2 \qquad (2.16)$$

The total energy associated with the wave can therefore be considered as travelling with this average density, as shown in Fig. 2.2, provided the amplitude remains constant. Equation (2.16) can be rewritten in the useful forms:

$$\bar{\rho}_{\text{total}}/\text{J m}^{-3} = 4{\cdot}427 \times 10^{-12}\,(E_{xo}/\text{V m}^{-1})^2 \qquad (2.17)$$

or

$$\bar{\rho}_{\text{total}}/\text{J m}^{-3} = 6{\cdot}283 \times 10^{-7}\,(H_{yo}/\text{A m}^{-1})^2 \qquad (2.18)$$

The amount of radiation energy which passes through a surface of unit area perpendicular to the propagation direction in one second can be obtained by multiplying the time-averaged total energy density $\bar{\rho}_{\text{total}}$ by the velocity of propagation. The resultant quantity is properly termed the irradiance of the surface or the surface flux density $\mathcal{I}$, and has units of W m^{-2}. Thus, expressed in terms of the electric field intensity, $\mathcal{I}$ is given by

$$\mathcal{I} = c\varepsilon_0\overline{E_x^2} = \tfrac{1}{2}c\varepsilon_0 E_{xo}^2 \qquad (2.19)$$

and in terms of the magnetic field intensity by

$$\mathcal{I} = c\mu_0\overline{H_y^2} = \tfrac{1}{2}c\mu_0 H_{yo}^2 \qquad (2.20)$$

These equations can be rewritten in the forms:

$$\mathcal{I}/\text{W m}^{-2} = 1{\cdot}327 \times 10^{-3}\,(E_{xo}/\text{V m}^{-1})^2 \qquad (2.21)$$

and

$$\mathcal{I}/\text{W m}^{-2} = 188{\cdot}4\,(H_{yo}/\text{A m}^{-1})^2 \qquad (2.22)$$

Since it is often necessary to calculate electric and magnetic field intensities from the irradiance, the following alternative forms of eqs. (2.21) and (2.22) will prove useful:

$$\log_{10}\{E_{xo}/\text{V m}^{-1}\} = 0{\cdot}5\log_{10}\{\mathcal{I}/\text{W m}^{-2}\} + 1{\cdot}439 \qquad (2.23)$$

and

$$\log_{10}\{H_{yo}/\text{A m}^{-1}\} = 0{\cdot}5\log_{10}\{\mathcal{I}/\text{W m}^{-2}\} - 1{\cdot}138 \qquad (2.24)$$

The irradiance of a small element of the wavefront of electromagnetic radiation at some point is a measure of what is often called the radiation intensity. In this connotation, the radiation intensity (time-averaged) at some point is proportional to the square of the amplitude of the electric field intensity. However, it is preferable to use the term irradiance ($\mathcal{I}$) to avoid confusion with the radiant intensity of a point source (I), which is the ratio of the power dΦ, in a given direction contained in a conical beam of solid angle dΩ about this direction, to the solid angle dΩ. Thus,

$$I = \frac{\text{d}\Phi}{\text{d}\Omega} \qquad (2.25)$$

and has units of W sr^{-1}. The power radiated by a point source into a conical beam of a given orientation in space and of solid angle dΩ is constant if I is constant. Thus, since $\text{d}\Omega = \text{d}A/r^2$, where d$A$ is the area of the cross-section of the cone perpendicular to the cone axis and at distance r from the point source, the irradiance produced by a point source decreases as r^{-2}. This may be contrasted with a parallel beam of electromagnetic radiation where the irradiance is the same at all points along the propagation direction.

An alternative approach to the calculation of the flow of energy involves the Poynting vector. The mutually perpendicular electric and magnetic field vectors are so oriented that their vector product $\boldsymbol{E} \times \boldsymbol{H}$ points in the direction of propagation, as shown in Fig. 2.1. This vector product is called the Poynting vector and is denoted by the symbol $\mathscr{S}$. Thus,

$$\mathscr{S} = \boldsymbol{E} \times \boldsymbol{H} \tag{2.26}$$

For electromagnetic waves, the Poynting vector may be regarded as representing a flow of energy per unit time, per unit area, in the direction of propagation of the wave. The magnitude of the Poynting vector thus corresponds to the irradiance and is a measure of the radiation intensity. For example, for the plane wave under consideration, the instantaneous value of the Poynting vector $\mathscr{S}_{\text{inst}}$ at a given z value is given by

$$\begin{aligned} \mathscr{S}_{\text{inst}} &= \boldsymbol{E} \times \boldsymbol{H} \\ &= E_x H_y (\boldsymbol{i} \times \boldsymbol{j}) = E_x H_y \boldsymbol{k} \\ &= c\varepsilon_0 E_x^2 \boldsymbol{k} = c\mu_0 H_y^2 \boldsymbol{k} \end{aligned} \tag{2.27}$$

where $\boldsymbol{i}$, $\boldsymbol{j}$, and $\boldsymbol{k}$ are unit vectors along the positive x-, y-, and z-directions. The time-averaged value over one cycle of the Poynting vector $\bar{\mathscr{S}}$ is given by

$$\bar{\mathscr{S}} = c\varepsilon_0 \overline{E_x^2} \boldsymbol{k} = \tfrac{1}{2} c\varepsilon_0 E_{x0}^2 \boldsymbol{k} = c\mu_0 \overline{H_y^2} \boldsymbol{k} = \tfrac{1}{2} c\mu_0 H_{y0}^2 \boldsymbol{k} \tag{2.28}$$

It is instructive to consider the peak values of the electric field intensity (i.e., E_{x0} values) associated with the irradiances that can be produced by various sources of radiation. With conventional sources, even those which might be described as powerful, only relatively low values of the irradiance can be achieved and the associated electric field intensities are quite small. For example, the irradiance produced by solar radiation at the earth's surface is about $1{\cdot}4 \times 10^3$ W m^{-2}, corresponding to an electric field intensity of 10^3 V m^{-1}. In lasers, the energy is propagated into an essentially non-diverging beam of small cross-section of the order of 10^{-5} m^2, which can be further reduced by focusing. Thus, even though the total power in a CW gas laser beam is at most only of the order of watts, irradiances of the order of 10^4 to 10^8 W m^{-2} can be achieved, depending on the type of laser and the extent to which the laser beam is concentrated by focusing. The corresponding electric field intensities are of the order of 3×10^3 to 3×10^5 V m^{-1}. A laser operating in the giant-pulse mode, with a pulse duration of the order of 2×10^{-8} s, energy of 2 J, and an unfocused beam of diameter 3×10^{-3} m, gives an irradiance of 10^{13} W m^{-2} and an electric field intensity of 10^8 V m^{-1}. A hundredfold reduction in beam radius increases the irradiance to 10^{17} W m^{-2} and thus the peak electric field intensity rises to 10^{10} V m^{-1}. The fields binding outer electrons in atoms and molecules are of the order of 10^{10}–10^{12} V m^{-1}, and so it is not surprising that a giant-pulse laser, when focused to a sufficiently small spot, causes ionization of air. Even larger irradiances, of the order of 10^{22} W m^{-2}, can be produced with giant-pulse laser systems using very short pulses and several stages of amplification. The electric field intensity then exceeds 10^{12} V m^{-1} and is sufficient to affect atomic nuclei; e.g., neutrons have been expelled from lithium deuteride with such lasers.

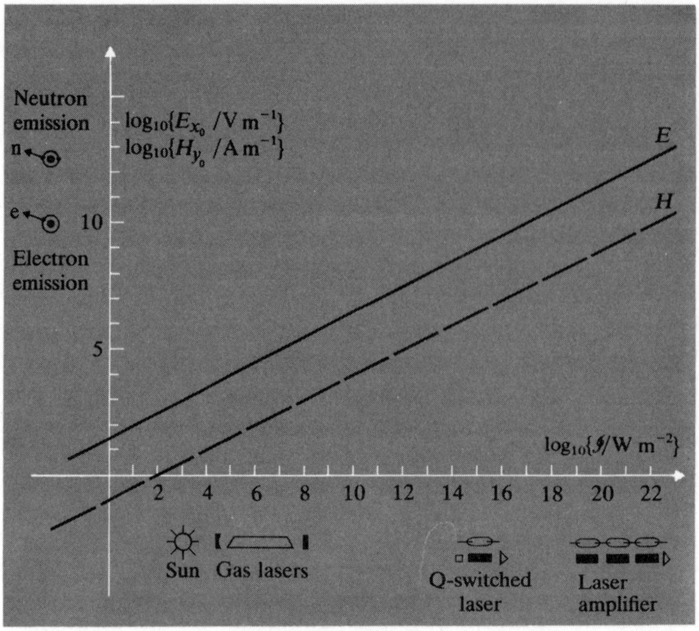

Fig. 2.3 Dependence of E_{x_0} and H_{y_0} on the irradiance $\mathscr{I}$ of a plane electromagnetic wave propagating along z

The magnetic fields associated with intense laser radiation are no less impressive. Whereas the magnetic field associated with solar radiation at the earth's surface is only $3\ \mathrm{A\,m^{-1}}$, at an irradiance of $10^{22}\ \mathrm{W\,m^{-2}}$ the magnetic field intensity exceeds $10^{10}\ \mathrm{A\,m^{-1}}$, corresponding to a magnetic flux density (or magnetic induction) in air of 10^8 gauss (10^4 tesla). For a plane electromagnetic wave propagating along z the dependence of E_{x0} and H_{y0} on the irradiance is shown graphically in Fig. 2.3.

2.3 States of polarization: monochromatic radiation

2.3.1 *Linear polarization*

The plane wave we have been considering propagates along the z-axis and has the electric vector parallel to the x-axis and the magnetic vector parallel to the y-axis. Such a wave, and the radiation associated with it, is often described as plane polarized, but this description is complete only if the plane is defined. Historically, the plane of polarization was taken to be the plane containing the direction of propagation and the magnetic vector. The direction of the magnetic vector was then referred to as the direction of polarization. However, the plane of polarization has also sometimes been defined in terms of the electric vector. Modern practice in terminology, which will be followed here, is to avoid the terms plane of polarization and direction of polarization. Instead, direction of vibration and plane of vibration are used to denote, respectively, the direction of a field vector and the plane containing the field vector and the direction of propagation, *the vector in question being specified in each case.* Thus, for the plane wave shown in Fig. 2.1, the direction of vibration of the electric vector is the x-direction and the plane of vibration of the electric vector is the xz-plane. This wave is said to be linearly polarized.

2.3.2 *Elliptical and circular polarization*

Waves with other states of polarization exist. They may be regarded as the result of combining together two plane waves of the same frequency propagating along the same direction, say the z-axis, with the direction of vibration of the electric vector in the x-direction for one wave and in the y-direction for the other wave. The resultant state of polarization depends on the amplitudes and phases of the two plane waves.

Using eq. (2.1), we may write for two such waves

$$E_x = E_{xo} \cos(\tau + \theta_x) = \mathrm{Re}\, E_{xo} \exp\{-i(\tau + \theta_x)\} \tag{2.29}$$

and

$$E_y = E_{yo} \cos(\tau + \theta_y) = \mathrm{Re}\, E_{yo} \exp\{-i(\tau + \theta_y)\} \tag{2.30}$$

where

$$\tau = \omega\left(t - \frac{z}{c}\right) \tag{2.31}$$

After some mathematical manipulation, it can be shown that at a given position along the z-axis, the locus of the points whose coordinates are E_x, E_y is, in general, an ellipse whose characteristics depend on the amplitudes, E_{xo} and E_{yo}, and $\Delta\theta = \theta_y - \theta_x$, the difference in phase of the two waves. In consequence, such waves are said to be elliptically polarized.

In general, the major and minor axes of the ellipse lie not along the initially chosen x- and y-directions but along a new set of rectangular axes, x' and y', where the x'-direction is the direction of the major axis of the ellipse and makes an angle ψ $(0 \leq \psi < \pi)$ with the x-direction (see Fig. 2.4). Then we may write for the relation

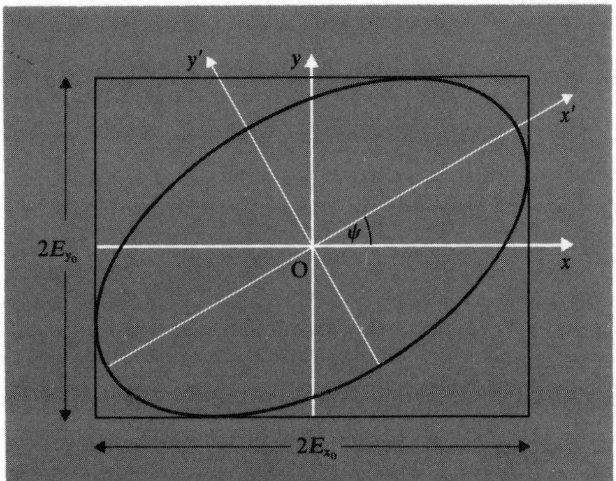

Fig. 2.4 The vibrational ellipse for the electric vector of an elliptically polarized wave propagating along the z axis, (out of the paper)

between the electric field components $E_{x'}$, $E_{y'}$ and the components E_x, E_y

$$E_{x'} = E_x \cos\psi + E_y \sin\psi \tag{2.32}$$

$$E_{y'} = -E_x \sin\psi + E_y \cos\psi \tag{2.33}$$

and for the time dependence of $E_{x'}$ and $E_{y'}$,

$$E_{x'} = E_{x'_0} \cos(\tau + \theta') = \text{Re}\, E_{x'_0} \exp\{-i(\tau + \theta')\} \tag{2.34}$$

$$E_{y'} = \mp E_{y'_0} \sin(\tau + \theta') = \text{Re}\,(\mp i) E_{y'_0} \exp\{-i(\tau + \theta')\} \tag{2.35}$$

where $E_{x'_0}$ and $E_{y'_0}$ are the semi-minor axes of the ellipse in the $x'y'$ axis system and $E_{x'_0} \geq E_{y'_0}$.

Following the traditional nomenclature, the polarization is said to be right-handed when, to an *observer looking in the direction from which the radiation is coming*, the end point of the electric vector would appear to describe the ellipse in the clockwise sense. Likewise, the polarization is said to be left-handed if, under the same conditions of observation, the end point of the electric vector appears to describe the ellipse anticlockwise. In eq. (2.35) the upper and lower signs refer to right and left elliptical polarization respectively as may be established by considering the values of $E_{x'}$ and $E_{y'}$ for $z = 0$ (a) when $\omega t = 0$ and (b) when $\omega t = \pi/2$ (i.e. a quarter of a period later).

The relationships between the various quantities in the two axis systems may be found by appropriate manipulation, and are most conveniently expressed using two auxiliary angles, ξ $(0 \leq \xi \leq \pi/2)$ and χ $(-\pi/4 \leq \chi \leq \pi/4)$ defined as follows:

$$\tan \xi = \frac{E_{yo}}{E_{xo}} \tag{2.36}$$

and

$$\tan \chi = \frac{\pm E_{y'_0}}{E_{x'_0}} \tag{2.37}$$

The numerical value of $\tan \chi$ represents the ratio of the axes of the ellipse and the upper and lower signs refer to right and left elliptical polarization respectively. (Note. $+\chi$ and $-i$ refer to right elliptical polarization). The required relationships are then

$$E_{xo}^2 + E_{yo}^2 = E_{x'_0}^2 + E_{y'_0}^2 \tag{2.38}$$

$$\tan 2\psi = \tan 2\xi \cos \Delta\theta \tag{2.39}$$

$$\sin 2\chi = \sin 2\xi \sin \Delta\theta \tag{2.40}$$

We see that, in general, the polarization ellipse can be characterized by three independent quantities: e.g., the amplitudes E_{xo} and E_{yo} and the phase difference $\Delta\theta$, or the major and minor axes $E_{x'_0}$ and $E_{y'_0}$ and the angle ψ which specifies the orientation of the ellipse, or the irradiance and the angles χ and ψ.

The various cases of elliptical polarization are shown in Fig. 2.5. Two special cases of elliptical polarization are of importance, namely, when the ellipse degenerates into a straight line or circle.

It can be seen from Fig. 2.5 that the ellipse will reduce to a straight line if

$$\Delta\theta = \theta_y - \theta_x = m\pi \; (m = 0, \pm 1, \pm 2, \ldots) \tag{2.41}$$

and then

$$\frac{E_y}{E_x} = (-1)^m \frac{E_{yo}}{E_{xo}} \tag{2.42}$$

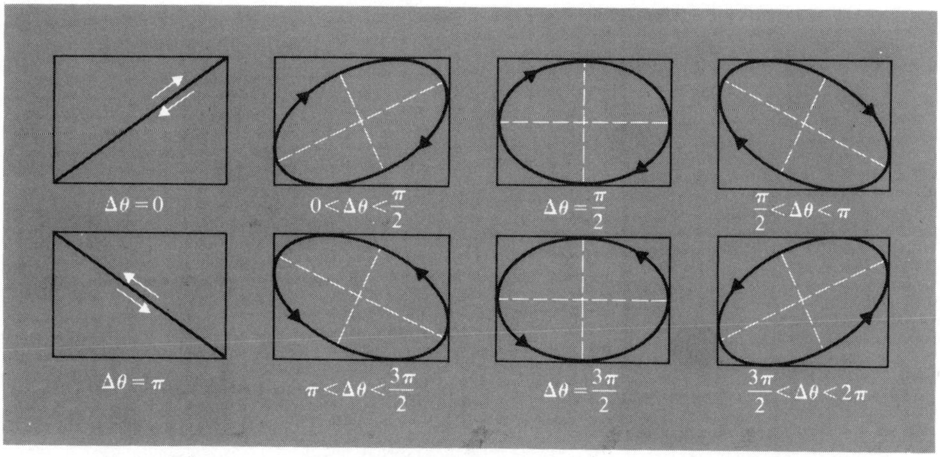

Fig. 2.5 Elliptical polarization with various values of the phase difference $\Delta\theta$

This is clearly a case of linear polarization. If the axis system is realigned so that, say, the x-axis lies along the line of vibration of the resultant electric vector, there is just one component of the electric vector, namely, E_x. There is then also only one component of the magnetic vector, the orthogonal component H_y. This, then, corresponds exactly to the linearly polarized wave we treated above. It can be seen that any linearly polarized wave can be considered to be the sum of two component waves propagating along the same direction, with their electric vectors in phase and linearly polarized in directions which are perpendicular to each other and the direction of propagation.

The ellipse degenerates into a circle if

$$E_{y_0} = E_{x_0} = E_0 \tag{2.43}$$

and

$$\Delta\theta = \theta_y - \theta_x = \frac{m\pi}{2}\,(m = \pm 1, \pm 3, \pm 5, \ldots) \tag{2.44}$$

We then have the case of circular polarization. The polarization is right-handed if

$$\sin \Delta\theta > 0 \tag{2.45}$$

so that

$$\Delta\theta = \frac{\pi}{2} + 2m\pi\,(m = 0, \pm 1, \pm 2, \ldots) \tag{2.46}$$

and thus

$$E_x = E_0 \cos(\tau + \theta_x) = \mathrm{Re}\, E_0 \exp\{-i(\tau + \theta_x)\} \tag{2.47}$$

and

$$E_y = E_0 \cos(\tau + \theta_x + \pi/2) = -E_0 \sin(\tau + \theta_x) = \mathrm{Re} - iE_0 \exp\{-i(\tau + \theta_x)\} \tag{2.48}$$

The polarization is left-handed if

$$\sin \Delta\theta < 0 \tag{2.49}$$

so that

$$\Delta\theta = -\pi/2 + 2m\pi \ (m = 0, \pm 1, \pm 2, \ldots) \tag{2.50}$$

and thus

$$E_x = E_0 \cos(\tau + \theta_x) = \text{Re } E_0 \exp\{-i(\tau + \theta_x)\} \tag{2.51}$$

and

$$E_y = E_0 \cos(\tau + \theta_x - \pi/2) = E_0 \sin(\tau + \theta_x) = \text{Re } iE_0 \exp\{-i(\tau + \theta_x)\} \tag{2.52}$$

Thus, in the exponential notation,

$$\frac{E_y}{E_x} = \exp\{-i(\pm\pi/2)\} = \mp i \tag{2.53}$$

where the upper and lower signs correspond to right and left circularly polarized radiation, respectively. More generally in this notation, for right-handed elliptical polarization the ratio E_y/E_x has a negative imaginary part, whereas for left-handed elliptical polarization the imaginary part is positive.

The electric vector of radiation in a general polarization state can be expressed as

$$\boldsymbol{E} = \tilde{\boldsymbol{E}}_0 \exp(-i\omega t) + \tilde{\boldsymbol{E}}_0^* \exp(i\omega t) \tag{2.54}$$

where $\boldsymbol{E}$ is always real but the amplitude $\tilde{\boldsymbol{E}}_0$ (conjugate complex $\tilde{\boldsymbol{E}}_0^*$) can be complex. For example eq. (2.54) yields $\boldsymbol{E} = E_0 \cos \omega t\, \boldsymbol{i}$ (linear polarization) if $\tilde{\boldsymbol{E}}_0 = \frac{1}{2}E_0\boldsymbol{i}$, and $\boldsymbol{E} = E_0(\cos \omega t\, \boldsymbol{i} - \sin \omega t\, \boldsymbol{j})$ (right circular polarization) if $\tilde{\boldsymbol{E}}_0 = \frac{1}{2}E_0(\boldsymbol{i} - i\boldsymbol{j})$.

2.3.3 *Stokes parameters*

It is convenient to characterize the states of polarization by Stokes parameters which all have the same dimensions. For a plane monochromatic wave there are four Stokes parameters and for a wave propagating along z, these are

$$S_0 = E_{xo}^2 + E_{yo}^2 \tag{2.55}$$

$$S_1 = E_{xo}^2 - E_{yo}^2 \tag{2.56}$$

$$S_2 = 2E_{xo}E_{yo} \cos \Delta\theta \tag{2.57}$$

$$S_3 = 2E_{xo}E_{yo} \sin \Delta\theta \tag{2.58}$$

Only three of these are independent quantities since they are related by the identity

$$S_0^2 = S_1^2 + S_2^2 + S_3^2 \tag{2.59}$$

For complex amplitudes, eqs. (2.55) to (2.58) take the form

$$S_0 = E_{xo}E_{xo}^* + E_{yo}E_{yo}^* \tag{2.60}$$

$$S_1 = E_{xo}E_{xo}^* - E_{yo}E_{yo}^* \tag{2.61}$$

$$S_2 = E_{xo}E_{yo}^* + E_{yo}E_{xo}^* \tag{2.62}$$

$$S_3 = i(E_{yo}E_{xo}^* - E_{xo}E_{yo}^*) \tag{2.63}$$

We see that S_0 is related to the irradiance of the radiation $\mathscr{I}$ as follows:

$$S_0 = \frac{2\mathscr{I}}{c\varepsilon_0} = 2\mu_0 c\mathscr{I} \tag{2.64}$$

S_1, S_2, and S_3 can be shown to be related to the angles ψ and χ as follows:

$$S_1 = S_0 \cos 2\chi \cos 2\psi \tag{2.65}$$

$$S_2 = S_0 \cos 2\chi \sin 2\psi \tag{2.66}$$

$$S_3 = S_0 \sin 2\chi \tag{2.67}$$

Hence

$$\mathscr{I} = \tfrac{1}{2} c\varepsilon_0 S_0 = \frac{S_0}{2\mu_0 c} \tag{2.68}$$

$$\psi = \tfrac{1}{2} \tan^{-1} (S_2/S_1) \tag{2.69}$$

$$\chi = \tfrac{1}{2} \tan^{-1} \frac{S_3}{(S_1^2 + S_2^2)^{1/2}} \tag{2.70}$$

It follows from eqs. (2.55) to (2.58) that for linearly polarized radiation ($\Delta\theta$ zero or an integral multiple of π), $S_3 = 0$; for right-handed circular polarized radiation ($E_{xo} = E_{yo}$, $\Delta\theta = +\pi/2$), $S_1 = S_2 = 0$, $S_3 = S_0$; and for left-handed circular polarized radiation ($E_{xo} = E_{yo}$, $\Delta\theta = -\pi/2$), $S_1 = S_2 = 0$, $S_3 = -S_0$.

Equations (2.65) to (2.67) indicate that a simple geometrical representation is possible for all states of polarization: S_1, S_2, and S_3 may be regarded as the Cartesian coordinates of a point P on a sphere of radius S_0 such that 2χ and 2ψ are the spherical angular coordinates of this point (see Fig. 2.6). Thus, to every possible state of polarization of a plane monochromatic wave of a given irradiance (S_0 constant) there corresponds one point on the sphere, and vice versa. This geometrical representation was first introduced by Poincaré and is called the Poincaré sphere. Since χ is positive

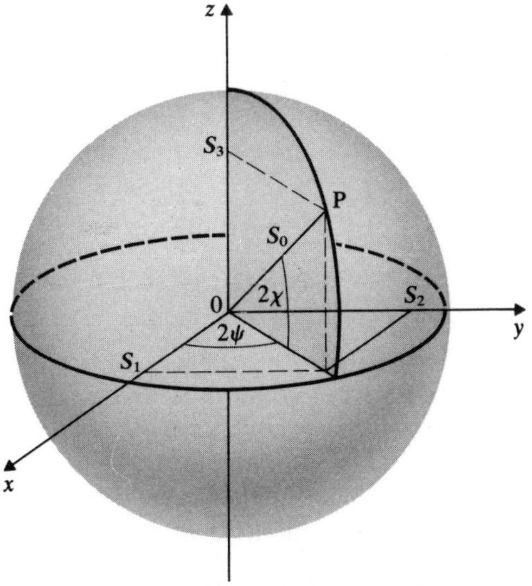

Fig. 2.6 Poincaré's representation of the state of polarization of a monochromatic wave (Poincaré's sphere)

or negative according to whether the polarization is right-handed or left-handed, right-handed polarization is represented by points on the sphere above the equatorial plane and left-handed polarization by points below this plane. Linearly polarized radiation ($\chi = 0$) is represented by points on the equatorial circumference. Right-handed circular polarization ($\chi = +\pi/4$) is represented by the north pole and left-handed circular polarization ($\chi = -\pi/4$) by the south pole.

2.4 States of polarization: quasi-monochromatic radiation

Strictly monochromatic radiation is always completely polarized; i.e., the end point of the electric (and also the magnetic) vector at each point in space moves periodically around an ellipse which may, in special cases, reduce to a circle or a straight line. In practice, we usually have to consider radiation that is only nearly monochromatic, containing frequencies in a small range centred on what is apparently a monochromatic frequency. Such waves are called quasi-monochromatic and can be represented by a superposition of strictly monochromatic waves of various frequencies.

Quasi-monochromatic radiation has an extra 'degree of freedom' in its permitted polarizations. At one extreme, the resultant electric vector can have the properties of a completely monochromatic wave and it is then said to be completely polarized. At the other extreme, the resultant electric vector can have no preferred directional properties and the radiation is said to be completely unpolarized or natural. In many instances, particularly with radiation produced by scattering, the radiation is partially polarized; i.e., it consists of a completely polarized part and a completely unpolarized part.

Natural light has the property that the intensity of its components in any direction perpendicular to the direction of propagation is the same; and, moreover, this intensity is not affected by any previous retardation of one of the rectangular components relative to another, into which the light may have been resolved. Natural light of irradiance $\mathcal{I}$ is equivalent to the superposition of any two *independent* linearly polarized waves each of irradiance $\mathcal{I}/2$ with their electric vectors vibrating in two mutually perpendicular directions at right angles to the direction of propagation. The independence of these two waves means that there exists no specific phase relationship between them. In consequence, the results of the interaction of natural light of irradiance $\mathcal{I}$ with a system can be calculated by considering the effects of two independent orthogonal electric field amplitudes E_{xo} and E_{yo}, where

$$\frac{1}{2}c\varepsilon_0 E_{xo}^2 = \frac{1}{2}c\varepsilon_0 E_{yo}^2 = \frac{\mathcal{I}}{2} \tag{2.71}$$

Natural light of irradiance $\mathcal{I}$ is also equivalent to two *independent* circularly polarized waves, one right-handed and the other left-handed, each of irradiance $\mathcal{I}/2$, or, indeed, to the superposition of any two *independent* orthogonal polarization states.

The polarization properties of such quasi-monochromatic radiation can be calculated following the procedure used in section 2.3.2, but eqs. (2.29) to (2.31) must be replaced by

$$E_x = E_{xo}(t) \cos\{\bar{\tau} + \theta_x(t)\} \tag{2.72}$$

$$E_y = E_{yo}(t) \cos\{\bar{\tau} + \theta_y(t)\} \tag{2.73}$$

and

$$\bar{\tau} = \bar{\omega}\left(t - \frac{z}{c}\right) \tag{2.74}$$

where $\bar{\omega}$ is the mean frequency of the radiation. In these equations the amplitudes and the phases are now time-dependent. As a consequence, for the Stokes parameters for a quasi-monochromatic wave, time-averaged quantities like $\overline{E_{xo}(t)^2}$ and $\overline{E_{xo}(t)E_{yo}(t)\cos\Delta\theta(t)}$ must be used in eqs. (2.55) to (2.58), and similarly for eqs. (2.60) to (2.63), where $\Delta\theta(t) = \theta_y(t) - \theta_x(t)$.

In general, the Stokes parameters are now four independent quantities and only in the special case of *complete* polarization does the relationship of eq. (2.59) exist. For completely unpolarized or natural radiation, $S_1 = S_2 = S_3 = 0$.

As already emphasized, complete polarization implies a strictly monochromatic wave for which the amplitudes and phases of the component waves are independent of time, and hence eqs. (2.55) to (2.58) or (2.60) to (2.63) define the Stokes parameters.

The Stokes parameters may be determined as follows: S_0 from the irradiance of the radiation; S_1 from the difference in irradiance of radiation transmitted by analysers that accept linear polarization with azimuth $\psi = 0$ and $\psi = \pi/2$, respectively; S_2 from the difference in irradiance of radiation transmitted by analysers which accept linear polarization with azimuths $\psi = \pi/4$ and $\psi = 3\pi/4$, respectively; and S_3 from the additional irradiance transmitted by a device that accepts right circularly polarized radiation over that transmitted by a device that accepts left circularly polarized radiation.

We now consider the decomposition of a partially polarized wave into mutually independent unpolarized and polarized portions,using Stokes parameters. Since, for an unpolarized wave, $S_1 = S_2 = S_3 = 0$, we can write the following sets of Stokes parameters for the polarized and unpolarized waves:

$$\text{Unpolarized wave: } S_0 - (S_1^2 + S_2^2 + S_3^2)^{1/2}, 0, 0, 0 \tag{2.75}$$

$$\text{Polarized wave: } (S_1^2 + S_2^2 + S_3^2)^{1/2}, S_1, S_2, S_3 \tag{2.76}$$

where S_0, S_1, S_2, and S_3 are the Stokes parameters of the undecomposed partially polarized radiation. We can now define a degree of polarization P of the original radiation as

$$P = \frac{\mathcal{I}_{\text{polarized}}}{\mathcal{I}_{\text{total}}} \tag{2.77}$$

Using eqs. (2.75) and (2.76), we obtain

$$P = \frac{(S_1^2 + S_2^2 + S_3^2)^{1/2}}{S_0} \tag{2.78}$$

For natural radiation $P = 0$ and for completely polarized radiation $P = 1$. We note also that in place of eq. (2.67) we now have

$$\sin 2\chi = \frac{S_3}{(S_1^2 + S_2^2 + S_3^2)^{1/2}} \tag{2.79}$$

but that the angle ψ is still defined by eq. (2.69).

The Stokes parameters provide a very convenient basis for a unified treatment of a radiation problem which embraces all polarization states. Such a general treatment can then be reduced to specific cases of polarization. For example, we note that an alternative representation of the four Stokes parameters for radiation of irradiance $\mathcal{I}$ and degree of polarization P is

$$S_0 = E_0^2 \tag{2.80}$$

$$S_1 = PE_0^2 \cos 2\chi \cos 2\psi \tag{2.81}$$

$$S_2 = PE_0^2 \cos 2\chi \sin 2\psi \tag{2.82}$$

$$S_3 = PE_0^2 \sin 2\chi \tag{2.83}$$

where

$$\mathcal{I} = \tfrac{1}{2} c\varepsilon_0 E_0^2 \tag{2.84}$$

Thus, if a problem like the scattering of radiation is treated for the general case of partially polarized incident radiation defined by the four Stokes parameters given by eqs. (2.80) to (2.84), the results for specific cases of polarization can be obtained by inserting into formulae resulting from the general treatment the appropriate values of P, χ, and ψ. For example, for linearly polarized radiation, $P=1$, $\chi=0$, and ψ is the angle made by the electric vector with the x-axis. For right circularly polarized radiation, $P=1$, $\chi=\pi/4$, and $\psi=0$, and for left circularly polarized radiation, $P=1$, $\chi=-\pi/4$, and $\psi=0$. For right elliptically polarized radiation, $P=1$, and ψ and χ determine the orientation and ellipticity of the ellipse. For the corresponding left elliptically polarized radiation, we have $P=1$, ψ, and $-\chi$. For natural light, $P=0$.

2.5 Change of polarization: depolarization ratios, reversal coefficients, and degrees of circularity

When electromagnetic radiation interacts with a system, there is often a change in the state of polarization. For example, incident natural radiation can produce scattered radiation which is either completely or partially polarized; changes in the state of polarization are also observed if the incident radiation is plane polarized or circularly polarized. These changes are important parameters in Rayleigh and Raman scattering because they may be correlated with the symmetry of the scattering species.

We now consider how quantitative expression can be given to such polarization changes. We first define a scattering plane as the plane containing the direction of propagation of the incident radiation and the direction of observation (provided their directions do not coincide). Then for plane polarized incident radiation with the electric vector *parallel* to the scattering plane we define a depolarization ratio $\rho_{\|}(\theta)$ for a direction of observation lying in the scatter plane and making an angle θ with the direction of propagation of the incident radiation as the intensity ratio

$$\rho_{\|}(\theta) = \frac{{}^{\|}I_{\perp}(\theta)}{{}^{\|}I_{\|}(\theta)} \tag{2.85}$$

where the superscript preceding I defines the relation of the incident electric vector to the scatter plane and the subscript following I defines the relation of the electric vector of the scattered radiation to the scatter plane. Similarly, if the electric vector of the incident radiation is perpendicular to the scattering plane, we can define a depolarization ratio $\rho_\perp(\theta)$ given by

$$\rho_\perp(\theta) = \frac{{}^{\perp}I_{\|}(\theta)}{{}^{\perp}I_{\perp}(\theta)} \tag{2.86}$$

and for incident natural radiation we can define a depolarization ratio $\rho_n(\theta)$ given by

$$\rho_n(\theta) = \frac{{}^{n}I_{\|}(\theta)}{{}^{n}I_{\perp}(\theta)} \tag{2.87}$$

Figure 2.7 illustrates how $\rho_{\|}(\pi/2)$ $\rho_\perp(\pi/2)$ and $\rho_n(\pi/2)$ are related to the scattered intensities for radiation incident along the z-axis of a Cartesian coordinate system and observation along the x or y directions.

Normally in Rayleigh and Raman scattering, the depolarization ratios as defined above have values such that $0 \leq \rho \leq 1$ for gases and liquids. If $\rho = 0$, the scattered radiation is completely polarized; if $\rho = 1$, the radiation is natural; and if $0 < \rho < 1$, the radiation is partially polarized. In certain special cases, which arise when the scattering system has an absorption band close to the frequency of the incident radiation, ρ values considerably greater than unity are observed. Such depolarization ratios are generally described as anomalous depolarization ratios. For scattering at 90° to the direction of the incident radiation, it happens that $\rho_{\|} = 1.0$, and hence for this geometry $\rho_{\|}$ can be equally well defined by $\rho_{\|} = I_{\|}/I_\perp$ and thus the same definition can be used for $\rho_{\|}$, $\rho_\perp$, and ρ_n. However, this definition leads to $\rho_{\|} > 1$ for certain geometries. The definition in eq. (2.85) reserves the term anomalous depolarization for situations where values of $\rho > 1$ arise only from scattering associated with the antisymmetric part of the scattering tensor (see chapter 5).) A special case is when ρ is infinity, and this is termed inverse polarization.

With incident circularly polarized radiation, a change in the handedness of the radiation can take place on scattering. We can define a reversal factor $\mathscr{P}(\theta)$ for the radiation scattered at an angle θ to the direction of the incident radiation by

$$\mathscr{P}(\theta) = \frac{{}^{\circledR}I_{\copyright}(\theta)}{{}^{\circledR}I_{\circledR}(\theta)} = \frac{{}^{\copyright}I_{\circledR}(\theta)}{{}^{\copyright}I_{\copyright}(\theta)} = \mathscr{P}(\theta+\pi)^{-1} \tag{2.88}$$

where the superscript preceding I defines the handedness of the incident radiation and the subscript following I, the handedness of the scattered radiation. $\mathscr{P}(\theta) = \mathscr{P}(\theta+\pi)^{-1}$ since a vector turning in one sense when viewed in the forward direction is seen to turn in the opposite sense when viewed in the backward direction.

Another useful property is the degree of circularity $\mathscr{C}(\theta)$ of the scattered radiation which is given by

$${}^{p_i}\mathscr{C}(\theta) = \frac{{}^{p_i}I_{\circledR}(\theta) - {}^{p_i}I_{\copyright}(\theta)}{{}^{p_i}I(\theta)_{\text{total}}} = -{}^{p_i}\mathscr{C}(\theta+\pi) \tag{2.89}$$

where ${}^{p_i}I(\theta)_{\text{total}}$ is the total intensity of the scattered radiation and p_i is the state of polarization of the incident radiation. It should be noted that ${}^{\circledR}\mathscr{C}(\theta) = -{}^{\copyright}\mathscr{C}(\theta)$.

The depolarization ratios, and $\mathscr{P}$ and $\mathscr{C}$, can also be expressed in terms of the Stokes parameters of the scattered radiation. Thus, for example since

$$\frac{{}^{p_i}I_{\|}(\pi/2)}{{}^{p_i}I_{\perp}(\pi/2)} = \frac{S_0(\pi/2) - S_1(\pi/2)}{S_0(\pi/2) + S_1(\pi/2)} \tag{2.90}$$

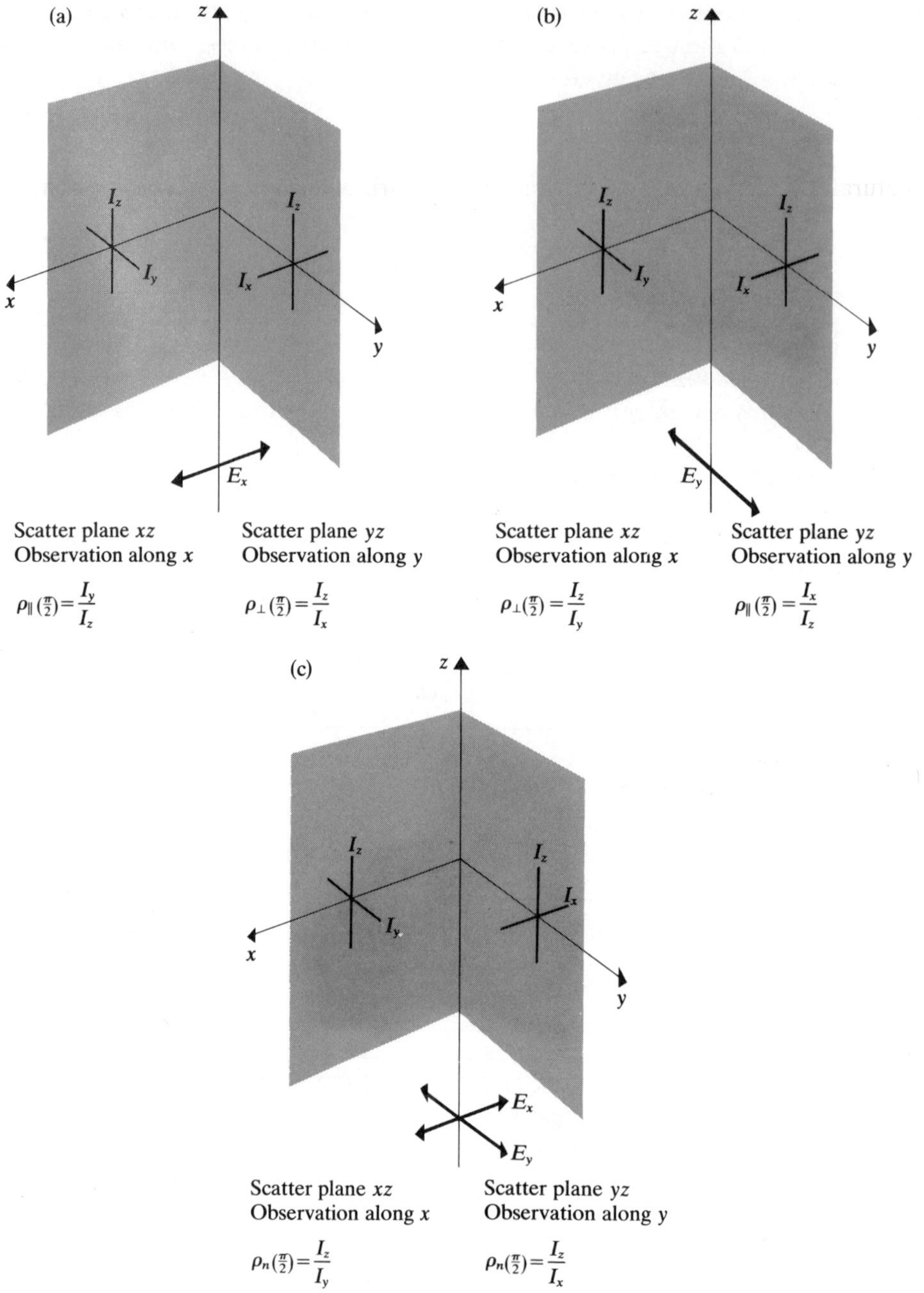

Fig. 2.7 Measurement of (a) $\rho_\parallel(\pi/2)$, (b) $\rho_\perp(\pi/2)$, and (c) $\rho_n(\pi/2)$ for illumination along the z-axis and observation along the x or y axes

$\rho_\parallel(\pi/2)$, $\rho_\perp(\pi/2)$ and $\rho_n(\pi/2)$ can be related to the Stokes parameters; similarly

$$\mathscr{P}(0) = \frac{S_0(0) - |S_3(0)|}{S_0(0) + |S_3(0)|} = \mathscr{P}(\pi)^{-1} \tag{2.91}$$

and

$$ {}^{p_i}\mathscr{C}(0) = \frac{S_3(0)}{S_0(0)} = -{}^{p_i}\mathscr{C}(\pi) \tag{2.92} $$

The Stokes parameters of the scattered radiation in eqs. (2.90) to (2.92) will of course depend on the state of polarization of the incident radiation.

We note that $\mathscr{P} \geqslant 0$ and in principle can range from 0 (no reversal) to ∞ (complete reversal). However ${}^{p_i}\mathscr{C}$ can range from $-1({}^{p_i}I_{Ⓡ} = 0;\ {}^{p_i}I_{Ⓛ} = {}^{p_i}I_{\text{total}})$ to $+1({}^{p_i}I_{Ⓛ} = 0;\ {}^{p_i}I_{Ⓡ} = {}^{p_i}I_{\text{total}})$; when ${}^{p_i}\mathscr{C} = 0$ the scattered radiation is natural $({}^{p_i}I_{Ⓡ} = {}^{p_i}I_{Ⓛ})$. For complete reversal, ${}^{Ⓡ}\mathscr{C} = -1$ but ${}^{Ⓛ}\mathscr{C} = 1$, and for no reversal ${}^{Ⓡ}\mathscr{C} = 1$ but ${}^{Ⓛ}\mathscr{C} = -1$. In practice the range of $\mathscr{P}$ and $\mathscr{C}$ values may be more restricted since symmetry does not always permit complete reversal on scattering.

2.6 Sources of electromagnetic radiation

The most important source of electromagnetic radiation is the oscillating electric dipole. Other sources of electromagnetic radiation include the oscillating magnetic dipole and higher order oscillating electric and magnetic multipoles as, for example, the oscillating electric quadrupole. The intensity of radiation from an oscillating electric dipole is usually several orders of magnitude greater than that from other kinds of oscillating multipoles, although in the molecular systems with which we are concerned it can sometimes happen, for reasons of symmetry, that there is no oscillating dipole. We shall only need to consider in detail the oscillating electric dipole as a source of electromagnetic radiation, since the major contributions to Rayleigh and Raman scattering arise from such sources. However, magnetic dipoles and electric quadrupoles assume importance for scattering from chiral systems, and will be considered briefly.

2.6.1 *The oscillating electric dipole as a source*

An electric dipole is formed when a pair of charges of equal magnitude and opposite sign are separated by a given distance. If $\boldsymbol{s}$ is a vector oriented from a charge $-q$ to a charge $+q$, and whose magnitude is the separation of the charges, the dipole moment vector $\boldsymbol{P}$ (magnitude P) is given by

$$ \boldsymbol{P} = q\boldsymbol{s} \tag{2.93} $$

We shall show that if such a dipole oscillates harmonically with circular frequency ω, then electromagnetic radiation of circular frequency ω is produced. First, we consider the ways in which molecular systems can give rise to such electric dipoles.

If the electron distribution is unsymmetrical in a molecule, it will have a so-called permanent dipole. During a vibration, the magnitude of this may change and there will result a dipole oscillating at the frequency of the vibration. If the electron distribution is symmetrical in a molecule and it has no permanent dipole in its equilibrium configuration, it is still possible for unsymmetrical vibrations to disturb the symmetry of the electron distribution and produce a dipole which again oscillates with the frequency of the vibration. Electric dipoles may also be induced in molecules by an external electric field, and such dipoles can oscillate and produce radiation. The frequencies of such dipoles are related to the frequency of the electric field *and* the frequency of the molecular vibration. If the electric field is a static one, the induced dipole oscillates with the frequency of the molecular vibration. If the electric field is an oscillating one, as for example in electromagnetic radiation, induced dipoles result which oscillate at the frequency of the electric field and also at frequencies which are beat frequencies between the electric field frequency and the molecular frequency.

We shall be particularly concerned with *induced* electric dipoles in this book, since they are the principal source of Rayleigh and Raman scattering. However, for the immediate purpose we can maintain generality since we can express the harmonic variation of *any* dipole oscillating with circular frequency ω by

$$\boldsymbol{P} = \boldsymbol{P}_0 \cos \omega t \tag{2.94}$$

where $\boldsymbol{P}_0$ is the amplitude vector of the *oscillating* dipole (magnitude P_0). It must be emphasized again that $\boldsymbol{P}_0$ (and hence $\boldsymbol{P}$) can be non-zero even if the system has no permanent dipole moment.

Using the standard procedures of electromagnetic theory, it is relatively straightforward to calculate the electric and magnetic field intensities as a function of time and position produced by such a harmonically oscillating dipole in free space. For the dipole orientation and coordinate system defined in Fig. 2.8, the electric and

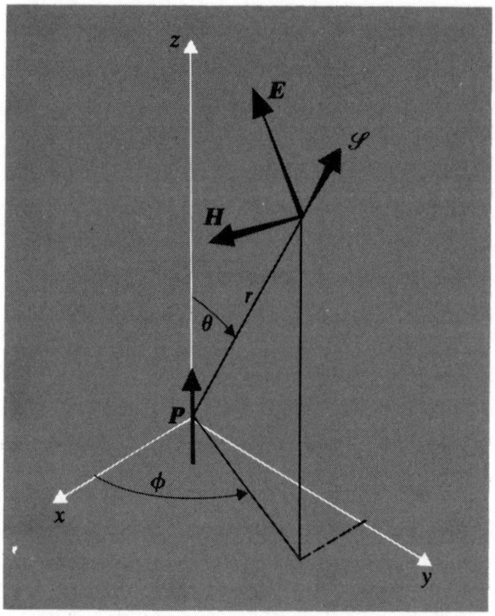

Fig. 2.8 Orientation of *E, H* and $\mathscr{S}$ vectors for an oscillating electric dipole oriented as shown ($r \gg \lambda$: phase angle zero). The unit polar vectors *i, j, k* at a given point lie along the directions of the vectors $\mathscr{S}$, $-\boldsymbol{E}$ and $-\boldsymbol{H}$ respectively

magnetic field intensities at some point with the polar coordinates r, θ, ϕ are given by

$$\boldsymbol{E} = \frac{-\omega^2 P \sin \theta\, \boldsymbol{j}}{4\pi\varepsilon_0 c^2 r} \tag{2.95}$$

and

$$\boldsymbol{H} = \frac{-\omega^2 P \sin \theta\, \boldsymbol{k}}{4\pi c r} \tag{2.96}$$

where

$$P = P_0 \cos\left\{\omega\left(t - \frac{r}{c}\right)\right\} \tag{2.97}$$

In deriving these formulae, it has been assumed that $s \ll r$, $s \ll \lambda$, and $r \gg \lambda$, where λ is the wavelength of the radiation of circular frequency ω. These conditions are readily satisfied for dipoles of molecular origin and radiation in the visible region, under practical conditions of observation.

It can be seen that both $\boldsymbol{E}$ and $\boldsymbol{H}$ propagate as waves with velocity c and vary inversely as the distance r. They both vary as ω^2, and so are proportional to the square of the frequency. As far as amplitude is concerned, they also have the same angular dependence which is determined by $\sin\theta$. Thus, the amplitudes of $\boldsymbol{E}$ and $\boldsymbol{H}$ have maximum values in the equatorial plane, are zero along the axis of the dipole, and have axial symmetry about the dipole axis as shown in Fig. 2.9, which is a plot of

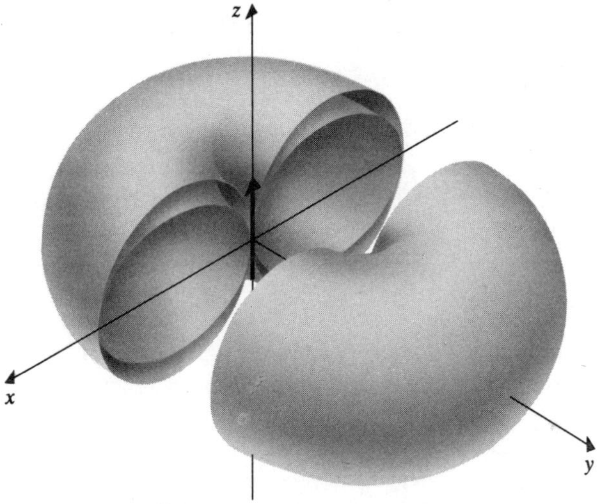

Fig. 2.9 Polar diagrams of $\sin\theta$ **(outer surface) and of** $\sin^2\theta$ **(inner surface) showing, respectively, the angular distributions of the amplitude of** $\boldsymbol{E}$ **or** $\boldsymbol{H}$ **and the magnitude of** $\bar{\mathscr{S}}$ **for an oscillating electric dipole situated at the origin. The radial distance from the origin to a surface is proportional to the magnitude of the corresponding quantity in that direction**

the amplitude of $\boldsymbol{E}$ (or $\boldsymbol{H}$) as a function of θ and ϕ. However, as far as *direction* is concerned, comparison of eqs. (2.95) and (2.96) shows that whereas $\boldsymbol{H}$ is azimuthal, $\boldsymbol{E}$ lies in a plane passing through the polar axis. The relative orientations of $\boldsymbol{E}$ and $\boldsymbol{H}$ at a particular point are shown in Fig. 2.8. The ratio of the magnitudes of $\boldsymbol{E}$ and $\boldsymbol{H}$ is $(\mu_0/\varepsilon_0)^{1/2}$.

It is clear from these considerations that the vectors $\boldsymbol{E}$ and $\boldsymbol{H}$ produced by an oscillating dipole are related to each other exactly as in a plane electromagnetic wave. An oscillating dipole of circular frequency ω is therefore a source of electromagnetic radiation of circular frequency ω.

Now that we have related the magnitude of the oscillating dipole to the electric field intensity of the radiation, the energy density and power per unit area of the radiation at the point (r, θ, ϕ) may be expressed in terms of the magnitude of the dipole. Using eq. (2.95) with eq. (2.14) gives, for the *instantaneous* total energy density,

$$\rho_{\text{total}} = \frac{\omega^4 P^2 \sin^2\theta}{16\pi^2 \varepsilon_0 c^4 r^2} = \frac{\pi^2 \tilde{\nu}^4 P^2 \sin^2\theta}{\varepsilon_0 r^2} \tag{2.98}$$

Using eqs. (2.95) and (2.97), eq. (2.16) gives, for the average total energy density,

$$\bar{\rho}_{\text{total}} = \frac{\omega^4 P_0^2 \sin^2\theta}{32\pi^2 \varepsilon_0 c^4 r^2} = \frac{\pi^2 \tilde{\nu}^4 P_0^2 \sin^2\theta}{2\varepsilon_0 r^2} \tag{2.99}$$

Substitution of eqs. (2.95) and (2.96) in eq. (2.26) gives, for the *instantaneous* value of the Poynting vector,

$$\mathscr{S}_{\text{inst}} = \frac{\omega^4 P^2 \sin^2\theta\, \boldsymbol{i}}{16\pi^2 \varepsilon_0 r^2 c^3} = \frac{\pi^2 c \tilde{\nu}^4 P^2 \sin^2\theta\, \boldsymbol{i}}{\varepsilon_0 r^2} \tag{2.100}$$

Using eqs. (2.95), (2.96) and (2.97), eq. (2.26) gives for the *average* value of the Poynting vector,

$$\bar{\mathscr{S}} = \frac{\omega^4 P_0^2 \sin^2\theta\, \boldsymbol{i}}{32\pi^2 \varepsilon_0 r^2 c^3} = \frac{\pi^2 c \tilde{\nu}^4 P_0^2 \sin^2\theta\, \boldsymbol{i}}{2\varepsilon_0 r^2} \tag{2.101}$$

The Poynting vector is entirely radial, as is shown in Fig 2.8. Since its angular dependence is determined by $\sin^2\theta$, its magnitude is zero along the axis of the dipole, is a maximum in the equatorial plane, and has axial symmetry about the dipole axis, as can be seen from Fig. 2.9.

The Poynting vector varies as r^{-2}, a condition which is necessary for conservation of energy, since under steady-state conditions the energy flow through any given solid angle must be the same for all r. The Poynting vector also depends on the fourth power of the frequency and on the square of the dipole moment.

We may use the expression (2.101) for the average value of the Poynting vector to calculate the average energy per second or average power from an oscillating dipole which passes through an area $\mathrm{d}A$ at distance r from the centre of the dipole where r makes an angle θ with the dipole direction (or, alternatively, the power into the corresponding solid angle $\mathrm{d}\Omega$).

For simplicity we denote the average power by $\mathrm{d}\Phi$ (rather than $\mathrm{d}\bar{\Phi}$). Thus

$$\mathrm{d}\Phi = |\bar{\mathscr{S}}|\, \mathrm{d}A \tag{2.102}$$

and substituting for $|\bar{\mathscr{S}}|$, using eq. (2.101), we obtain

$$\mathrm{d}\Phi = \frac{\omega^4 P_0^2 \sin^2\theta\, \mathrm{d}A}{32\pi^2 \varepsilon_0 r^2 c^3} = \frac{\pi^2 c \tilde{\nu}^4 P_0^2 \sin^2\theta\, \mathrm{d}A}{2\varepsilon_0 r^2} \tag{2.103}$$

$$\frac{\mathrm{d}A}{r^2} = \mathrm{d}\Omega \tag{2.104}$$

where $\mathrm{d}\Omega$ is an element of solid angle, we may also write

$$\mathrm{d}\Phi = \frac{\omega^4 P_0^2 \sin^2\theta\, \mathrm{d}\Omega}{32\pi^2 \varepsilon_0 c^3} = \frac{\pi^2 c \tilde{\nu}^4 P_0^2 \sin^2\theta\, \mathrm{d}\Omega}{2\varepsilon_0} \tag{2.105}$$

To obtain the total power radiated by the dipole we must integrate over $\mathrm{d}\Omega$. Thus,

$$\Phi = \frac{\omega^4 P_0^2}{32\pi^2 \varepsilon_0 c^3} \int \sin^2\theta \,\mathrm{d}\Omega \tag{2.106}$$

and putting

$$\mathrm{d}\Omega = \sin\theta \,\mathrm{d}\theta \,\mathrm{d}\phi \tag{2.107}$$

we have

$$\Phi = \frac{\omega^4 P_0^2}{32\pi^2 \varepsilon_0 c^3} \int_0^{2\pi} \int_0^{\pi} \sin^3\theta \,\mathrm{d}\theta \,\mathrm{d}\phi \tag{2.108}$$

The integral has the value $8\pi/3$ and hence

$$\Phi = \frac{\omega^4 P_0^2}{12\pi\varepsilon_0 c^3} = \frac{4\pi^3 c\tilde{\nu}^4 P_0^2}{3\varepsilon_0} \tag{2.109}$$

It follows from eqs. (2.25) and (2.105) that the radiant intensity I of the dipole in a particular direction is given by

$$I = \frac{\mathrm{d}\Phi}{\mathrm{d}\Omega} = \frac{\omega^4 P_0^2 \sin^2\theta}{32\pi^2 \varepsilon_0 c^3} = \frac{\pi^2 c\tilde{\nu}^4 P_0^2 \sin^2\theta}{2\varepsilon_0} \tag{2.110}$$

This corresponds to what is often described, rather loosely, as the 'intensity' of a source in a given direction.

It is important to realize that measurements of radiated power usually involve collection over a finite solid angle. Since $\mathrm{d}\Phi$ is angle-dependent, the total power $\Delta\Phi$ in a given finite solid angle $\Delta\Omega$ is then given by integrating eq. (2.105) over the appropriate range of θ and ϕ values. That is,

$$\begin{aligned}\Delta\Phi &= \frac{\omega^4 P_0^2}{32\pi^2 \varepsilon_0 c^3} \int_{\theta-\Delta\theta}^{\theta+\Delta\theta} \int_{\phi-\Delta\phi}^{\phi+\Delta\phi} \sin^3\theta \,\mathrm{d}\theta \,\mathrm{d}\phi \\ &= \frac{\pi^2 c\tilde{\nu}^4 P_0^2}{2\varepsilon_0} \int_{\theta-\Delta\theta}^{\theta+\Delta\theta} \int_{\phi-\Delta\phi}^{\phi+\Delta\phi} \sin^3\theta \,\mathrm{d}\theta \,\mathrm{d}\phi\end{aligned} \tag{2.111}$$

This angular dependence can lead to a rather complicated angular dependence of $\Delta\Phi$ in certain situations in light scattering, particularly where contributions to the scattering from two dipoles at right angles to each other have to be considered.

Equations (2.103), (2.105), (2.109), (2.110), and (2.111) can be expressed in the useful forms:

$$\mathrm{d}\Phi = 1.6709 \times 10^{28} \tilde{\nu}^4 \sin^2\theta P_0^2 \frac{\mathrm{d}A}{r^2} \tag{2.112}$$

$$\mathrm{d}\Phi = 1.6709 \times 10^{28} \tilde{\nu}^4 \sin^2\theta P_0^2 \,\mathrm{d}\Omega \tag{2.113}$$

$$I = \frac{\mathrm{d}\Phi}{\mathrm{d}\Omega} = 1.6709 \times 10^{28} \tilde{\nu}^4 \sin^2\theta P_0^2 \tag{2.114}$$

$$\Phi = 1.3998 \times 10^{29} \tilde{\nu}^4 P_0^2 \tag{2.115}$$

$$\Delta\Phi = 1.6709 \times 10^{28} \tilde{\nu}^4 P_0^2 \int_{\theta-\Delta\theta}^{\theta+\Delta\theta} \int_{\phi-\Delta\phi}^{\phi+\Delta\phi} \sin^3\theta \,\mathrm{d}\theta \,\mathrm{d}\phi \tag{2.116}$$

where Φ is in W, I in W sr^{-1}, $\tilde{\nu}$ in cm^{-1}, P_0 in C m, and $\mathrm{d}\Omega = \mathrm{d}A/r^2$ in sr.

In scattering, where the radiation originates from dipoles *induced* by the electric field of the radiation, it is convenient to introduce the concept of *scattering cross-section.* If the scattering is induced by a beam of monochromatic radiation with irradiance $\mathcal{I}$, where $\mathcal{I} = \frac{1}{2}\varepsilon_0 c E_0^2$ and E_0 is the amplitude of the electric field intensity, it will be shown subsequently that for normal (linear) scattering the amplitude of the induced dipole moment is proportional to E_0. It follows from eq. (2.109) that the total power Φ radiated by an induced dipole is proportional to E_0^2. Thus, the ratio of the total scattered power to the irradiance of the incident radiation will be independent of E_0^2 and have the dimensions of area. This ratio is termed the scattering cross-section σ and has units of m^2. Thus

$$\sigma = \frac{\Phi}{\mathcal{I}} \tag{2.117}$$

Clearly, σ measures the rate at which energy is removed from the incident beam by scattering, relative to the rate at which energy crosses a unit area perpendicular to the direction of propagation of the incident beam. Since σ has the dimensions of area, we may say that a system with scattering cross-section σ scatters all the radiation incident on the area σ. The differential cross-section $d\sigma/d\Omega$ relates to the power scattered into a given element of solid angle $d\Omega$; i.e.,

$$\frac{d\sigma}{d\Omega} = \frac{d\Phi/d\Omega}{\mathcal{I}} \tag{2.118}$$

where $d\Phi/d\Omega$ is defined by eq. (2.110). The differential cross-section has units of $m^2\ sr^{-1}$. Where the scattering arises from an assembly of molecules and has a finite wavenumber range, σ and $d\sigma/d\Omega$ are usually related to a single molecule and a unit wavenumber interval (e.g. σ in m^2 molecule^{-1} cm^{-1}).

2.6.2 *The oscillating magnetic dipole as a source*

By analogy with an electric dipole, a separation of a pair of magnetic poles of opposite sign will produce a magnetic dipole. If m is the magnetic pole magnitude, then the magnetic dipole vector $\boldsymbol{M}$ is given by

$$\boldsymbol{M} = \mathrm{m}\boldsymbol{s} \tag{2.119}$$

where $\boldsymbol{s}$ has the same significance as in eq. (2.93). This definition of a magnetic dipole in terms of magnetic poles preserves the analogy with electric dipoles, but is only a convenient artifice. In atomic and molecular systems, magnetic dipoles arise from the motion of charges. If a particle of charge $-q$ and mass m moves with momentum $\boldsymbol{p}$ in a circle of radius vector $\boldsymbol{r}$, then it can be shown that the magnetic moment is given, in the classical picture, by

$$\boldsymbol{M} = -\left(\frac{q}{2m}\right)\boldsymbol{r}\times\boldsymbol{p} \tag{2.120}$$

This situation is illustrated in Fig. 2.10, where we see that the magnetic moment is antiparallel to the angular momentum. An oscillating magnetic dipole is also a source of electromagnetic radiation, and it can be shown that provided $r \gg \lambda \gg s$ the electric

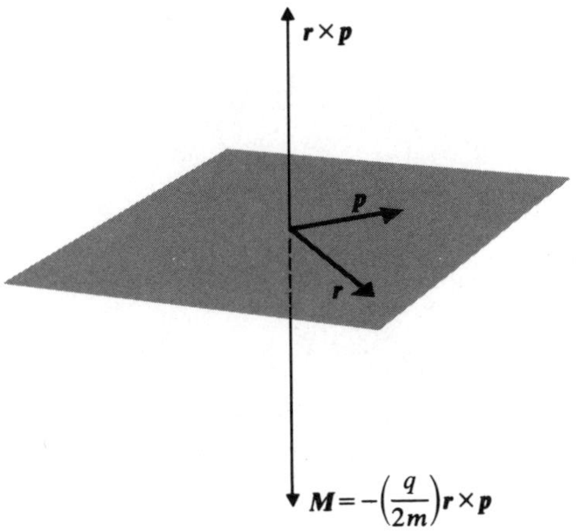

Fig. 2.10 Magnetic moment and angular momentum

field intensity $\boldsymbol{E}$ is given by

$$\boldsymbol{E}=\left(\frac{\mu_0}{\varepsilon_0}\right)^{1/2}\frac{M_0\omega^2}{4\pi rc^2}\cos\left\{\omega\left(t-\frac{r}{c}\right)\right\}\sin\theta\,\boldsymbol{k} \tag{2.121}$$

the magnetic field intensity $\boldsymbol{H}$ by

$$\boldsymbol{H}=\frac{-M_0\omega^2}{4\pi rc^2}\cos\left\{\omega\left(t-\frac{r}{c}\right)\right\}\sin\theta\,\boldsymbol{j} \tag{2.122}$$

and the time-averaged Poynting vector $\bar{\mathscr{S}}$ by

$$\bar{\mathscr{S}}=\frac{\mu_0 M_0^2\omega^4}{32\pi^2 r^2 c^3}\sin^2\theta\,\boldsymbol{i} \tag{2.123}$$

where M_0 is the amplitude of the oscillating magnetic dipole. Comparing eqs. (2.121) and (2.122) with eqs. (2.95), (2.96), and (2.97), we see that

$$\boldsymbol{E}_{\text{magnetic dipole}}=\frac{-M_0}{P_0}\mu_0\boldsymbol{H}_{\text{electric dipole}} \tag{2.124}$$

and

$$\boldsymbol{H}_{\text{magnetic dipole}}=\frac{M_0}{P_0}\varepsilon_o\boldsymbol{E}_{\text{electric dipole}} \tag{2.125}$$

It can be seen from eqs. (2.121) and (2.122) that, for magnetic dipole radiation, $\boldsymbol{E}$ is azimuthal and $\boldsymbol{H}$ lies in a plane passing through the polar axis, in contrast to the situation for an electric dipole. Also, the sign of $\boldsymbol{E}$ for the magnetic dipole is opposite to that of $\boldsymbol{H}$ for the electric dipole.

The relative orientations of $\boldsymbol{H}$, $\boldsymbol{E}$, and $\mathscr{S}$ for magnetic dipole radiation are shown in Fig. 2.11.

2.6.3 *The oscillating electric quadrupole as a source*

We may conveniently illustrate the properties of an electric quadrupole by considering the special case of a *linear* quadrupole. A linear quadrupole is made up of two

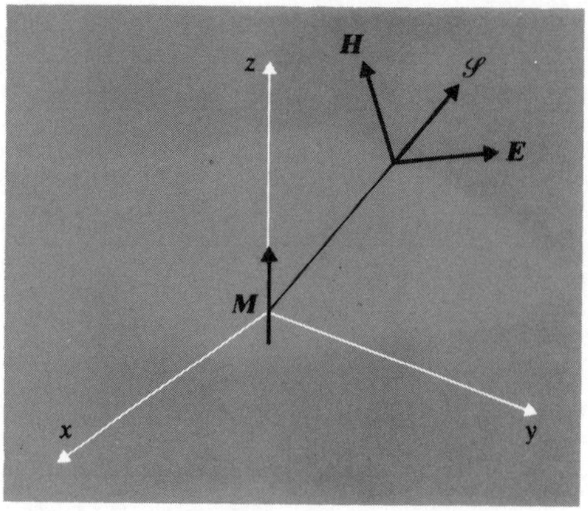

Fig. 2.11 Relative orientations of *H, E*, and $\mathscr{S}$ for a magnetic dipole along the *z*-axis

dipoles of opposite polarity arranged in line to give three charges $+q$, $-2q$, and $+q$, as in Fig. 2.12. Although the electric dipole moment of such a charge distribution is zero, the quadrupole moment Θ_{zz} is not, and is given by

$$\begin{aligned}\Theta_{zz} &= \sum qz^2 \\ &= 2qs^2\end{aligned} \tag{2.126}$$

The electric and magnetic field intensities of such a quadrupole may be calculated by combining the fields of the two constituent dipoles.

The following expressions result for $\boldsymbol{E}$ and $\boldsymbol{H}$ provided $r \gg \lambda \gg s$:

$$\boldsymbol{E} = \frac{\Theta_{zz_0}\omega^3}{4\pi\varepsilon_0 rc^3} \sin\left\{\omega\left(t - \frac{r}{c}\right)\right\} \sin\theta \cos\theta \, \boldsymbol{j} \tag{2.127}$$

and

$$\boldsymbol{H} = \frac{\Theta_{zz_0}\omega^3}{4\pi rc^2} \sin\left\{\omega\left(t - \frac{r}{c}\right)\right\} \sin\theta \cos\theta \, \boldsymbol{k} \tag{2.128}$$

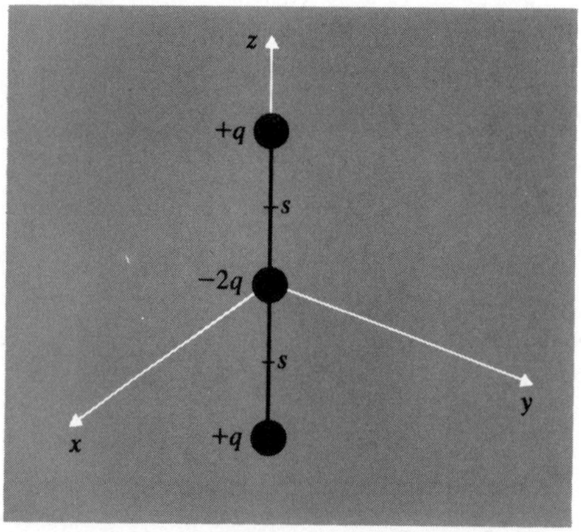

Fig. 2.12 Linear electric quadrupole formed of two dipoles of opposite polarity, one above the other. The dipole centred at $-s/2$ has a moment of $-qs$ and the dipole centred at $+s/2$ has a moment of $+qs$

and Θ_{zzo} is the amplitude of the oscillating electric quadrupole. ***E*** and ***H*** for electric quadrupole radiation have some features in common with ***E*** and ***H*** for electric dipole radiation. The relative orientations of ***E*** and ***H*** and the dependence on r^{-1} are the same in both cases. However, the angular distribution is different. For electric quadrupole radiation, both ***E*** and ***H*** are zero, not only along the axis of the quadrupole ($\theta = 0$ or π) where neither constituent dipole radiates but also along the equator ($\theta = \pi/2$) where the fields of the constituent dipoles are equal and opposite; the maximum field intensity occurs along the surface of a cone at 45° to the z-axis. A plot of the amplitude of ***E*** and ***H*** is shown in Fig. 2.13.

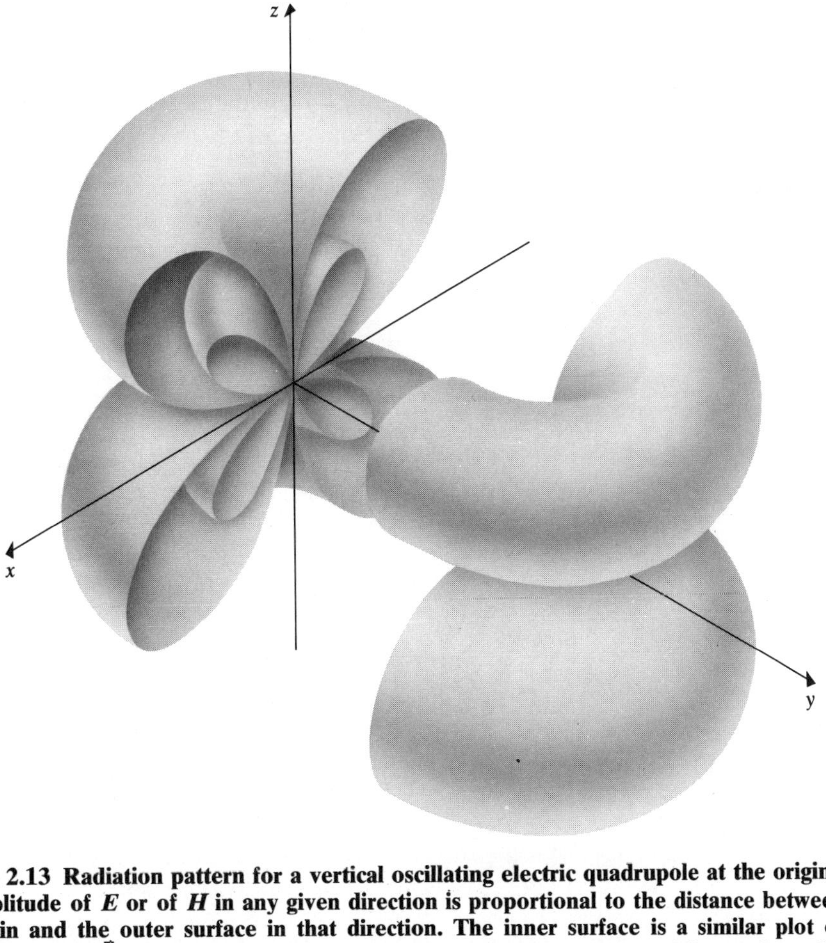

Fig. 2.13 Radiation pattern for a vertical oscillating electric quadrupole at the origin. The amplitude of *E* or of *H* in any given direction is proportional to the distance between the origin and the outer surface in that direction. The inner surface is a similar plot of the magnitude of $\bar{\mathscr{S}}$. There is no field along the axis or along the equator of the quadrupole. The maximum field intensity occurs along the surface of a cone at 45° to the axis

The dependence on ω is also different. For the electric quadrupole, ***E*** and ***H*** depend on ω^3.

The time-averaged Poynting vector $\bar{\mathscr{S}}$ is given by

$$\bar{\mathscr{S}} = \frac{\Theta_{zzo}^2 \omega^6}{32\pi^2 \varepsilon_0 r^2 c^5} \sin^2\theta \cos^2\theta \, \boldsymbol{i} \tag{2.129}$$

$\bar{\mathscr{S}}$ is again radial and a plot of the magnitude of $\bar{\mathscr{S}}$ is included in Fig. 2.13. $\bar{\mathscr{S}}$ depends on ω^6.

A more general definition of the electric quadrupole is given by

$$\Theta_{lm} = \tfrac{1}{2}\sum_i q_i(3l_i m_i - r_i^2 \delta_{lm}) \tag{2.130}$$

where l, m can be x, y or z, and the i-th charge q_i is at a distance r_i from the origin. δ_{lm} is a component of the Kronecker delta and has the value unity if $l = m$ and zero if $l \neq m$.

2.7 Propagation in air

The formulae derived so far relate to electromagnetic radiation in free space. In air, the values of the wavelength (and hence the wavenumber) and the velocity of propagation of radiation are slightly different from the values in free space. If u is the velocity of propagation in air we may write

$$\tilde{\nu}_{air} = \frac{1}{\lambda_{air}} = \nu\left(\frac{1}{u}\right) \tag{2.131}$$

in contrast to

$$\tilde{\nu}_{vac} = \frac{1}{\lambda_{vac}} = \nu\left(\frac{1}{c}\right) \tag{2.132}$$

The velocities of propagation in air and free space are related as follows

$$n = \frac{c}{u} \tag{2.133}$$

where n is the refractive index. Thus

$$\tilde{\nu}_{vac} = \frac{1}{n}(\tilde{\nu}_{air}) = \frac{1}{n}\left(\frac{1}{\lambda_{air}}\right) \tag{2.134}$$

Both n and u vary slightly with the frequency of the radiation. Consequently whereas $\tilde{\nu}_{vac}$ is related to the frequency ν by a constant proportionality factor, $\tilde{\nu}_{air}$ is not. Thus, strictly, only $\tilde{\nu}_{vac}$ values may be compared. Since, in Raman spectroscopy, wavelengths are measured in air, appropriate corrections[1] must be made to obtain $\tilde{\nu}_{vac}$.

In this book when characterizing radiation used for excitation of Raman spectra we quote λ_{air} (in Å and nm) *but* $\tilde{\nu}_{vac}$ (in cm^{-1}), unless otherwise stated.

References

1. G. Strey, Spectrochim. Acta, **25A**, 163, 1969.

3 A classical treatment of Rayleigh and vibrational Raman scattering

"*Se non è vero, è molto ben trovato.*"

Giordano Bruno

3.1 Introduction

As we have seen in chapter 2, according to the classical theory of electromagnetic radiation, electric and magnetic multipoles oscillating with a given frequency radiate electromagnetic radiation of that frequency. Light scattering phenomena may be given a classical explanation in terms of the electromagnetic radiation produced by such multipoles *induced* in a scattering system by the electric and magnetic fields of the incident radiation.

This chapter will be devoted mainly to such a classical treatment. Although this treatment cannot account for all aspects of Rayleigh and Raman scattering, it will serve as a gentle introduction to more rigorous treatments and enable the newcomer to the subject to become familiar with the mathematical techniques needed to describe the directional properties of light scattering. The reader need not fear that the classical edifice we are about to erect will be totally demolished subsequently under the hammer of quantum mechanics. For molecular vibrations, many of the results are still valid when quantum considerations are introduced, and even where there are changes these may be readily grafted on to the classical results. (Any classical formulae which will be modified subsequently are specially marked to warn against their misuse. See, for example, page 56.) In the case of molecular rotations this is not so, because classical theory does not ascribe specific discrete rotational frequencies to molecules. We shall therefore exclude rotational Raman spectra from our considerations in this chapter.

3.2 The induced electric dipole: some general considerations

For the majority of systems, we need consider only an induced electric dipole moment $\boldsymbol{P}$ which is related to the electric field $\boldsymbol{E}$ of the radiation by the power series

$$\boldsymbol{P} = \boldsymbol{P}^{(1)} + \boldsymbol{P}^{(2)} + \boldsymbol{P}^{(3)} + \dots \tag{3.1}$$

where

$$\boldsymbol{P}^{(1)} = \boldsymbol{\alpha} \cdot \boldsymbol{E} \tag{3.2}$$

$$\boldsymbol{P}^{(2)} = \tfrac{1}{2} \boldsymbol{\beta} : \boldsymbol{E}\boldsymbol{E} \tag{3.3}$$

and

$$\boldsymbol{P}^{(3)} = \tfrac{1}{6}\boldsymbol{\gamma} : \boldsymbol{EEE} \tag{3.4}$$

In these equations, $\boldsymbol{P}$ and $\boldsymbol{E}$ are vectors with units of C m and V m^{-1}, respectively, and $\boldsymbol{\alpha}$, $\boldsymbol{\beta}$, and $\boldsymbol{\gamma}$, which relate $\boldsymbol{P}$ and $\boldsymbol{E}$, are tensors since, in general, the direction of the induced dipole is not the same as that of the electric field producing it: i.e., in general, each component of $\boldsymbol{P}^{(1)}$ is a different linear combination of the components of $\boldsymbol{E}$, each component of $\boldsymbol{P}^{(2)}$ is a different linear combination of the components of $\boldsymbol{EE}$, and so on. In the classical picture, eqs. (3.2) to (3.4) are to be understood to imply that the tensors are *real and symmetric*. The full significance of these tensor relationships will emerge as we proceed.

$\boldsymbol{\alpha}$ is termed the polarizability tensor and is a second-rank tensor whose components have units of CV^{-1} m^2; $\boldsymbol{\beta}$ is the hyperpolarizability tensor and is a third-rank tensor whose components have units of CV^{-2} m^3; $\boldsymbol{\gamma}$ is termed the second hyperpolarizability tensor and is a fourth-rank tensor whose components have units of C V^{-3} m^4. The polarizabilities $\boldsymbol{\alpha}$, $\boldsymbol{\beta}$, and $\boldsymbol{\gamma}$ can be regarded as measures of the ease with which electrons can be displaced to produce an electric dipole under the action of an electric field. The tensors $\boldsymbol{\alpha}$, $\boldsymbol{\beta}$, and $\boldsymbol{\gamma}$ can be time-dependent, and hence the time dependence of the induced dipoles can be different from that of the electric field terms $\boldsymbol{E}$, $\boldsymbol{EE}$, etc.

Typical orders of magnitude for components of $\boldsymbol{\alpha}$, $\boldsymbol{\beta}$, and $\boldsymbol{\gamma}$ are as follows: α, 10^{-40} C V^{-1} m^2; β, 10^{-50} C V^{-2} m^3; and γ, 10^{-61} C V^{-3} m^4. For such values, if the contribution to the induced dipole from $\boldsymbol{P}^{(2)}$ is to reach one per cent of that from $\boldsymbol{P}^{(1)}$, an electric field intensity of the order of 10^9 V m^{-1} is required; and the electric field intensity must rise to about 10^{10} V m^{-1} for the contribution from $\boldsymbol{P}^{(3)}$ to be one per cent of that from the $\boldsymbol{P}^{(1)}$ term. It will be recalled from chapter 2 that, as far as electromagnetic radiation is concerned, such electric field intensities are reached only in focused giant-pulse laser beams. Since Rayleigh and Raman scattering are observed quite readily with very much lower electric field intensities, we may expect to be able to explain Rayleigh and Raman scattering in terms of the linear induced dipole $\boldsymbol{P}^{(1)}$ only, and this will indeed prove to be the case.

The first-order term $\boldsymbol{P}^{(1)}$ will be found to have the correct frequency dependence for Rayleigh and Raman scattering. The second-order term $\boldsymbol{P}^{(2)}$ does not include any induced dipoles whose frequencies correspond to normal Rayleigh and Raman scattering, but it can give rise to non-linear Rayleigh and Raman scattering, termed hyper Rayleigh and hyper Raman scattering, which have a different frequency dependence. Although the third-order term $\boldsymbol{P}^{(3)}$ does include induced dipoles with the correct frequency dependence for Rayleigh and Raman scattering, the contributions from such dipoles is quite negligible. $\boldsymbol{P}^{(3)}$ also gives rise to yet another kind of non-linear Rayleigh and Raman scattering, known as second hyper Rayleigh and Raman scattering, which have a still different frequency dependence.

Non-linear light scattering phenomena will be considered in the last chapter. Apart from this, the remainder of this book will be devoted to normal Rayleigh and Raman scattering arising from linear induced dipoles. While for the majority of systems the linear induced dipole is described by equations of the form of eq. (3.2) involving a *real symmetric* tensor, this does not suffice in the following cases: non-transparent systems where the frequency of the incident radiation is close to a vibronic absorption frequency; systems involving chiral molecules; and systems perturbed by other fields which may be external or intra-molecular in origin. Although such systems represent a minority of the systems studied by Raman spectroscopy, they are by no means unimportant. Their treatment involves the consideration of polarizability tensors which can be complex and non-symmetrical, and also of additional contributions to the scattered radiation from tensors involving magnetic dipole and electric quadrupole terms. Such matters will be deferred until the quantum mechanical treatment of light scattering is considered.

3.3 The real symmetric polarizability tensor

3.3.1 *Directional properties*

We must now consider the significance of eq. (3.2) in which $\boldsymbol{P}$ and $\boldsymbol{E}$ are vectors and $\boldsymbol{\alpha}$ is a tensor. This equation implies that the magnitudes of the components of $\boldsymbol{P}$ are related to the magnitudes of the components of $\boldsymbol{E}$ by the following three linear equations:

$$\left.\begin{aligned} P_x &= \{\alpha_{xx}E_x + \alpha_{xy}E_y + \alpha_{xz}E_z\} \\ P_y &= \{\alpha_{yx}E_x + \alpha_{yy}E_y + \alpha_{yz}E_z\} \\ P_z &= \{\alpha_{zx}E_x + \alpha_{zy}E_y + \alpha_{zz}E_z\} \end{aligned}\right\} \tag{3.5}$$

The nine coefficients α_{ij} are called the components of the polarizability tensor $\boldsymbol{\alpha}$. We may write these nine components in an ordered array as follows:

$$\begin{matrix} \alpha_{xx} & \alpha_{xy} & \alpha_{xz} \\ \alpha_{yx} & \alpha_{yy} & \alpha_{yz} \\ \alpha_{zx} & \alpha_{zy} & \alpha_{zz} \end{matrix} \tag{3.6}$$

and this constitutes a representation of the tensor $\boldsymbol{\alpha}$. Since we are considering a real symmetric polarizability tensor, $\alpha_{xy} = \alpha_{yx}$, $\alpha_{xz} = \alpha_{zx}$, and $\alpha_{yz} = \alpha_{zy}$, and the tensor has a maximum of six distinct components, all of which are real.

It must be emphasized that the values of the tensor components for a given system will depend on the choice of coordinate system. Of course, for a given system fixed in space, the application of an electric field along a particular direction produces an induced dipole whose magnitude and direction is unique, but its *components* will be determined by the choice of coordinate system. It can be appreciated that, as in so many problems of this nature, a particular choice of coordinate system can lead to an especially simple form for the tensor. We shall return to this point subsequently. For the moment, we shall proceed as generally as possible with the sole restriction that the tensor is symmetric.

The set of linear equations (3.5) can be written in matrix form:

$$\begin{bmatrix} P_x \\ P_y \\ P_z \end{bmatrix} = \begin{bmatrix} \alpha_{xx} & \alpha_{xy} & \alpha_{xz} \\ \alpha_{yx} & \alpha_{yy} & \alpha_{yz} \\ \alpha_{zx} & \alpha_{zy} & \alpha_{zz} \end{bmatrix} \begin{bmatrix} E_x \\ E_y \\ E_z \end{bmatrix} \tag{3.7}$$

Here the components of the vectors $\boldsymbol{E}$ and $\boldsymbol{P}$ form column matrices and the components of the polarizability tensor $\boldsymbol{\alpha}$ constitute the elements of a square matrix.

If we adopt the notation $[P]$ and $[E]$ for the column matrices and $[\alpha]$ for the square matrix, eq. (3.7) may be written in the form

$$[P]=[\alpha][E] \tag{3.8}$$

Since the polarizability tensor is symmetric, the matrix will also be symmetric. It is clear that the matrix $[\alpha]$ is intimately related to the tensor $\boldsymbol{\alpha}$. However, it must be emphasized that the matrix $[\alpha]$ is not itself a tensor. Although tensors may always be written in matrix form, if so desired, the elements of a matrix do not need to transform in the same manner as tensor components.

It can be readily seen from eq. (3.5) that the direction of the induced dipole will usually be different from that of the electric field, since, in general, each component of $\boldsymbol{P}$ is determined by contributions from all three components of $\boldsymbol{E}$. Even if there is only one component of $\boldsymbol{E}$, it is possible to produce three components of $\boldsymbol{P}$. The several diagrams in Fig. 3.1 show how the direction of the induced dipole $\boldsymbol{P}$ is

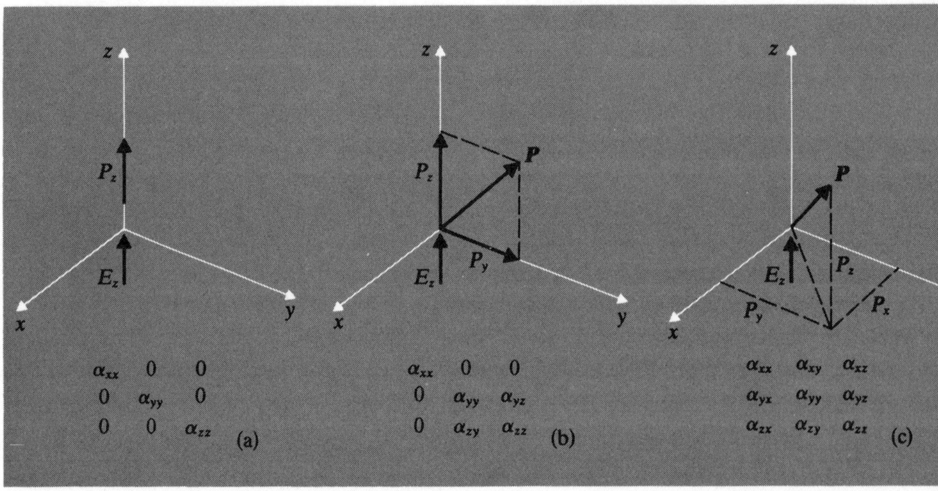

Fig. 3.1 Dependence of direction of induced dipole $\boldsymbol{P}$ on the form of the polarizability tensor: (a) diagonal tensor ($\alpha_{ij}=0$); (b) $\alpha_{xy}=\alpha_{xz}=0$; and (c) no non-zero components

determined by the form of the polarizability tensor for systems subject to an electric field along the z-axis. Only if α_{xz} and α_{yz} are zero will P_x and P_y be zero; and then, provided $\alpha_{zz}\neq 0$, the only non-zero component of the induced dipole will be P_z, which is coincident in direction with the incident field E_z. In all other cases, the direction of the induced dipole will not coincide with the direction of the electric field. It is clear that the pattern of entries in a tensor is of prime importance in determining the directional relationship between $\boldsymbol{P}$ and $\boldsymbol{E}$. This pattern is itself determined by the symmetry properties of the system involved.

3.3.2 *The polarizability ellipsoid*

The polarizability tensor may be given a graphical representation by use of the polarizability ellipsoid. Consider the equation

$$\alpha_{xx}x^2+\alpha_{yy}y^2+\alpha_{zz}z^2+2\alpha_{xy}xy+2\alpha_{yz}yz+2\alpha_{zx}zx=1 \tag{3.9}$$

This is an equation of an ellipsoid which has its centre at the origin of the coordinate system, but whose axes are not coincident with the axes of the coordinate system. Such a polarizability ellipsoid is depicted in Fig. 3.2(a). It can be shown that a line drawn from the origin to a point on the surface of the ellipsoid with coordinates x, y, z, has a length equal to $\alpha_E^{-1/2}$, where α_E is the polarizability in that direction. This

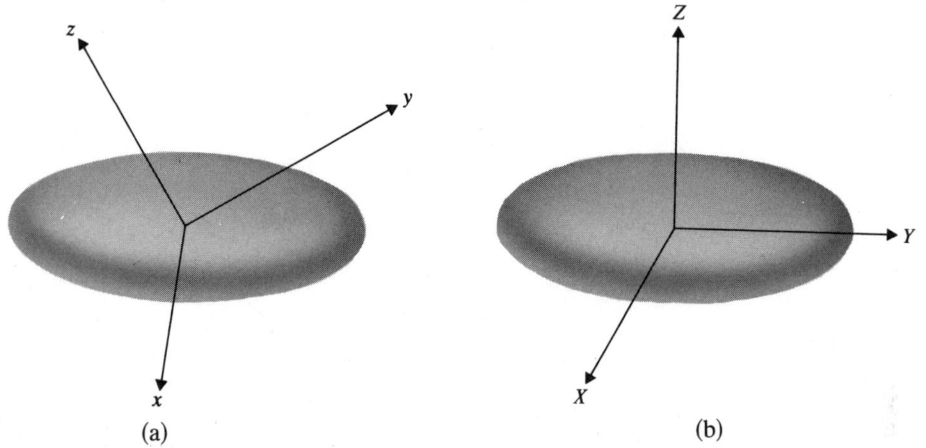

Fig. 3.2 The polarizability ellipsoid: (a) principal axes not coincident with system axes; and (b) principal axes coincident with system axes

means that, if an electric field $\boldsymbol{E}$ is applied along this direction, then the *component* P_E of $\boldsymbol{P}$ in this direction is given by

$$P_E = \alpha_E |\boldsymbol{E}| \tag{3.10}$$

Along the principal axes a, b, and c of the ellipsoid, and only along these directions, the *direction* of $\boldsymbol{P}$ is the same as the direction of $\boldsymbol{E}$. Thus, for these directions, we may write

$$\left.\begin{aligned} P_a &= \alpha_a E_a \\ P_b &= \alpha_b E_b \\ P_c &= \alpha_c E_c \end{aligned}\right\} \tag{3.11}$$

where α_a, α_b, α_c are the polarizabilities along the principal axes a, b, and c.

Consequently, if the coordinate system is rotated in such a way that the new Cartesian axes X, Y, Z coincide with the principal axes of the polarizability ellipsoid, then the polarizability tensor takes on a much simpler form in which $\alpha_{XY} = \alpha_{YZ} = \alpha_{ZX} = 0$. In this new coordinate system, the equations (3.11) become

$$\left.\begin{aligned} P_X &= \alpha_{XX} E_X \\ P_Y &= \alpha_{YY} E_Y \\ P_Z &= \alpha_{ZZ} E_Z \end{aligned}\right\} \tag{3.12}$$

and the array of polarizability tensor components is now

$$\begin{matrix} \alpha_{XX} & 0 & 0 \\ 0 & \alpha_{YY} & 0 \\ 0 & 0 & \alpha_{ZZ} \end{matrix} \tag{3.13}$$

where α_{XX}, α_{YY}, and α_{ZZ} are called the principal values of the polarizability.

The matrix $[\alpha]$ is now

$$\begin{bmatrix} \alpha_{XX} & 0 & 0 \\ 0 & \alpha_{YY} & 0 \\ 0 & 0 & \alpha_{ZZ} \end{bmatrix} \tag{3.14}$$

and so has diagonal form. The equation for the polarizability ellipsoid becomes

$$\alpha_{XX}X^2+\alpha_{YY}Y^2+\alpha_{ZZ}Z^2=1 \tag{3.15}$$

and its semi-minor axes have the values $\alpha_{XX}{}^{-1/2}$, $\alpha_{YY}{}^{-1/2}$, and $\alpha_{ZZ}{}^{-1/2}$ (note that this means that the largest axis of the ellipsoid corresponds to the smallest principal value of the polarizability, and vice versa). Two special cases are worth noting. If two of the principal values of the polarizability are equal, say, $\alpha_{XX}=\alpha_{YY}<\alpha_{ZZ}$, then the polarizability ellipsoid is a solid of revolution generated by rotating about the Z-axis the ellipse defined by

$$\alpha_{XX}X^2+\alpha_{ZZ}Z^2=1 \tag{3.16}$$

Such an ellipsoid is shown in Fig. 3.2(b). If the three principal values of the polarizability are equal, the polarizability ellipsoid becomes a sphere of radius $\alpha_{XX}{}^{-1/2}=\alpha_{YY}{}^{-1/2}=\alpha_{ZZ}{}^{-1/2}$.

3.3.3 *Rotation of axes*

The relationship between the tensor components in one Cartesian system x, y, z and another Cartesian system x', y', z' with the same origin but a different orientation in space may be shown to be

$$\alpha_{xy}=\sum_{x'y'}\alpha_{x'y'}\cos(xx')\cos(yy') \tag{3.17}$$

where the direction cosine, $\cos(xx')$, is the cosine of the angle between the x-axis and x'-axis, and so on, and the summation is over all possible pairs of Cartesian axes in the $x'y'z'$ system, viz., $x'x'$, $y'y'$, $z'z'$, $x'y'$, $y'x'$, $x'z'$, $z'x'$, $y'z'$, and $z'y'$.

Although the values of individual components of the polarizability tensor change on rotation of the axis system, certain combinations of them remain invariant. For the symmetric tensor (3.6) these are the mean polarizability a defined by

$$a=\tfrac{1}{3}(\alpha_{xx}+\alpha_{yy}+\alpha_{zz}) \tag{3.18}$$

and the anisotropy γ defined by

$$\gamma^2=\tfrac{1}{2}\{(\alpha_{xx}-\alpha_{yy})^2+(\alpha_{yy}-\alpha_{zz})^2+(\alpha_{zz}-\alpha_{xx})^2+6(\alpha_{xy}^2+\alpha_{yz}^2+\alpha_{zx}^2)\} \tag{3.19}$$

The invariance of these combinations under rotation may be established using eq. (3.17) and the properties of direction cosines.

In the special case where the coordinate axes coincide with the principal axes of the ellipsoid, we have

$$a = \tfrac{1}{3}(\alpha_{XX} + \alpha_{YY} + \alpha_{ZZ}) \tag{3.20}$$

$$\gamma^2 = \tfrac{1}{2}\{(\alpha_{XX} - \alpha_{YY})^2 + (\alpha_{YY} - \alpha_{ZZ})^2 + (\alpha_{ZZ} - \alpha_{XX})^2\} \tag{3.21}$$

If two of the principal values of the polarizability are equal, say $\alpha_{YY} = \alpha_{ZZ}$, then we have

$$|\gamma| = |\alpha_{XX} - \alpha_{YY}| \tag{3.22}$$

and if all three values are equal then $\gamma = 0$. These considerations enable us to see that γ, as its name implies, is indeed a measure of the anisotropy of the polarizability. The invariant a is clearly a measure of the 'average' polarizability.

3.3.4 *Space averages*

So far, we have considered the system subject to the electric field to be space-fixed. If the system is freely rotating, then it is necessary to average over all orientations taken up by the system with respect to the electric field. Since the power of the scattered radiation from a freely rotating induced dipole will be determined by the space-averaged value of the square of the induced dipole, we shall require the average of the squares of the tensor components over all orientations of the system-fixed axes relative to a set of space-fixed axes which we shall here denote as x, y, z. The necessary results may be obtained by starting with eq. (3.17) where x', y', z' are regarded as the system-fixed axes, squaring and averaging over the appropriate cosine terms. From such calculations we find that for the symmetric tensor (3.6):

$$\overline{\alpha_{xx}^2} = \overline{\alpha_{yy}^2} = \overline{\alpha_{zz}^2} = \frac{45a^2 + 4\gamma^2}{45} \tag{3.23}$$

$$\overline{\alpha_{yx}^2} = \overline{\alpha_{yz}^2} = \overline{\alpha_{zx}^2} = \frac{\gamma^2}{15} \tag{3.24}$$

$$\overline{\alpha_{xx}\alpha_{yy}} = \overline{\alpha_{yy}\alpha_{zz}} = \overline{\alpha_{zz}\alpha_{xx}} = \frac{45a^2 - 2\gamma^2}{45} \tag{3.25}$$

and all other terms, i.e., those which involve any subscript once only, like $\overline{\alpha_{xx}\alpha_{xy}}$, and so on, are zero. It should not be surprising that these space averages are expressed in terms of the invariants of the polarizability tensor.

3.3.5 *Tensor decomposition: the isotropic and symmetric tensors*

The symmetric polarizability tensor $\boldsymbol{\alpha}$ can always be written as the sum of two other symmetric tensors: an isotropic tensor $\boldsymbol{\alpha}_{\text{iso}}$ and an anisotropic tensor $\boldsymbol{\alpha}_{\text{aniso}}$. These tensors have the following forms:

$$\boldsymbol{\alpha}_{\text{iso}}: \quad \begin{matrix} a & 0 & 0 \\ 0 & a & 0 \\ 0 & 0 & a \end{matrix} \tag{3.26}$$

where a is given by eq. (3.18), and

$$\boldsymbol{a}_{\text{aniso}}: \quad \begin{matrix} \alpha_{xx}-a & \alpha_{xy} & \alpha_{xz} \\ \alpha_{yx} & \alpha_{yy}-a & \alpha_{yz} \\ \alpha_{zx} & \alpha_{zy} & \alpha_{zz}-a \end{matrix} \tag{3.27}$$

where $\alpha_{xy}=\alpha_{yx}$, $\alpha_{yz}=\alpha_{zy}$, and $\alpha_{zx}=\alpha_{xz}$. It is clear that

$$\boldsymbol{a}=\boldsymbol{a}_{\text{iso}}+\boldsymbol{a}_{\text{aniso}} \tag{3.28}$$

The traces of the products of these tensors with themselves are readily shown to be as follows:

$$\text{Trace}\,\{\boldsymbol{a}_{\text{iso}}:\boldsymbol{a}_{\text{iso}}\}=3a^2 \tag{3.29}$$

$$\text{Trace}\,\{\boldsymbol{a}_{\text{aniso}}:\boldsymbol{a}_{\text{aniso}}\}=\tfrac{2}{3}\gamma^2 \tag{3.30}$$

where γ^2 is given by eq. (3.19).

It will transpire that for a real symmetric polarizability tensor the intensity of the scattered radiation can always be expressed as the sum of two contributions, an isotropic part (involving only a^2), and an anisotropic part (involving only γ^2). This separation of the scattered radiation into isotropic and anisotropic parts is convenient since the two parts have different angular dependences. Rather unfortunately, Placzek[1] refers to the anisotropic part of the scattering as quadrupole scattering, because it has the same symmetry properties as scattering from a quadrupole moment. Placzek's choice of nomenclature is confusing and will not be adopted here.

3.4 Classical theory of Rayleigh and Raman scattering

The classical theory enables us to treat quantitatively a number of facets of light scattering, particularly the frequency dependence, the directional and polarization properties, and some aspects of the selection rules. We shall now consider each of these topics in turn in relation to vibrations in molecular systems.

3.4.1 *Frequency dependence*

We shall now consider the interaction of a molecular system with the harmonically oscillating electric field associated with electromagnetic radiation of circular frequency ω_0. Initially we shall consider the scattering system to be one molecule which is free to vibrate, but does not rotate; i.e., the molecule is space-fixed in the equilibrium configuration, but the nuclei may vibrate about their equilibrium positions.

It is to be expected that, in general, the polarizability will be a function of the nuclear coordinates. The variation of the polarizability with vibrations of the molecule can be expressed by expanding each component α_{ij} of the polarizability tensor in a Taylor series with respect to the normal coordinates of vibration, as follows:

$$\alpha_{ij}=(\alpha_{ij})_0+\sum_k\left(\frac{\partial\alpha_{ij}}{\partial Q_k}\right)_0 Q_k+\tfrac{1}{2}\sum_{k,l}\left(\frac{\partial^2\alpha_{ij}}{\partial Q_k\partial Q_l}\right)_0 Q_kQ_l+\ldots \tag{3.31}$$

where $(\alpha_{ij})_0$ is the value of α_{ij} at the equilibrium configuration, Q_k, Q_l, ... are normal coordinates of vibration associated with vibrational frequencies ω_k, ω_l, ..., and the summations are over all normal coordinates. The subscript '0' on the derivatives indicates that these are to be taken at the equilibrium configurations. We shall, for the time being, neglect the terms which involve powers of Q higher than the first. This approximation is sometimes referred to as the electrical harmonic approximation. We shall also initially fix our attention on one normal mode of vibration Q_k. We may then write eq. (3.31) in the special form:

$$(\alpha_{ij})_k = (\alpha_{ij})_0 + (\alpha'_{ij})_k Q_k \tag{3.32}$$

where

$$(\alpha'_{ij})_k = \left(\frac{\partial \alpha_{ij}}{\partial Q_k}\right)_0 \tag{3.33}$$

The $(\alpha'_{ij})_k$ are components of a new tensor which we shall call the derived polarizability tensor, since all its components are polarizability derivatives with respect to a given normal coordinate. The properties of the derived tensor components $(\alpha'_{ij})_k$ differ in some respects from those of the equilibrium tensor components $(\alpha_{ij})_0$, as we shall see subsequently. We shall represent the derived polarizability tensor associated with the normal mode k by $\boldsymbol{\alpha}'_k$. If we define $\boldsymbol{\alpha}_k$ and $\boldsymbol{\alpha}_0$ to be tensors with components $(\alpha_{ij})_k$ and $(\alpha_{ij})_0$ respectively, then since eq. (3.32) is valid for all tensor components:

$$\boldsymbol{\alpha}_k = \boldsymbol{\alpha}_0 + \boldsymbol{\alpha}'_k Q_k \tag{3.34}$$

where Q_k, the k-th normal coordinate, is a scalar quantity multiplying all components of $\boldsymbol{\alpha}'_k$. Assuming simple harmonic motion, i.e., mechanical harmonicity, the time dependence of Q_k is given by

$$Q_k = Q_{k_0} \cos(\omega_k t + \delta_k) \tag{3.35}$$

where Q_{k_0} is the normal coordinate amplitude and δ_k is a phase factor. Inserting eq. (3.35) into eq. (3.34) gives the time dependence of the polarizability tensor resulting from the k-th molecular vibration

$$\boldsymbol{\alpha}_k = \boldsymbol{\alpha}_0 + \boldsymbol{\alpha}'_k Q_{k_0} \cos(\omega_k t + \delta_k) \tag{3.36}$$

Now, under the influence of electromagnetic radiation of circular frequency ω_0, the linear induced electric dipole will be given by a form of eq. (3.2)

$$\boldsymbol{P}^{(1)} = \boldsymbol{\alpha}_k \cdot \boldsymbol{E} \tag{3.37}$$

where $\boldsymbol{P}^{(1)}$ is the linear induced dipole vector at some time t, $\boldsymbol{\alpha}_k$ is the time-dependent polarizability tensor defined in eq. (3.36), and $\boldsymbol{E}$ is the electric field vector at time t. Since the variation of the electric field intensity with time is given by

$$\boldsymbol{E} = \boldsymbol{E}_0 \cos \omega_0 t \tag{3.38}$$

we may rewrite eq. (3.37) as

$$\boldsymbol{P}^{(1)} = \boldsymbol{\alpha}_k \cdot \boldsymbol{E}_0 \cos \omega_0 t \tag{3.39}$$

where $\cos \omega_0 t$ is a scalar quantity which multiplies every term in the linear equations implicit in eq. (3.39). Introducing the relation (3.36) for $\boldsymbol{a}_k$ we obtain

$$\begin{aligned}\boldsymbol{P}^{(1)} &= \boldsymbol{a}_0 \cdot \boldsymbol{E}_0 \cos \omega_0 t \\ &\quad + \boldsymbol{a}'_k \cdot \boldsymbol{E}_0 Q_{k_0} \cos \omega_0 t \cos (\omega_k t + \delta_k)\end{aligned} \tag{3.40}$$

Using the trigonometric identity

$$\cos A \cos B = \tfrac{1}{2}\{\cos (A+B) + \cos (A-B)\} \tag{3.41}$$

the second term in eq. (3.40) may be reformulated and we may then write $\boldsymbol{P}^{(1)}$ in the form

$$\boldsymbol{P}^{(1)} = \boldsymbol{P}^{(1)}(\omega_0) + \boldsymbol{P}^{(1)}(\omega_0 - \omega_k) + \boldsymbol{P}^{(1)}(\omega_0 + \omega_k) \tag{3.42}$$

where

$$\boldsymbol{P}^{(1)}(\omega_0) = \boldsymbol{P}_0^{(1)}(\omega_0)\, \{\cos \omega_0 t\} \tag{3.43}$$

with

$$\boldsymbol{P}_0^{(1)}(\omega_0) = \boldsymbol{a}_0 \cdot \boldsymbol{E}_0 \tag{3.44}$$

$$\boldsymbol{P}^{(1)}(\omega_0 - \omega_k) = \boldsymbol{P}_0^{(1)}(\omega_0 - \omega_k)\, \{\cos (\omega_0 - \omega_k)t - \delta_k\} \tag{3.45}$$

with

$$\boldsymbol{P}_0^{(1)}(\omega_0 - \omega_k) = \tfrac{1}{2} Q_{k_0} \boldsymbol{a}'_k \cdot \boldsymbol{E}_0 \tag{3.46}$$

and

$$\boldsymbol{P}^{(1)}(\omega_0 + \omega_k) = \boldsymbol{P}_0^{(1)}(\omega_0 + \omega_k)\, \{\cos (\omega_0 + \omega_k)t + \delta_k\} \tag{3.47}$$

with

$$\boldsymbol{P}_0^{(1)}(\omega_0 + \omega_k) = \tfrac{1}{2} Q_k \boldsymbol{a}'_k \cdot \boldsymbol{E}_0 \tag{3.48}$$

In these equations, the cosine functions define the *frequencies* of the induced dipoles and are scalar quantities which multiply every term in the linear equations implicit in eqs. (3.43), (3.45) and (3.48); the *amplitudes* of the induced dipoles are given by the vectors $\boldsymbol{P}_0^{(1)}(\omega_0)$, $\boldsymbol{P}_0^{(1)}(\omega_0 - \omega_k)$, and $\boldsymbol{P}_0^{(1)}(\omega_0 + \omega_k)$. By redefining $\boldsymbol{E}_0$, we can omit the factor $\frac{1}{2}$ from eqs. (3.46) and (3.48), and we shall do this subsequently.

We see that the linear induced dipole $\boldsymbol{P}^{(1)}$ has three distinct frequency components: $\boldsymbol{P}^{(1)}(\omega_0)$, which gives rise to radiation at ω_0, and so accounts for Rayleigh scattering; $\boldsymbol{P}^{(1)}(\omega_0 - \omega_k)$, which gives rise to radiation at $\omega_0 - \omega_k$, and so accounts for Stokes Raman scattering; and $\boldsymbol{P}^{(1)}(\omega_0 + \omega_k)$, which gives rise to radiation at $\omega_0 + \omega_k$, and so accounts for anti-Stokes Raman scattering. It should be noted that whereas the induced dipole $\boldsymbol{P}^{(1)}(\omega_0)$ has the same phase as that of the incident field, the induced dipoles $\boldsymbol{P}^{(1)}(\omega_0 \pm \omega_k)$ are shifted in phase relative to the incident field by $\pm\delta_k$. The quantity $\pm\delta_k$ defines the phase of the normal vibration Q_k relative to the field and will be different for different molecules. The consequences of the different phase relationship for Rayleigh and Raman scattering will be discussed later.

The time dependence of the electric field of circular frequency ω_0 and the linear induced dipoles $\boldsymbol{P}^{(1)}$ which it can produce are compared in Fig. 3.3(a) and (b). In Fig. 3.3(a), the scattering molecule is not vibrating ($\omega_k = 0$) and the total induced dipole then has only one frequency component $\boldsymbol{P}^{(1)}(\omega_0)$. In Fig. 3.3(b), the scattering molecule is vibrating with circular frequency ω_k and the total induced dipole is

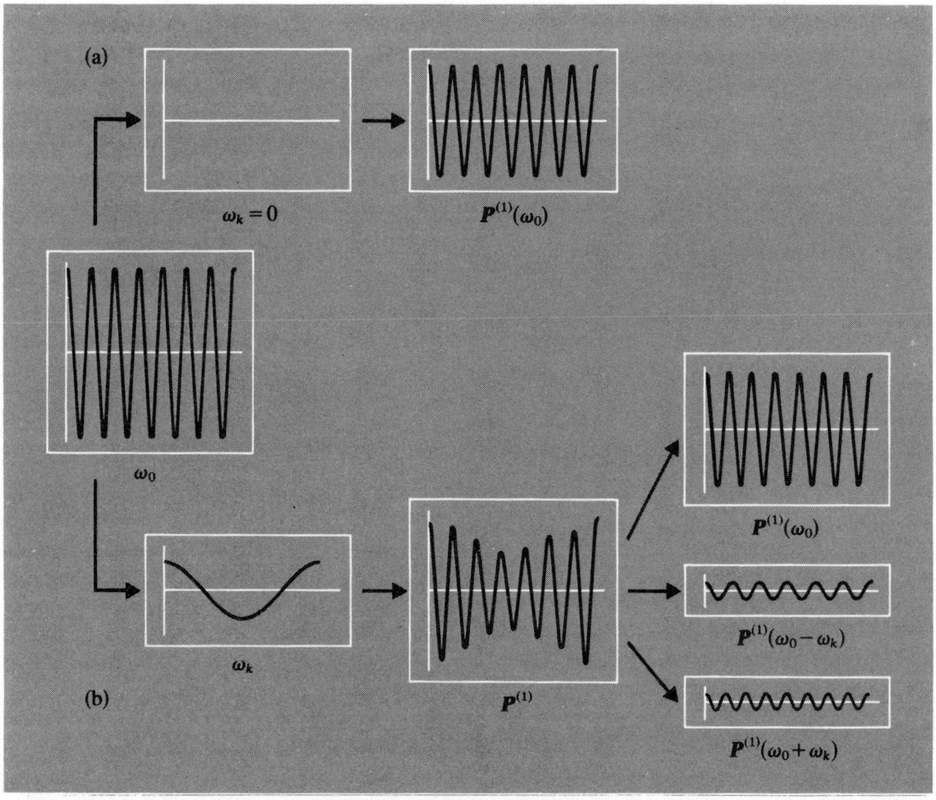

Fig. 3.3 Time dependence of the linear induced dipoles $P^{(1)}$ produced by electromagnetic radiation of frequency ω_0: (a) scattering molecule not vibrating $\omega_k = 0 : P^{(1)} = P^{(1)}(\omega_0)$; and (b) scattering molecule vibrating with frequency ω_k: $P^{(1)} = P^{(1)}(\omega_0) + P^{(1)}(\omega_0 - \omega_k) + P^{(1)}(\omega_0 + \omega_k)$

resolvable into three frequency components, $P^{(1)}(\omega_0)$, $P^{(1)}(\omega_0 - \omega_k)$, and $P^{(1)}(\omega_0 + \omega_k)$.

From these relatively simple mathematical manipulations, there emerges a useful qualitative picture of the mechanisms of Rayleigh and Raman scattering in terms of classical radiation theory. Rayleigh scattering arises from the dipole oscillating at ω_0 induced in the molecule by the electric field of the incident radiation, which itself oscillates at ω_0. Raman scattering arises from the dipoles oscillating at $\omega_0 \pm \omega_k$, which are produced when the dipole oscillating at ω_0 is modulated by the molecule oscillating at frequency ω_k. The necessary coupling between the nuclear motions and the electric field is provided by the electrons whose rearrangement with nuclear motion impose a harmonic variation on the polarizability. Alternatively, we may say that the frequencies observed in Raman scattering are beat frequencies between the radiation frequency ω_0 and the molecular frequency ω_k.

3.4.2 *Directional properties and polarization for scattering from a space-fixed molecule*

The linear equations implicit in eq. (3.44) which relates the amplitude of the induced dipole associated with Rayleigh scattering to the incident electric field are

$$\left.\begin{aligned} P^{(1)}_{xo}(\omega_0) &= \{(\alpha_{xx})_0 E_{xo} + (\alpha_{xy})_0 E_{yo} + (\alpha_{xz})_0 E_{zo}\} \\ P^{(1)}_{yo}(\omega_0) &= \{(\alpha_{yx})_0 E_{xo} + (\alpha_{yy})_0 E_{yo} + (\alpha_{yz})_0 E_{zo}\} \\ P^{(1)}_{zo}(\omega_0) &= \{(\alpha_{zx})_0 E_{xo} + (\alpha_{zy})_0 E_{yo} + (\alpha_{zz})_0 E_{zo}\} \end{aligned}\right\} \tag{3.49}$$

The $(\alpha_{ij})_0$ are components of the polarizability tensor $\boldsymbol{\alpha}_0$ for the molecule in its equilibrium position. This tensor has the properties discussed in section 3.3.

The linear equations implicit in eqs. (3.46) and (3.48) which relate the amplitude of the induced dipoles associated with Stokes and anti-Stokes Raman scattering to the incident electric field are

$$\left.\begin{aligned} P^{(1)}_{xo}(\omega_0 \pm \omega_k) &= \{(\alpha'_{xx})_k E_{xo} + (\alpha'_{xy})_k E_{yo} + (\alpha'_{xz})_k E_{zo}\} Q_{ko} \\ P^{(1)}_{yo}(\omega_0 \pm \omega_k) &= \{(\alpha'_{yx})_k E_{xo} + (\alpha'_{yy})_k E_{yo} + (\alpha'_{yz})_k E_{zo}\} Q_{ko} \\ P^{(1)}_{zo}(\omega_0 \pm \omega_k) &= \{(\alpha'_{zx})_k E_{xo} + (\alpha'_{zy})_k E_{yo} + (\alpha'_{zz})_k E_{zo}\} Q_{ko} \end{aligned}\right\} \tag{3.50}$$

In these equations, the $(\alpha'_{ij})_k$ are components of a tensor $\boldsymbol{\alpha}'_k$, the derived polarizability tensor for the k-th normal mode. This tensor differs in several respects from the equilibrium polarizability tensor $\boldsymbol{\alpha}_0$. Firstly, the principal axes of the derived polarizability ellipsoid need not coincide with those of the equilibrium polarizability ellipsoid. Thus, even if the axis system has been chosen so that eqs. (3.20) and (3.21) apply for the calculation of the invariants of the equilibrium polarizability tensor, it will usually be necessary to revert to the more general formulae given by eqs. (3.18) and (3.19) for the calculation of the invariants of the derived polarizability tensor. Secondly, the components of the derived polarizability tensor can be positive (meaning that the polarizability increases as the normal coordinate increases during a vibration) or negative (meaning that the polarizability decreases as the normal coordinate increases during a vibration), whereas the components of the polarizability tensor $\boldsymbol{\alpha}_0$ can only be positive. An important consequence of this is that $(a')_k$, the mean value of the derived polarizability tensor, can be zero, whereas $(a)_0$, the mean value of the equilibrium polarizability, can never be zero. The anisotropy $(\gamma')_k$ of the derived polarizability tensor $\boldsymbol{\alpha}'_k$ can be zero, as, of course, can that of the equilibrium polarizability tensor $(\gamma)_0$. The difference in the behaviour of $\boldsymbol{\alpha}_0$ and $\boldsymbol{\alpha}'_k$ leads to important differences in the directional properties and states of polarization of Rayleigh and Raman scattering, as we shall soon see. Finally, it should be noted that since components of $\boldsymbol{\alpha}'_k$ can be negative, the derived polarizability cannot always be represented by a real ellipsoid. Despite these differences, the space averages of the squares of the components of the derived polarizability tensor are given by formulae analogous to eqs. (3.23), (3.24), and (3.25), but involving the squares of the invariants of the derived polarizability tensor $(a')^2_k$ and $(\gamma')^2_k$.

We are now in a position to consider the directional and polarization properties of Rayleigh and Raman scattering. We shall treat, first, scattering from a single molecule with a fixed orientation in space. Although this represents a situation not found in practice, we shall consider it in some detail since many aspects of the treatment can be extended readily to other, more realistic cases. The additional considerations that arise for an assembly of molecules and random molecular orientations will then be examined.

We shall consider the single molecule to be at the origin of a space-fixed Cartesian coordinate system x, y, z. This coordinate system will be the reference frame for definition of the directions of the incident and scattered radiation. Since the molecule has a fixed orientation, this coordinate system can also serve as the reference frame for the components of the polarizability tensors of the molecule. The illumination will be along the positive z-axis and we consider, first, observation along the x-axis, so that xz is the scattering plane (see Fig. 3.4(a)).

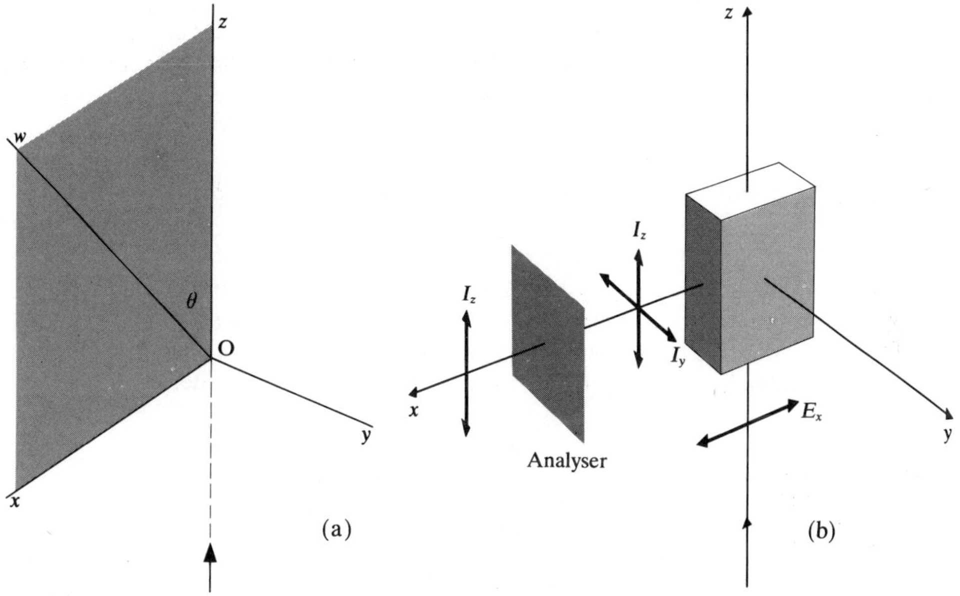

Fig. 3.4(a) Raman scattering from a gas or liquid : illumination and observation geometry

Fig. 3.4(b) Raman scattering from a crystal : geometry for $z(xz)x$ measurement

To calculate the power scattered into a solid angle $d\Omega$ about the x-direction, we need the y and z components of the amplitude of the linear induced dipole, with the correct frequency dependence for the scattered radiation in question. Thus, for Rayleigh scattering we need $P^{(1)}_{yo}(\omega_0)$ and $P^{(1)}_{zo}(\omega_0)$ and for Raman scattering $P^{(1)}_{yo}(\omega_0 \pm \omega_k)$ and $P^{(1)}_{zo}(\omega_0 \pm \omega_k)$.

For Rayleigh scattering produced by incident plane polarized radiation with $E_y = E_z = 0$ and $E_x \neq 0$ (i.e., polarized *parallel* to the scattering plane), the appropriate components of the amplitudes of the induced dipole are given by

$$P^{(1)}_{yo}(\omega_0) = (\alpha_{yx})_0 E_{xo} \tag{3.51}$$

$$P^{(1)}_{zo}(\omega_0) = (\alpha_{zx})_0 E_{xo} \tag{3.52}$$

The relationship between the amplitude of the induced dipole and the power $d\Phi$ scattered into a small solid angle $d\Omega$ about a direction perpendicular to the dipole axis is given by eq. (2.105). Alternatively, the corresponding radiant intensity I is given by eq. (2.110). We recall that the scattered power and radiant intensity are axially symmetrical about the dipole axis. Thus, we may write

$$
{}^{\|}I_{\perp}(\pi/2) = \frac{{}^{\|}\mathrm{d}\Phi_{\perp}(\pi/2)}{\mathrm{d}\Omega} = k'_{\omega}\omega_0^4(\alpha_{yx})_0^2 E_{x_0}^2 \\
= k'_{\tilde{\nu}}\tilde{\nu}_0^4(\alpha_{yx})_0^2 E_{x_0}^2 \tag{3.53}
$$

where the symbol ${}^{\|}I_{\perp}(\pi/2)$ is to be read sequentially as follows: incident radiation plane polarized *parallel* ($\|$) to the scatter plane, *radiant intensity* (I) of scattered radiation plane polarized *perpendicular* ($\perp$) to the scatter plane and propagating along a direction in the scatter plane making an *angle* ($\pi/2$) to the direction of propagation of the incident radiation. The symbol ${}^{\|}\mathrm{d}\Phi_{\perp}(\pi/2)$ is to be interpreted similarly for the scattered power $\mathrm{d}\Phi$. This information defines I and $\mathrm{d}\Phi$ uniquely. With the addition of the following symbols the notation can be extended to cover incident and scattered radiation with other polarization characteristics and other directions of observation: Ⓡ, right circularly polarized radiation; Ⓛ, left circularly polarized radiation; n, natural (unpolarized radiation); p_i polarization state (unspecified) of incident radiation; p_s polarization state (unspecified) of scattered radiation; and θ, angle made by the direction of observation (in the scatter plane) with the direction of propagation of the incident radiation. The absence of a subscript after I means that the total radiation is observed without any analyser. The absence of a superscript before I means that the incident radiation is in a general polarization state.

Similarly,

$$
{}^{\|}I_{\|}(\pi/2) = \frac{{}^{\|}\mathrm{d}\Phi_{\|}(\pi/2)}{\mathrm{d}\Omega} = k'_{\omega}\omega_0^4(\alpha_{zx})_0^2 E_{x_0}^2 \\
= k'_{\tilde{\nu}}\tilde{\nu}_0^4(\alpha_{zx})_0^2 E_{x_0}^2 \tag{3.54}
$$

where

$$
k'_{\omega} = \frac{1}{32\pi^2\varepsilon_0 c^3} \tag{3.55}
$$

and

$$
k'_{\tilde{\nu}} = \frac{\pi^2 c}{2\varepsilon_0} \tag{3.56}
$$

Using eq. (2.19), we may write eqs. (3.53) and (3.54) in terms of the irradiance $\mathcal{I}$ of the incident radiation as follows:

$$
{}^{\|}I_{\perp}(\pi/2) = \frac{{}^{\|}\mathrm{d}\Phi_{\perp}(\pi/2)}{\mathrm{d}\Omega} = k_{\tilde{\nu}}\tilde{\nu}_0^4(\alpha_{yx})_0^2\mathcal{I} \tag{3.57}
$$

$$
{}^{\|}I_{\|}(\pi/2) = \frac{{}^{\|}\mathrm{d}\Phi_{\|}(\pi/2)}{\mathrm{d}\Omega} = k_{\tilde{\nu}}\tilde{\nu}_0^4(\alpha_{zx})_0^2\mathcal{I} \tag{3.58}
$$

where

$$
k_{\tilde{\nu}} = \frac{\pi^2}{\varepsilon_0^2} \tag{3.59}
$$

Subsequently, we shall normally only quote formulae for scattered radiant intensities and powers in terms of $\mathscr{I}$ and $\tilde{\nu}_0$.

The total radiant intensity scattered along the x-axis is given by

$$\begin{aligned} {}^{\|}I(\pi/2) &= {}^{\|}I_{\perp}(\pi/2) + {}^{\|}I_{\|}(\pi/2) \\ &= k_{\tilde{\nu}}\tilde{\nu}_0^4\{(\alpha_{yx})_0^2 + (\alpha_{zx})_0^2\}\mathscr{I} \end{aligned} \tag{3.60}$$

Similarly, for Rayleigh scattering produced by plane polarized radiation with $E_x = E_z = 0$ and $E_y \neq 0$ (i.e., polarized *perpendicular* to the scattering plane), the appropriate components of the amplitudes of the induced dipoles are given by

$$P_{y\mathrm{o}}^{(1)}(\omega_0) = (\alpha_{yy})_0 E_{y\mathrm{o}} \tag{3.61}$$

$$P_{z\mathrm{o}}^{(1)}(\omega_0) = (\alpha_{zy})_0 E_{y\mathrm{o}} \tag{3.62}$$

Hence

$${}^{\perp}I_{\perp}(\pi/2) = \frac{{}^{\perp}\mathrm{d}\Phi_{\perp}(\pi/2)}{\mathrm{d}\Omega} = k_{\tilde{\nu}}\tilde{\nu}_0^4(\alpha_{yy})_0^2\mathscr{I} \tag{3.63}$$

$${}^{\perp}I_{\|}(\pi/2) = \frac{{}^{\perp}\mathrm{d}\Phi_{\|}(\pi/2)}{\mathrm{d}\Omega} = k_{\tilde{\nu}}\tilde{\nu}_0^4(\alpha_{zy})_0^2\mathscr{I} \tag{3.64}$$

and the total scattered radiant intensity is given by

$${}^{\perp}I(\pi/2) = \frac{{}^{\perp}\mathrm{d}\Phi(\pi/2)}{\mathrm{d}\Omega} = k_{\tilde{\nu}}\tilde{\nu}_0^4\{(\alpha_{yy})_0^2 + (\alpha_{zy})_0^2\}\mathscr{I} \tag{3.65}$$

In the case of Rayleigh scattering produced by incident natural radiation with $E_z = 0, E_x = E_y \neq 0$ but with no fixed phase relationship between them, we can obtain the scattered radiant intensities by *adding* the intensities from the appropriate dipoles induced by E_x and E_y. Thus, from eqs. (3.57), (3.63), and (2.71),

$${}^{n}I_{\perp}(\pi/2) = \frac{{}^{n}\mathrm{d}\Phi_{\perp}(\pi/2)}{\mathrm{d}\Omega} = k_{\tilde{\nu}}\tilde{\nu}_0^4\{(\alpha_{yx})_0^2 + (\alpha_{yy})_0^2\}\mathscr{I}/2 \tag{3.66}$$

and, from eqs. (3.58), (3.64), and (2.71),

$${}^{n}I_{\|}(\pi/2) = \frac{{}^{n}\mathrm{d}\Phi_{\|}(\pi/2)}{\mathrm{d}\Omega} = k_{\tilde{\nu}}\tilde{\nu}_0^4\{(\alpha_{zx})_0^2 + (\alpha_{zy})_0^2\}\mathscr{I}/2 \tag{3.67}$$

where $\mathscr{I}$ is now the irradiance of the natural radiation. The total scattered radiant intensity ${}^{n}I(\pi/2)$, is given by the sum of eqs. (3.66) and (3.67).

We consider, now, observation along a direction Ow lying in the xz-plane and making an angle θ with the z-axis, the direction of illumination (Fig. 3.4(a)). The induced dipoles P_z and P_x both make contributions to the radiant intensity of the scattered radiation observed along Ow and polarized with the electric vector in the xz-plane. P_z makes a contribution proportional to $P_z^2 \sin^2\theta$ and P_x makes a contribution proportional to $P_x^2 \cos^2\theta$. The scattered radiation observed along Ow and polarized with the electric vector perpendicular to the xz-plane is, however, proportional simply to P_y^2, that is ${}^{p_i}I_{\perp}(\theta) = {}^{p_i}I_{\perp}(\pi/2)$.

Thus, for incident radiation plane polarized with $E_y = E_z = 0$ and $E_x \neq 0$, we have

$$
\begin{aligned}
{}^{\|}I_{\|}(\theta) &= \frac{{}^{\|}\mathrm{d}\Phi_{\|}(\theta)}{\mathrm{d}\Omega} \\
&= k_{\tilde{\nu}}\tilde{\nu}_0^4\{(\alpha_{zx})_0^2 \sin^2\theta + (\alpha_{xx})_0^2 \cos^2\theta\}\mathscr{I}
\end{aligned}
\tag{3.68}
$$

with ${}^{\|}I_{\perp}(\theta)$ given by eq. (3.57); for incident radiation plane polarized with $E_x = E_z = 0$ and $E_y \neq 0$, we have

$$
\begin{aligned}
{}^{\perp}I_{\|}(\theta) &= \frac{{}^{\perp}\mathrm{d}\Phi_{\|}(\theta)}{\mathrm{d}\Omega} \\
&= k_{\tilde{\nu}}\tilde{\nu}_0^4\{(\alpha_{zy})_0^2 \sin^2\theta + (\alpha_{xy})_0^2 \cos^2\theta\}\mathscr{I}
\end{aligned}
\tag{3.69}
$$

with ${}^{\perp}I_{\perp}(\theta)$ given by eq. (3.63); and for natural incident radiation with $E_x = E_y \neq 0$, $E_z = 0$, and irradiance $\mathscr{I}$ we have

$$
\begin{aligned}
{}^{n}I_{\|}(\theta) &= \frac{{}^{n}\mathrm{d}\Phi_{\|}(\theta)}{\mathrm{d}\Omega} \\
&= k_{\tilde{\nu}}\tilde{\nu}_0^4\{[(\alpha_{zx})_0^2 + (\alpha_{zy})_0^2] \sin^2\theta + [(\alpha_{xx})_0^2 + (\alpha_{xy})_0^2] \cos^2\theta\}\mathscr{I}/2
\end{aligned}
\tag{3.70}
$$

with ${}^{n}I_{\perp}(\theta)$ given by eq. (3.66).

In the case of Raman scattering, the corresponding expressions for the powers and radiant intensities may be obtained by combining appropriate forms of eq. (3.50) with eqs. (2.105) and (2.110). For example, for incident radiation plane polarized with $E_x = E_z = 0$ and $E_y \neq 0$ and observation along x,

$$
{}^{\perp}I_{\perp}(\pi/2) = \frac{{}^{\perp}\mathrm{d}\Phi_{\perp}(\pi/2)}{\mathrm{d}\Omega} = k_{\tilde{\nu}}(\tilde{\nu}_0 \pm \tilde{\nu}_k)^4(\alpha'_{yy})_k^2 Q_{k\mathrm{o}}^2\mathscr{I} \tag{3.71*}
$$

$$
{}^{\perp}I_{\|}(\pi/2) = \frac{{}^{\perp}\mathrm{d}\Phi_{\|}(\pi/2)}{\mathrm{d}\Omega} = k_{\tilde{\nu}}(\tilde{\nu}_0 \pm \tilde{\nu}_k)^4(\alpha'_{zy})_k^2 Q_{k\mathrm{o}}^2\mathscr{I} \tag{3.72*}
$$

and

$$
{}^{\perp}I(\pi/2) = \frac{{}^{\perp}\mathrm{d}\Phi(\pi/2)}{\mathrm{d}\Omega} = k_{\tilde{\nu}}(\tilde{\nu}_0 \pm \tilde{\nu}_k)^4\{(\alpha'_{yy})_k^2 + (\alpha'_{zy})_k^2\} Q_{k\mathrm{o}}^2\mathscr{I} \tag{3.73*}
$$

where the + sign refers to anti-Stokes and the − sign to Stokes Raman scattering.

It is evident from a comparison of eqs. (3.71), (3.72), and (3.73) with eqs. (3.63), (3.64), and (3.65) that the formulae for Raman scattering can be obtained from the corresponding formulae for Rayleigh scattering by replacing $\tilde{\nu}_0^4$ by $(\tilde{\nu}_0 \pm \tilde{\nu}_k)^4$, by replacing the components of the equilibrium polarizability tensor by the corresponding components of the derived polarizability tensor, and by including the amplitude factor $Q_{k\mathrm{o}}^2$. It will therefore not be necessary to give the formulae for Raman scattering under other conditions of irradiation and observation.

For Rayleigh scattering where the incident vector is parallel to the scattering plane (xz) and observation is along x, we have, from eqs. (2.85), (3.57), and (3.58), *provided* that the observation of the scattered power for each polarized component is made

* Quantum mechanics will lead to a modification of these formulae, replacing $Q_{k\mathrm{o}}^2$ by another amplitude factor. Consequently, these formulae have been given a "grey" designation!

with the same incident irradiance and over the same collection angle, which must be small,

$$\rho_{\|}(\pi/2)=\frac{(\alpha_{yx})_0^2}{(\alpha_{zx})_0^2} \tag{3.74}$$

For the case where the incident vector is perpendicular to the scattering plane we have, from eqs. (2.86), (3.63), and (3.64),

$$\rho_{\perp}(\pi/2)=\frac{(\alpha_{zy})_0^2}{(\alpha_{yy})_0^2} \tag{3.75}$$

and for incident natural radiation we have, from eqs. (2.87), (3.66), and (3.67),

$$\rho_n(\pi/2)=\frac{\{(\alpha_{zx})_0^2+(\alpha_{zy})_0^2\}}{\{(\alpha_{yx})_0^2+(\alpha_{yy})_0^2\}} \tag{3.76}$$

Corresponding expressions may be obtained for Raman scattering by replacing the components of the equilibrium polarizability tensor by the corresponding components of the derived polarizability tensor.

3.4.3 *Directional properties and polarization for scattering from freely rotating molecules*

The above equations were obtained for a single scattering molecule, fixed in space. However, in any experimental study, scattering will be observed from a relatively large number of molecules and their orientations will not necessarily be fixed. We shall now consider how, in particular cases, these factors may be taken into account. We must note first, however, that there is an important difference between Rayleigh and Raman scattering from an assembly of molecules. This difference arises because, as we saw earlier, Rayleigh scattering is in phase with the incident radiation, whereas Raman scattering bears an arbitrary phase relation to the incident radiation. Thus, for Rayleigh scattering, interference between the scattering from different molecules is possible. In the idealized case of a perfect crystal at absolute zero, this interference could result in the Rayleigh scattering averaging to zero. In all other cases, the random motions of the scattering molecules prevent this. The relationship between the Rayleigh scattering from one molecule and the Rayleigh scattering from $\mathcal{N}$ molecules thus depends on the state of matter involved. For solids and liquids it can be rather complicated; only in the case of a gas at low pressure is the scattering from individual centres simply additive. The strong dependence of Rayleigh scattering on the state of matter is illustrated by the fact that Rayleigh scattering from water is only 200 times stronger than from air, even though water contains 1200 times more molecules in the same volume! For Raman scattering, the 'built-in' arbitrary phase factor ensures that the scattering from individual centres is always additive. We note also that whereas forward Raman scattering ($\theta=0$ in Fig. 3.4a) is meaningful, forward Rayleigh scattering is not, since interference occurs between the forward scattered waves and the transmitted incident waves which have the same frequency.

We now proceed to examine, in more detail, two particular cases of scattering from an assembly of scattering molecules: Rayleigh and Raman scattering from a gas at low pressure and Raman scattering from a crystal.

In a gas at low pressure the molecules are freely rotating. To find the power and radiant intensity scattered by $\mathcal{N}$ molecules, it is therefore necessary to calculate for one molecule the square of the amplitude of the induced moment for each orientation, average over all orientations, introduce the resultant average into eqs. (2.105) and (2.110), and multiply by $\mathcal{N}$. We shall illustrate the procedure by considering in detail the case of Raman scattering where illumination is along the z-axis with plane polarized radiation such that $E_x = E_z = 0, E_y \neq 0$ and observation is along the x-axis; the scatter plane is thus xz (Fig. 3.4(a)). By analogy with eqs. (3.71), (3.72), and (3.73), we have for the k-th normal mode

$$^{\perp}I_{\perp}(\pi/2) = \frac{^{\perp}\mathrm{d}\Phi_{\perp}(\pi/2)}{\mathrm{d}\Omega} = k_{\tilde{\nu}}\mathcal{N}(\tilde{\nu}_0 \pm \tilde{\nu}_k)^4 \overline{(\alpha'_{yy})_k^2} Q_{k_0}^2 \mathcal{I} \tag{3.77}$$

$$^{\perp}I_{\parallel}(\pi/2) = \frac{^{\perp}\mathrm{d}\Phi_{\parallel}(\pi/2)}{\mathrm{d}\Omega} = k_{\tilde{\nu}}\mathcal{N}(\tilde{\nu}_0 \pm \tilde{\nu}_k)^4 \overline{(\alpha'_{zy})_k^2} Q_{k_0}^2 \mathcal{I} \tag{3.78}$$

and

$$^{\perp}I(\pi/2) = \frac{^{\perp}\mathrm{d}\Phi(\pi/2)}{\mathrm{d}\Omega} = k_{\tilde{\nu}}\mathcal{N}(\tilde{\nu}_0 \pm \tilde{\nu}_k)^4 \{\overline{(\alpha'_{yy})_k^2} + \overline{(\alpha'_{zy})_k^2}\} Q_{k_0}^2 \mathcal{I} \tag{3.79}$$

The space averages of the squares of the derived polarizability components may be expressed in terms of the invariants $(a')_k$ and $(\gamma')_k$ of the derived polarizability tensor associated with the k-th normal mode, using appropriate forms of eqs. (3.23) and (3.24), as discussed on page 52. Thus, eqs. (3.77), (3.78), and (3.79) become

$$^{\perp}I_{\perp}(\pi/2) = \frac{^{\perp}\mathrm{d}\Phi_{\perp}(\pi/2)}{\mathrm{d}\Omega} = k_{\tilde{\nu}}\mathcal{N}(\tilde{\nu}_0 \pm \tilde{\nu}_k)^4 \left\{\frac{45(a')_k^2 + 4(\gamma')_k^2}{45}\right\} Q_{k_0}^2 \mathcal{I} \tag{3.80}$$

$$^{\perp}I_{\parallel}(\pi/2) = \frac{^{\perp}\mathrm{d}\Phi_{\parallel}(\pi/2)}{\mathrm{d}\Omega} = k_{\tilde{\nu}}\mathcal{N}(\tilde{\nu}_0 \pm \tilde{\nu}_k)^4 \left\{\frac{(\gamma')_k^2}{15}\right\} Q_{k_0}^2 \mathcal{I} \tag{3.81}$$

$$^{\perp}I(\pi/2) = \frac{^{\perp}\mathrm{d}\Phi(\pi/2)}{\mathrm{d}\Omega} = k_{\tilde{\nu}}\mathcal{N}(\tilde{\nu}_0 \pm \tilde{\nu}_k)^4 \left\{\frac{45(a')_k^2 + 7(\gamma')_k^2}{45}\right\} Q_{k_0}^2 \mathcal{I} \tag{3.82}$$

Thus,

$$\text{Raman: } \rho_{\perp}(\pi/2) = \frac{(3\gamma')_k^2}{45(a')_k^2 + 4(\gamma')_k^2} \tag{3.83}$$

and, similarly,

$$\text{Raman: } \rho_{\parallel}(\pi/2) = 1 \text{ (provided } (\gamma')_k \neq 0) \tag{3.84}$$

$$\text{Raman: } \rho_n(\pi/2) = \frac{6(\gamma')_k^2}{45(a')_k^2 + 7(\gamma')_k^2} \tag{3.85}$$

$$\text{Rayleigh: } \rho_{\parallel}(\pi/2) = 1 \text{ (provided } (\gamma)_0 \neq 0) \tag{3.86}$$

$$\text{Rayleigh: } \rho_{\perp}(\pi/2) = \frac{3(\gamma)_0^2}{45(a)_0^2 + 4(\gamma)_0^2} \tag{3.87}$$

$$\text{Rayleigh: } \rho_n(\pi/2) = \frac{6(\gamma)_0^2}{45(a)_0^2 + 7(\gamma)_0^2} \tag{3.88}$$

It can be seen that $\rho_{\perp}(\pi/2)$ can be determined directly by measuring the ratio of scattered intensities ${}^{\perp}I_{\|}(\pi/2)/{}^{\perp}I_{\perp}(\pi/2)$. However for experimental reasons a different set of measurements is often used. For the illumination–observation geometry we have been considering, the procedure is as follows: (a) illuminate along the z-axis using radiation plane polarized with $E_x \neq 0$, $E_y = E_z = 0$, and observe the total radiant intensity scattered along the x-axis, i.e., ${}^{\|}I_{\perp}(\pi/2) + {}^{\|}I_{\|}(\pi/2)$; (b) illuminate along the z-axis using radiation plane polarized with $E_y \neq 0$, $E_x = E_z = 0$, and again observe along the x-axis, i.e., ${}^{\perp}I_{\perp}(\pi/2) + {}^{\perp}I_{\|}(\pi/2)$. Then the ratio $\{{}^{\|}I_{\perp}(\pi/2) + {}^{\|}I_{\|}(\pi/2)\}/\{{}^{\perp}I_{\perp}(\pi/2) + {}^{\perp}I_{\|}(\pi/2)\}$ yields $\rho_n(\pi/2)$ (Raman) or $\rho_n(\pi/2)$ (Rayleigh) defined by eqs. (3.85) and (3.88).

We shall now consider the case of circularly polarized radiation incident along the positive z-axis and calculate ${}^{®}I_{®}(0)$, ${}^{©}I_{©}(0)$, ${}^{®}I_{©}(0)$, ${}^{©}I_{®}(0)$ and $\mathscr{P}(0)$ for Rayleigh and Raman scattering in the forward direction from an assembly of $\mathcal{N}$ freely rotating molecules.

As shown in chapter 2, we can represent incident *right* circularly polarized radiation of frequency ω_0 by the superposition of two waves

$$E_{x_0} \exp\{-i\omega_0 t\} \text{ and } -iE_{y_0} \exp\{-i\omega_0 t\} \tag{3.89}$$

and incident *left* circularly polarized radiation by the superposition of

$$E_{x_0} \exp\{-i\omega_0 t\} \text{ and } iE_{y_0} \exp\{-i\omega_0 t\} \tag{3.90}$$

where $E_{x_0} = E_{y_0}$.

In this notation, the Rayleigh scattering will have a time dependence given by $\exp\{-i\omega_0 t\}$ and the Raman scattering a time dependence given by $\exp\{-i(\omega_0 \pm \omega_k)t \pm i\delta_k\}$, where the + sign corresponds to anti-Stokes and the − sign to Stokes Raman scattering. The amplitudes of the circularly polarized components of the scattered radiation will be given by $P_{x_0} + iP_{y_0}$ for the left circularly polarized component and by $P_{x_0} - iP_{y_0}$ for the right circularly polarized component where the superscript (1) on P has been omitted for simplicity.

Now, for one molecule in a specific orientation it follows from eqs. (3.49) and (3.50) that

$$P_{x_0} = \alpha_{xx} E_{x_0} + \alpha_{xy} E_{y_0} \tag{3.91}$$

$$iP_{y_0} = i\alpha_{xy} E_{x_0} + i\alpha_{yy} E_{y_0} \tag{3.92}$$

where to maintain generality α_{xx} represents either $(\alpha_{xx})_0$ for Rayleigh scattering or $(\alpha'_{xx})_k Q_{k_0}$ for Raman scattering and so on. Hence

$$P_{x_0} + iP_{y_0} = (\alpha_{xx} + i\alpha_{xy})E_{x_0} + (\alpha_{xy} + i\alpha_{yy})E_{y_0} \tag{3.93}$$

$$P_{x_0} - iP_{y_0} = (\alpha_{xx} - i\alpha_{xy})E_{x_0} + (\alpha_{xy} - i\alpha_{yy})E_{y_0} \tag{3.94}$$

Using the identity

$$ax + by = \left(\frac{a - ib}{2}\right)(x + iy) + \left(\frac{a + ib}{2}\right)(x - iy) \tag{3.95}$$

eqs. (3.93) and (3.94) may be rearranged to give

$$
\begin{aligned}
(P_0)_{\text{Ⓛ}} &= P_{x_0} + iP_{y_0} \\
&= \tfrac{1}{2}(\alpha_{xx}+\alpha_{yy})(E_{x_0}+iE_{y_0}) + \tfrac{1}{2}(\alpha_{xx}-\alpha_{yy}+2i\alpha_{xy})(E_{x_0}-iE_{y_0})
\end{aligned} \tag{3.96}
$$

$$
\begin{aligned}
(P_0)_{\text{Ⓡ}} &= P_{x_0} - iP_{y_0} \\
&= \tfrac{1}{2}(\alpha_{xx}-\alpha_{yy}-2i\alpha_{xy})(E_{x_0}+iE_{y_0}) + \tfrac{1}{2}(\alpha_{xx}+\alpha_{yy})(E_{x_0}-iE_{y_0})
\end{aligned} \tag{3.97}
$$

Equation (3.96) gives the contribution to $(P_0)_{\text{Ⓛ}}$, the amplitude of the left circularly polarized scattered radiation from right circularly polarized incident radiation $(E_{x_0}-iE_{y_0})$ and from left circularly polarized incident radiation $(E_{x_0}+iE_{y_0})$. Similarly, eq. (3.97) gives the corresponding contributions to $(P_0)_{\text{Ⓡ}}$, the amplitude of the right circularly polarized scattered radiation.

To obtain the scattered radiant intensities we must now form the products $(P_0)_{\text{Ⓡ}}(P_0)^*_{\text{Ⓡ}}$ and $(P_0)_{\text{Ⓛ}}(P_0)^*_{\text{Ⓛ}}$, average over all orientations and multiply by $\mathcal{N}$. We should note particularly that because of the phase relationship between E_x and E_y (and hence P_x and P_y) it was necessary to combine the induced dipole moments before squaring and averaging. This procedure should be contrasted with that used when considering scattering produced by incident natural radiation (see page 55).

For the evaluation of $\overline{(P_0)_{\text{Ⓡ}}(P_0)^*_{\text{Ⓡ}}}$ and $\overline{(P_0)_{\text{Ⓛ}}(P_0)^*_{\text{Ⓛ}}}$ we note that

$$\overline{(\alpha_{xx}+\alpha_{yy})^2} = \overline{\alpha^2_{xx}} + \overline{\alpha^2_{yy}} + 2\overline{\alpha_{xx}\alpha_{yy}} \tag{3.98}$$

and, using eqs. (3.23) to (3.25), we find that

$$\overline{\tfrac{1}{4}(\alpha_{xx}+\alpha_{yy})^2} = \frac{45a^2+\gamma^2}{45} \tag{3.99}$$

Similarly, the other space average is given by

$$\overline{\tfrac{1}{4}(\alpha_{xx}-\alpha_{yy}+2i\alpha_{xy})(\alpha_{xx}-\alpha_{yy}-2i\alpha_{xy})} = \frac{2\gamma^2}{15} \tag{3.100}$$

Thus for Rayleigh scattering the radiant intensities are

$$^{\text{Ⓡ}}I_{\text{Ⓛ}}(0) = {}^{\text{Ⓛ}}I_{\text{Ⓡ}}(0) = k_{\tilde{\nu}}\mathcal{N}\tilde{\nu}_0^4 \frac{2(\gamma)_0^2}{15}\mathcal{I} \tag{3.101}$$

and

$$^{\text{Ⓡ}}I_{\text{Ⓡ}}(0) = {}^{\text{Ⓛ}}I_{\text{Ⓛ}}(0) = k_{\tilde{\nu}}\mathcal{N}\tilde{\nu}_0^4 \left\{\frac{45(a)_0^2+(\gamma)_0^2}{45}\right\}\mathcal{I} \tag{3.102}$$

$\mathcal{I} = \frac{1}{2}c\varepsilon_0\{E^2_{x_0}+E^2_{y_0}\}$, is the irradiance of the incident circularly polarized radiation. Similarly for Raman scattering

$$^{\text{Ⓡ}}I_{\text{Ⓛ}}(0) = {}^{\text{Ⓛ}}I_{\text{Ⓡ}}(0) = k_{\tilde{\nu}}\mathcal{N}(\tilde{\nu}_0 \pm \tilde{\nu}_k)^4 \frac{2(\gamma')_k^2}{15} Q^2_{k_0}\mathcal{I} \tag{3.103}$$

and

$$^{\text{Ⓡ}}I_{\text{Ⓡ}}(0) = {}^{\text{Ⓛ}}I_{\text{Ⓛ}}(0) = k_{\tilde{\nu}}\mathcal{N}(\tilde{\nu}_0 \pm \tilde{\nu}_k)^4 \left\{\frac{45(a')_k^2+(\gamma')_k^2}{45}\right\} Q^2_{k_0}\mathcal{I} \tag{3.104}$$

Thus, for Rayleigh scattering, we obtain using eq. (2.88)

$$\mathscr{P}(0)=\frac{6(\gamma)_0^2}{45(a)_0^2+(\gamma)_0^2}=\mathscr{P}(\pi)^{-1} \tag{3.105}$$

For incident circularly polarized radiation using eq. (2.89) the degree of circularity of Rayleigh scattering is given by

$$^{\circledR}\mathscr{C}(0)=\frac{45(a)_0^2-5(\gamma)_0^2}{45(a)_0^2+7(\gamma)_0^2}=-^{\circledL}\mathscr{C}(0) \tag{3.106}$$

For Raman scattering, it follows that

$$\mathscr{P}(0)=\frac{6(\gamma')_k^2}{45(a')_k^2+(\gamma')_k^2}=\mathscr{P}(\pi)^{-1} \tag{3.107}$$

and

$$^{\circledR}\mathscr{C}(0)=\frac{45(a')_k^2-5(\gamma')_k^2}{45(a')_k^2+7(\gamma')_k^2}=-^{\circledL}\mathscr{C}(0) \tag{3.108}$$

A little manipulation will show that for Rayleigh or Raman scattering

$$\mathscr{P}(0)=\frac{2\rho_\perp(\pi/2)}{1-\rho_\perp(\pi/2)} \tag{3.109}$$

or

$$\rho_\perp(\pi/2)=\frac{\mathscr{P}(0)}{2+\mathscr{P}(0)} \tag{3.110}$$

that

$$\mathscr{P}(0)=\frac{\rho_n(\pi/2)}{1-\rho_n(\pi/2)} \tag{3.111}$$

or

$$\rho_n(\pi/2)=\frac{\mathscr{P}(0)}{1+\mathscr{P}(0)} \tag{3.112}$$

and that

$$^{\circledR}\mathscr{C}(0)=\frac{1-3\rho_\perp(\pi/2)}{1+\rho_\perp(\pi/2)} \tag{3.113}$$

or

$$\rho_\perp(\pi/2)=\frac{1-^{\circledR}\mathscr{C}(0)}{3+^{\circledR}\mathscr{C}(0)} \tag{3.114}$$

3.4.4 *Symmetry and depolarization ratios, reversal coefficients, and degrees of circularity*

Comparison of the expressions for depolarization ratios, reversal coefficients, and degrees of circularity for Rayleigh and Raman scattering reveals some important differences. We recall that the mean value of the equilibrium polarizability tensor $(a)_0$ can never be zero, since all molecular systems are polarizable to some extent; the anisotropy $(\gamma)_0$ will, however, be zero for isotropic systems. Thus, for Rayleigh scattering

$$0 \leqslant \rho_\perp(\pi/2) < \tfrac{3}{4} \tag{3.115}$$

$$0 \leqslant \rho_n(\pi/2) < \tfrac{6}{7} \tag{3.116}$$

$$0 \leqslant \mathscr{P}(0) < 6 \tag{3.117}$$

and

$$-\tfrac{5}{7} < {}^{\circledR}\mathscr{C}(0) \leqslant 1 \tag{3.118}$$

By contrast for Raman scattering, either the mean value $(a')_k$ or the anisotropy $(\gamma')_k$ of the derived polarizability tensor can be zero. Thus, for Raman scattering

$$0 \leqslant \rho_\perp(\pi/2) \leqslant \tfrac{3}{4} \tag{3.119}$$

$$0 \leqslant \rho_n(\pi/2) \leqslant \tfrac{6}{7} \tag{3.120}$$

$$0 \leqslant \mathscr{P}(0) \leqslant 6 \tag{3.121}$$

and

$$-\tfrac{5}{7} \leqslant {}^{\circledR}\mathscr{C}(0) \leqslant 1 \tag{3.122}$$

Conventionally, a Raman line is said to be *depolarized* (*dp*) when $\rho_\perp(\pi/2) = \frac{3}{4}$ (or $\rho_n(\pi/2) = \frac{6}{7}$); *polarized* (*p*) when $0 < \rho_\perp(\pi/2) < \frac{3}{4}$ (or $0 < \rho_n(\pi/2) < \frac{6}{7}$); and *completely polarized* when $\rho_\perp(\pi/2) = 0$ (or $\rho_n(\pi/2) = 0$). The circular polarization is said to be *completely unreversed* when $\mathscr{P}(0) = 0$ (or ${}^{\circledR}\mathscr{C}(0) = 1$ or ${}^{\copyright}\mathscr{C}(0) = -1$) and *partly reversed* when $0 < \mathscr{P}(0) \leqslant 6$ (or $-\frac{5}{7} \leqslant {}^{\circledR}\mathscr{C}(0) < 1$ or $-1 < {}^{\copyright}\mathscr{C}(0) \leqslant \frac{5}{7}$).

The values of depolarization ratios, reversal coefficients, and degrees of circularity in Raman scattering are determined by the symmetry properties of the derived polarizability tensor, which itself reflects the symmetry of the mode of vibration. Thus, measurements of these quantities in Raman scattering can provide evidence for the assignment of vibrational modes to symmetry classes. However, the symmetry information obtained from such measurements from randomly oriented molecules is somewhat circumscribed. It will be seen later that it only enables us to distinguish between those vibrations which preserve the equilibrium molecular symmetry (e.g. $\rho_\perp(\pi/2) < \frac{3}{4}$, $\mathscr{P}(0) < 6$) and those which do not (e.g. $\rho_\perp(\pi/2) = \frac{3}{4}$, $\mathscr{P}(0) = 6$). There can be more than one class of vibrations in a molecule for which $\rho_\perp(\pi/2) = \frac{3}{4}$ (or $\mathscr{P}(0) = 6$), and these cannot be distinguished just from such measurements. However, although all such vibrations have in common the properties that $(a')_k = 0$ and $(\gamma')_k \neq 0$, vibrations with different symmetries will, in general, have different patterns of entries in their derived polarizability tensors, and so can be distinguished if this pattern can be determined.

It can be seen from eqs. (3.84) and (3.86) that since $\rho_\parallel(\pi/2)$ is always unity for Rayleigh and Raman scattering it yields no useful information. In the special case of

Rayleigh scattering from systems with $(\gamma)_0 = 0$ and Raman scattering from a vibration with $(\gamma')_k = 0$, formulae (3.84) and (3.86) do not apply, since there is then no scattered radiation along the x-axis.

Although it is possible to measure not only intensities but also ρ, $\mathscr{P}$, and $\mathscr{C}$ for scattered radiation, it must be emphasized that these are not all independent quantities. The scattering from an assembly of randomly oriented molecules is determined by $(a)_0^2$ and $(\gamma)_0^2$ for Rayleigh scattering and $(a')_k^2$ and $(\gamma')_k^2$ for Raman scattering; and these two quantities can be determined, for example, from one intensity measurement and one depolarization ratio measurement. In particular experimental situations it may be more convenient to combine an intensity measurement with a measurement of $\mathscr{P}$ or $\mathscr{C}$.

It is interesting to note that, in Raman scattering, if $\rho_\perp(\pi/2) > \frac{1}{3}$, ${}^{®}\mathscr{C}(0)$ will be negative and hence ${}^{®}I_{®}(0) - {}^{®}I_{Ⓛ}(0)$ will be negative. Thus if the Raman spectrum is recorded in the form of ${}^{®}I_{®}(0) - {}^{®}I_{Ⓛ}(0)$ an elegant distinction may be made between strongly polarized bands ($\rho_\perp(\pi/2) < \frac{1}{3}$) for which ${}^{®}I_{®}(0) - {}^{®}I_{Ⓛ}(0)$ is positive and weakly polarized and depolarized bands ($\rho_\perp(\pi/2) > \frac{1}{3}$) for which ${}^{®}I_{®}(0) - {}^{®}I_{Ⓛ}(0)$ is negative.

3.4.5 *Caveat*

It must also be remembered that the formulae we have obtained for Rayleigh and Raman scattering in an assembly of molecules have been derived on the assumption that all orientations are equally possible. Strictly, then, these equations apply only to gases at low pressure. They are usually applied to liquids, but in such systems the assumption of no preferred orientation may not always be correct.

3.4.6 *Raman scattering from a crystal*

We now turn to the case of Raman scattering from a crystal (see Fig. 3.4(b)). Here we have an assembly of essentially space-fixed molecules or ions. If the crystal is illuminated along the z-axis with plane polarized radiation for which $E_x \neq 0$, $E_y = E_z = 0$, and we observe the Stokes Raman scattering which is polarized with the electric vector perpendicular to the x-axis and in the zx-plane, and is delivered into a small solid angle $d\Omega$ about the x-axis, then the scattered power associated with the k-th vibration is proportional to $\{P_{z\text{o}}^{(1)}\}^2$, where the label $(\omega_0 - \omega_k)$ has been omitted for brevity. It follows from the equations of (3.50) that

$$\{P_{z\text{o}}^{(1)}\}^2 \propto (\alpha'_{zx})_k^2 E_{x\text{o}}^2 \tag{3.123}$$

Similarly, if the scattering polarized with the electric vector perpendicular to the x-axis and in the yx-plane is observed, then the scattered power is proportional to $\{P_{y\text{o}}^{(1)}\}^2$, and, from the equations of (3.50),

$$\{P_{y\text{o}}^{(1)}\}^2 \propto (\alpha'_{yx})_k^2 E_{x\text{o}}^2 \tag{3.124}$$

Thus, the relative magnitudes (but not the signs) of $(\alpha'_{zx})_k$ and $(\alpha'_{yx})_k$ can be determined. It can easily be seen that with illumination along the z-axis with plane polarized radiation for which $E_y \neq 0$, $E_x = E_z = 0$, $\{P_{z\text{o}}^{(1)}\}^2$ will be proportional to $(\alpha'_{zy})_k^2$ and $\{P_{y\text{o}}^{(2)}\}^2$ proportional to $(\alpha'_{yy})_k^2$. Provided the relative values of $E_{x\text{o}}^2$ and $E_{y\text{o}}^2$ are known, the magnitudes of the four tensor components, $(\alpha'_{zx})_k$, $(\alpha'_{yx})_k$, $(\alpha'_{zy})_k$, $(\alpha'_{yy})_k$, can be put on the same scale and compared. To obtain the relative magnitudes of the remaining two distinct tensor components, $(\alpha'_{xx})_k$ and $(\alpha'_{zz})_k$, it is necessary to make observations of the scattering when the crystal is illuminated along other axes. It is clear from the equations of (3.50) that illumination geometries that permit illumination with $E_x \neq 0$ and observation of $\{P^{(1)}_{x0}\}^2$, and illumination with $E_z \neq 0$ and observation of $\{P^{(1)}_{z0}\}^2$ will give the required information.

Porto[2] has proposed a notation for describing polarization data from single crystals. This involves four symbols (appropriately chosen from x, y, z or other axis labels) which define the propagation direction of the incident radiation, the direction of the electric vector of the incident radiation, the direction of the electric vector of the scattered radiation being examined, and the direction of propagation of the scattered radiation; the second and third symbols are placed in brackets. Hence, the observations relevant to eq. (3.123) can be expressed in the Porto notation as $z(xz)x$, and the observations relevant to eq. (3.124) as $z(xy)x$. It can be seen that the two symbols in brackets define the component of the scattering tensor being measured.

It must be emphasized that in such measurements it is the derived polarizability tensor components of species in the crystal relative to axes fixed in relation to the crystal that are being determined. These space-fixed axes will not generally be coincident with the molecular axes and an appropriate transformation is necessary to obtain the molecular tensor components. It is usually satisfactory to assume that these tensor components are not seriously different from those of the free molecule.

The symmetry of a scattering molecule and its vibrations determines which components of the derived polarizability tensor will be zero, and how many distinct non-zero components the tensor will contain. Generally, the pattern of entries in the derived polarizability tensor is characteristic of a particular symmetry class. Thus, the experimental determination of the magnitudes of tensor components from the directional properties of Raman scattering from oriented single crystals can afford a valuable method of determining the symmetry properties of molecules and their normal modes of vibration. This method is much more informative than measurement of depolarization ratios. It is illustrated in more detail in Appendix III with a worked example.

3.4.7 *The relation of classical theory to quantum theory*

We shall see in chapter 4 that on the basis of a partial quantum theory treatment some modifications to the expressions for the scattered power $d\Phi$ and radiant intensity I arise for Raman scattering, but not for Rayleigh scattering. These modifications do not alter the general form of the relationships so far obtained; we have merely to replace the classical amplitude of vibration Q_{k_0} in the expansion of the polarizability by a quantum mechanical amplitude factor (see eqs. (3.71) to (3.73), (3.77) to (3.79), (3.80) to (3.82) and (3.103) and (3.104) and the footnote on page 56). However, these factors cancel in the ratios of scattered powers and radiant intensities that lead to depolarization ratios, reversal coefficients, and degrees of circularity, and thus the classical theory gives the same results for polarization properties as the partial quantum mechanical treatment of chapter 4.

In chapter 5 a more complete quantum mechanical treatment is given, and this leads to new additional contributions to Rayleigh and Raman scattering. However, these only arise when the system absorbs ω_0 or is subjected to magnetic perturbation, and they can be grafted on to the classical–quantum formulae since the effects are simply additive. Thus, our excursion into classical theory is of lasting value, as we indicated at the beginning of this chapter.

3.5 Selection rules for fundamental vibrations

3.5.1 *General considerations*

We noted in chapter 1 that although a simple picture based on energy transfer between the incident radiation and the molecule could account for the pattern of frequencies observed in Raman scattering, it gave no indication of which transitions would participate in Raman scattering. The classical radiation treatment we have just given provides a basis for an answer to this important question. It is immediately clear from the equations of (3.50) that the induced dipoles responsible for Raman scattering associated with the molecular frequency ω_k will be zero unless at least one of the components of the derived polarizability tensor $(\alpha'_{ij})_k$ is non-zero. We recall that

$$(\alpha'_{ij})_k = \left(\frac{\partial \alpha_{ij}}{\partial Q_k}\right)_0$$

is the derivative of the ij component of the polarizability tensor with respect to the normal coordinate of vibration Q_k, taken at the equilibrium position. Thus, the condition for Raman activity is that, for at least one component of the polarizability tensor, a plot of that component against the normal coordinate must have a non-zero gradient at the equilibrium position. The corresponding condition for infrared activity, according to classical theory, is that at least one of the dipole moment component derivatives with respect to the normal coordinate Q_k, taken at the equilibrium position, should be non-zero. This means that, for at least one of the dipole moment components, a plot of that component against the normal coordinate must have a non-zero gradient at the equilibrium position.

In principle we can now determine the infrared and Raman activity of vibrations in particular molecules. However, the selection rules in their classical form, although appearing deceptively simple in concept, become progressively more difficult to apply as the complexity of the molecule increases. This is particularly true for Raman scattering because of the tensor nature of the polarizability. The ensuing discussion of some specific cases will illustrate this.

3.5.2 *Diatomic molecules*

We shall consider first a homonuclear diatomic molecule A_2 which has just one mode of vibration. Such a molecule has no permanent dipole moment in the equilibrium position because of the symmetry of the electron distribution. This symmetry does not change with small changes in the internuclear separation, and so the dipole remains zero during a vibration and hence the derivative is zero. The vibration is therefore infrared inactive. We turn now to the question of Raman activity. Clearly the molecule has a non-zero polarizability, and we may represent this by a polarizability ellipsoid so oriented at equilibrium that it has one principal axis along the bond direction and its other two principal axes at right angles to the bond direction. Such an ellipsoid is defined by a maximum of three polarizability components. However, since we are considering a σ bond, the polarizability will be the same in all directions at right angles to the bond, and the polarizability tensor is then defined by just two components which we may designate as $\alpha_{\parallel}$, the polarizability along the bond, and $\alpha_{\perp}$, the polarizability at right angles to the bond. For a given internuclear separation the mean polarizability a is then given by $a = \frac{1}{3}(\alpha_{\parallel} + 2\alpha_{\perp})$ and the anisotropy γ by

$\gamma = \alpha_{\parallel} - \alpha_{\perp}$. We have now to ask how these polarizability components change during the vibration of the A_2 molecule. It is not as easy as many texts would imply to deduce conclusively the properties of $(\partial\alpha_{\parallel}/\partial Q)_0$ and $(\partial\alpha_{\perp}/\partial Q)_0$ from *a priori* considerations of how electron distributions are affected by a change of internuclear distance. Let us consider the specific case of the hydrogen molecule H_2. We can see that as the internuclear separation tends to zero the polarizability tends towards that of a helium atom, and as the internuclear separation tends to infinity the polarizability tends to that of two hydrogen atoms. Since we know that the polarizability of two hydrogen atoms is greater than that of a helium atom, we can infer that, in general terms, the polarizability does change with internuclear separation. However, these limiting cases do not define the form of the variation of the polarizability with internuclear separation, and in principle the derivative at the equilibrium internuclear separation could be positive, negative, or zero (see Fig. 3.5(a)). Also, since atoms have isotropic

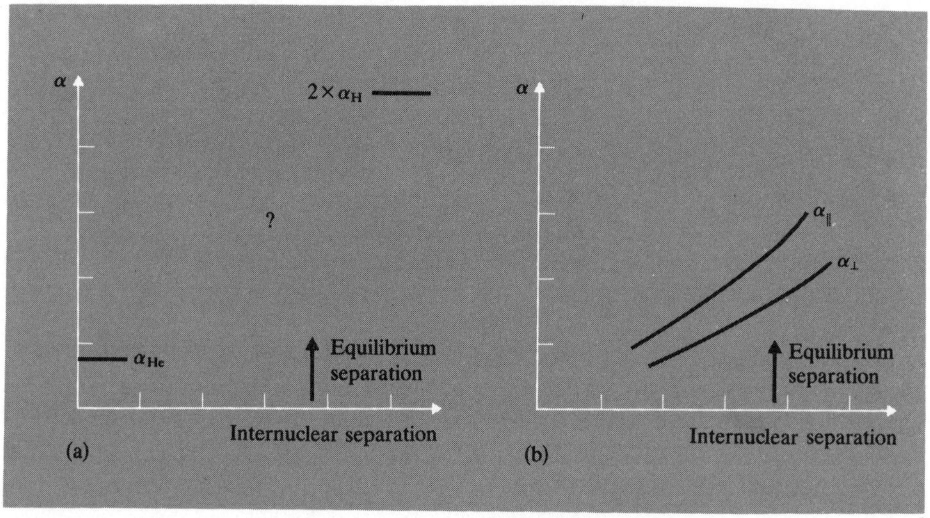

Fig. 3.5 Variation of polarizability with internuclear distance in the H_2 molecule (schematic): (a) limiting values for zero and infinite internuclear distances; and (b) $\alpha_{\parallel}$ and $\alpha_{\perp}$ around the equilibrium separation (after Bell and Long[3])

polarizabilities with $a \neq 0$ and $\gamma = 0$, whereas near the equilibrium position in H_2 we have $a \neq 0$ and $\gamma \neq 0$, the forms of the variation of a and γ with internuclear separation in H_2 must be different. For the fullest information, we can only appeal to quantum mechanical calculations to ascertain the variation of the polarizability components with internuclear separation in the neighbourhood of the equilibrium position. Such calculations[3] (see Fig. 3.5(b)) show that both $(\partial\alpha_{\parallel}/\partial Q)_0$ and $(\partial\alpha_{\perp}/\partial Q)_0$ are non-zero and positive. Further, the calculations show that $(\partial\alpha_{\parallel}/\partial Q)_0 > (\partial\alpha_{\perp}/\partial Q)_0$ and so $(\partial\gamma/\partial Q)_0$ is also non-zero and positive. Thus, in H_2 the vibration will be Raman active. (Note that only in those special cases where the general definition of γ^2 in eq. (3.21) can be reduced to a linear form, as in eq. (3.22), can we write $\gamma' = (\partial\gamma/\partial Q)_0 = (\partial\alpha_{\parallel}/\partial Q)_0 - (\partial\alpha_{\perp}/\partial Q)_0$.)

These results for H_2 may be used to guide us towards some generalizations for other homonuclear diatomic molecules. In such molecules, we may also reasonably expect $(\partial\alpha_{\parallel}/\partial Q)_0$ and $(\partial\alpha_{\perp}/\partial Q)_0$ to be non-zero and of different magnitudes and hence $(\partial a/\partial Q)_0$ and $(\partial \gamma/\partial Q)_0$ to be non-zero, but the signs and relative magnitudes need not necessarily follow the pattern in H_2. Thus, the vibrations of A_2 diatomic molecules will be Raman active.

We next consider the case of a heteronuclear diatomic molecule AB which also has just one mode of vibration. The arguments given above for polarizability changes in A_2 molecules can be expected to apply to AB molecules, and thus the vibration will be Raman active. The molecule AB will necessarily have a permanent dipole moment because there will be an asymmetry in its electron distribution. Since the dipole moment must be zero for both infinitely large and zero internuclear separations, the normal form of the variation of the dipole moment component along the bond direction with internuclear distance will be as shown in Fig. 3.6. The components of the dipole at right angles to the bond direction are, of course, always zero.

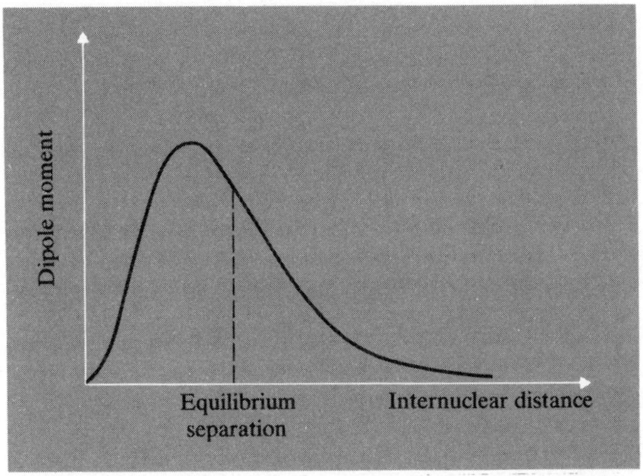

Fig. 3.6 Variation of dipole moment with internuclear distance in a diatomic molecule AB (here the dipole moment derivative at the equilibrium separation is negative)

For infrared activity, the maximum dipole moment must occur at an internuclear distance different from the equilibrium distance so that the derivative at the equilibrium position is non-zero. This is the case for all heteronuclear diatomic molecules and thus the vibration in AB molecules will be infrared active. The form of the dipole moment/internuclear distance curve will, however, vary from one molecule to another, and thus the magnitude and sign of the derivative will also vary considerably. The A_2 and AB cases are compared in Fig. 3.7(a) and (b).

3.5.3 *Polyatomic molecules*

In polyatomic molecules, if we regard the total dipole moment as made up of contributions from individual bond dipoles (at least to a first approximation), then we may regard each heteronuclear bond as having a non-zero bond dipole derivative at the equilibrium position and combine such derivatives vectorially to determine infrared activities in particular modes of vibration, taking into account the relative phases of the motions in each bond.

Similarly, we may regard the total molecular polarizability as made up of contributions from individual bond polarizabilities (at least to a first approximation), and

	(a)	(b)
Molecule	A——A	A——B
Mode of vibration	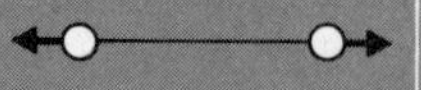	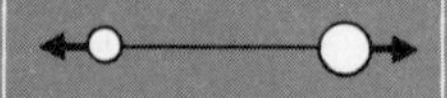
Variation of polarizability with normal coordinate (schematic)	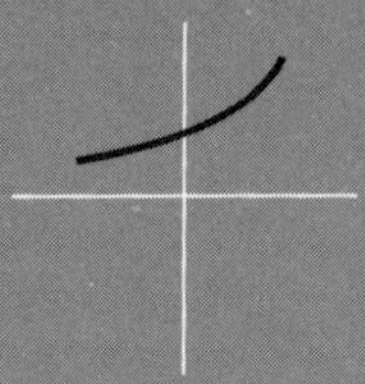	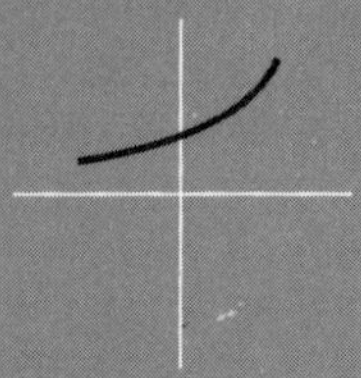
Polarizability derivative	$\neq 0$	$\neq 0$
Raman activity	Yes	Yes
Variation of dipole moment with normal coordinate (schematic)	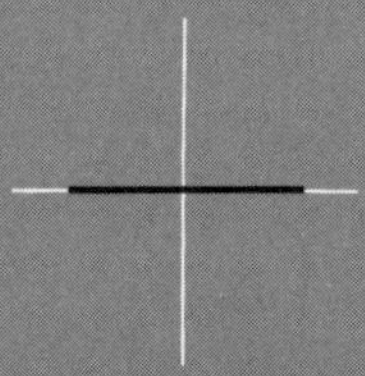	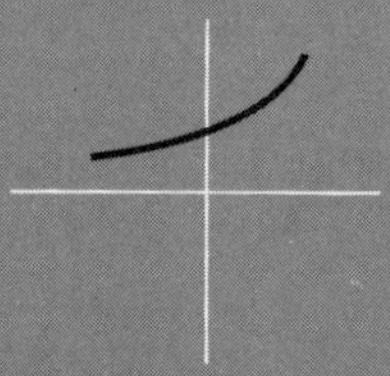
Dipole moment derivative	$= 0$	$\neq 0$
Infrared activity	No	Yes

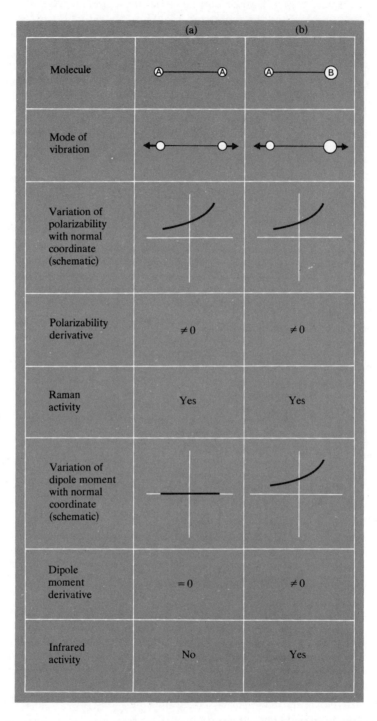

Fig. 3.7 Comparison of polarizability and dipole moment variations in the neighbourhood of the equilibrium position and vibrational Raman and infrared activities for (a) an A_2 and (b) an AB molecule

assume that these bond polarizabilities show the same qualitative behaviour as in H_2 (at least for σ bonds). However, in polyatomic molecules the forms of the vibrations are relatively complicated and involve, for example, the stretching and compression of more than one bond. The Raman activity of such a vibration depends on the components of the overall derived polarizability tensor which is formed by tensor addition of individual derived bond polarizability tensors, taking into account the relative phases of the motions in each bond. It is profitable to consider only very simple polyatomic molecules as examples.

We examine first a linear symmetric molecule ABA. Such a molecule has four modes of vibration: a symmetric stretching mode Q_1, an antisymmetric stretching mode Q_2, and two bending modes Q_{3a} and Q_{3b} which form a degenerate pair and have the same frequency of vibration. These vibrations are illustrated in Fig. 3.8; the z axis is the bond axis. This ABA molecule has no permanent dipole because of the symmetry of the electron distribution. For the symmetric stretching mode Q_1, in which both $A—B$ bonds are simultaneously stretched in one phase and simultaneously compressed in the other phase, the non-zero bond dipole derivatives in the two $A—B$ bonds always act in opposition and cancel each other exactly; the vibration is therefore infrared inactive. On the other hand, for this vibration the non-zero bond polarizability derivatives in the two $A—B$ bonds are additive and the vibration is Raman active. Specifically, since the axes of the molecular polarizability ellipsoid do not change in this vibration, we have for the molecular polarizability derivatives $(\partial\alpha_{xx}/\partial Q_1)_0=(\partial\alpha_{yy}/\partial Q_1)_0$ and $(\partial\alpha_{zz}/\partial Q_1)_0$ non-zero; and $(\partial\alpha_{xy}/\partial Q_1)_0=(\partial\alpha_{yz}/\partial Q_1)_0=(\partial\alpha_{zx}/\partial Q_1)_0=0$. Thus, $(\partial a/\partial Q_1)_0$ and $(\partial\gamma/\partial Q_1)_0$ are both non-zero.

The situation is quite different for the antisymmetric stretching mode Q_2, in which one AB bond is stretched (or compressed) as the other AB bond is compressed (or stretched). The non-zero bond dipole derivatives are additive and thus this vibration is infrared active. On the other hand, the bond polarizability derivatives cancel each other and we have for the molecular polarizability derivatives $(\partial\alpha_{xx}/\partial Q_2)_0=(\partial\alpha_{yy}/\partial Q_2)_0=(\partial\alpha_{zz}/\partial Q_2)_0=(\partial\alpha_{xy}/\partial Q_2)_0=(\partial\alpha_{yz}/\partial Q_2)_0=(\partial\alpha_{zx}/\partial Q_2)_0=0$, and so $(\partial a/\partial Q_2)_0$ and $(\partial\gamma/\partial Q_2)_0$ are both zero. Thus, the Q_2 mode is Raman inactive.

For the degenerate bending modes Q_{3a} and Q_{3b}, as far as infrared activity is concerned, it is readily seen that the dipole moment derivatives at right angles to the molecular axis are non-zero and thus these vibrations are infrared active. However, all six molecular polarizability derivatives are zero for both modes. Thus, $(\partial a/\partial Q_{3a})_0=(\partial a/\partial Q_{3b})_0=(\partial\gamma/\partial Q_{3a})_0=(\partial\gamma/\partial Q_{3b})_0=0$, and both modes are Raman inactive. The general forms of the dipole moment and polarizability changes with the various normal coordinates are included in Fig. 3.8.

Finally, we consider the case of a non-linear ABA molecule. Such a molecule has three modes of vibration: Q_1, a symmetric stretching mode; Q_2, a symmetric bending mode; and Q_3, an antisymmetric stretching mode. The forms of these modes and the axis system are given in Fig. 3.9. We may perhaps leave it to the reader to endeavour to convince himself that: (a) Q_1 is infrared active and also Raman active with $(\partial\alpha_{xx}/\partial Q_1)_0$, $(\partial\alpha_{yy}/\partial Q_1)_0$, and $(\partial\alpha_{zz}/\partial Q_1)_0$ non-zero and $(\partial\alpha_{xy}/\partial Q_1)_0=(\partial\alpha_{yz}/\partial Q_1)_0=(\partial\alpha_{zx}/\partial Q_1)_0=0$, and hence both $(\partial a/\partial Q_1)_0$ and $(\partial\gamma/\partial Q_1)_0$ non-zero; (b) Q_2 is also both infrared active and Raman active with the same dipole and polarizability components non-zero. The case of Q_3 calls for a little more explanation. It is quite easy to see that Q_3 is infrared active; Q_3 is also Raman active. This

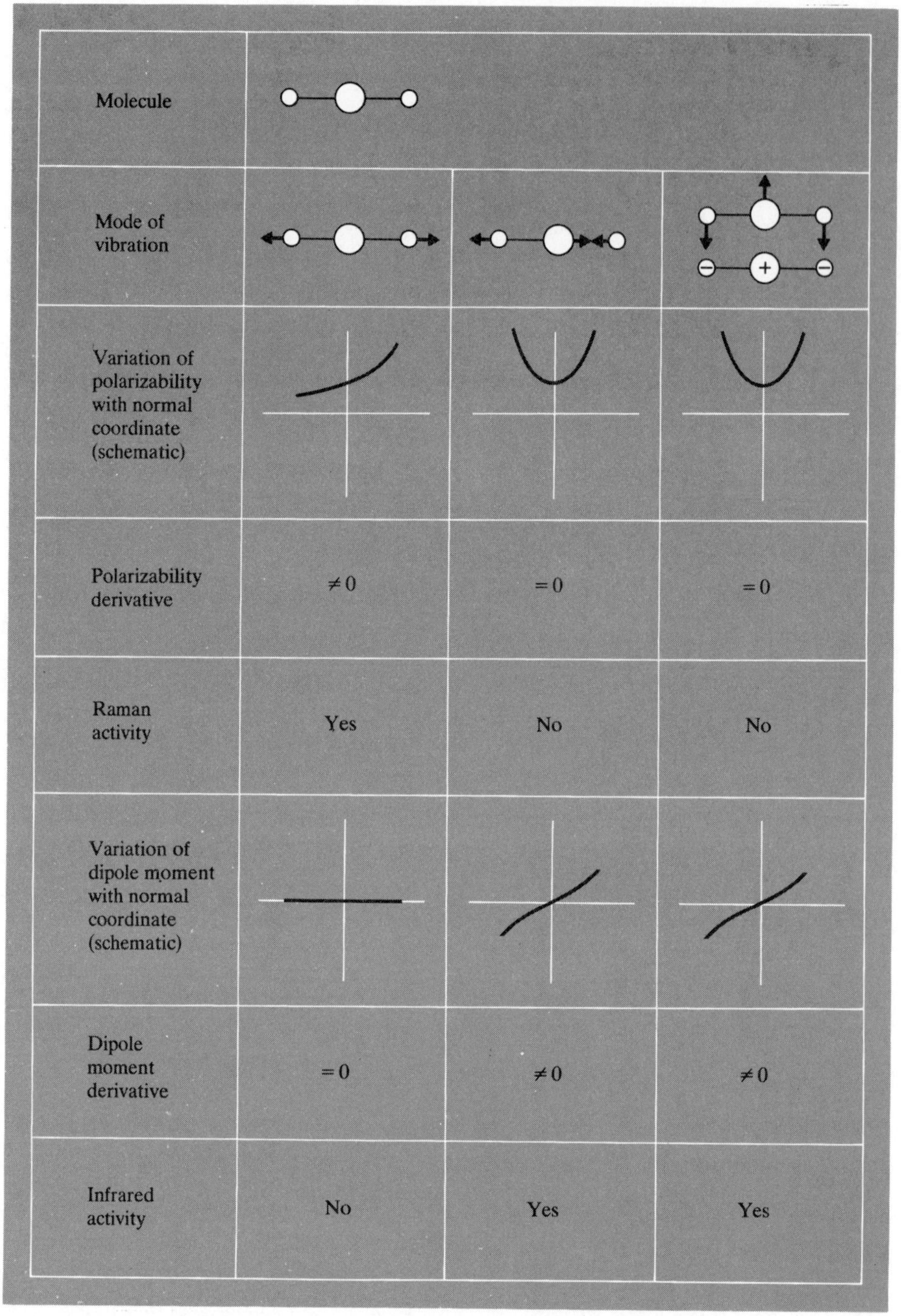

Fig. 3.8 Polarizability and dipole moment variations in the neighbourhood of the equilibrium position and vibrational Raman and infrared activities for a linear *ABA* molecule

activity arises because the space-fixed z- and y-axes no longer remain axes of the polarizability ellipsoid during the whole of the vibration. Thus, although α_{yz} is zero, in the equilibrium configuration $(\partial\alpha_{yz}/\partial Q_3)_0$ is non-zero. This is the only non-zero component of the derived polarizability for this mode, and hence $(\partial a/\partial Q_3)_0=0$ although $(\partial\gamma/\partial Q_3)_0$ is non-zero. Figure 3.9 shows the forms of the dipole moment and polarizability changes with the various normal coordinates.

Comparison of the vibrational activities in the molecules considered above shows that, for those molecules with a centre of symmetry, those vibrations which are Raman active are infrared inactive, and vice versa. This can be shown to be a general rule which is often termed the rule of mutual exclusion. It can form the basis for distinguishing between two alternative configurations of a molecule; e.g., between a linear and a non-linear configuration for an ABA molecule. In many other cases, it is also possible to distinguish between alternative configurations of a molecule by comparing the predicted numbers of vibrational modes which are Raman and/or infrared active. This has proved to be a valuable method of structural elucidation. Its main limitation is that, whereas observation of a band is proof of its activity, the converse is not necessarily true. Some bands, although permitted in principle, may fail to be observed, either because they are inherently weak or because of limitations in the experimental technique. A careful study of the literature will teach the spectroscopist the importance of tempered judgement and guarded optimism!

Broadly speaking, with molecules of relatively low symmetry, all, or nearly all, vibrations are both infrared and Raman active. In some molecules, especially those of high symmetry, some vibrations may be both infrared and Raman inactive; but such modes are often active in light scattering spectra of non-linear origin, as, for example, in hyper Raman spectra (see chapter 8).

These qualitative considerations will have served to indicate the importance of the symmetry of a molecule and its vibrational modes in determining infrared and Raman activity. It will also have emerged that the qualitative arguments used here would be difficult to apply to more complicated molecules. It would be difficult to infer with certainty the symmetry properties of the modes of vibration, let alone the behaviour of the dipole moment and the polarizability. Fortunately, there exist mathematical procedures which enable the symmetry properties of vibrational modes, dipole moment, and polarizability derivatives for molecules to be predicted for an assumed equilibrium configuration of the nuclei. We shall consider how such procedures operate after we have discussed quantum mechanical approaches to light scattering phenomena. It should be noted in advance that these new procedures are completely independent of *special* assumptions regarding mechanical and electrical anharmonicity. They are derived solely on the basis of the symmetry of the vibrations and the electrical properties of the molecules, and consequently have wide validity. However, these symmetry arguments tell us nothing about the intensity with which a vibrational band will appear in infrared or Raman spectra. Once again we emphasize that a vibration, while formally active, may be so weak as to be undetectable in practice.

3.6 Selection rules for overtones and combinations

In the foregoing we have been concerned entirely with fundamental vibrations; indeed, in the approximation of mechanical and electrical harmonicity, it would

Molecule	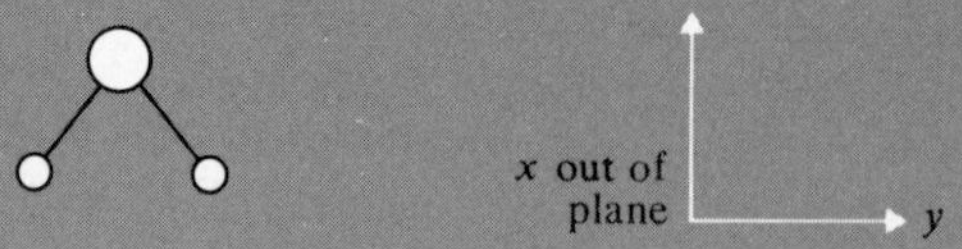		
Mode of vibration	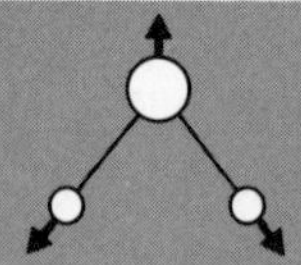	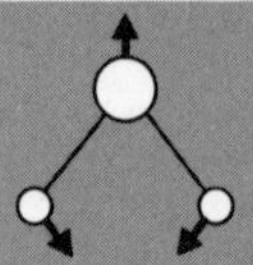	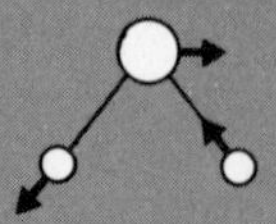
Variation of polarizability with normal coordinate (schematic)	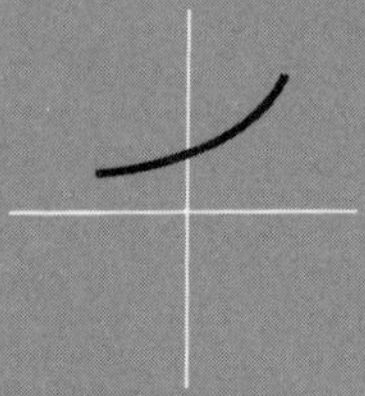	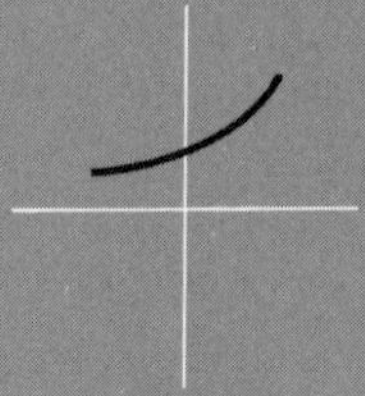	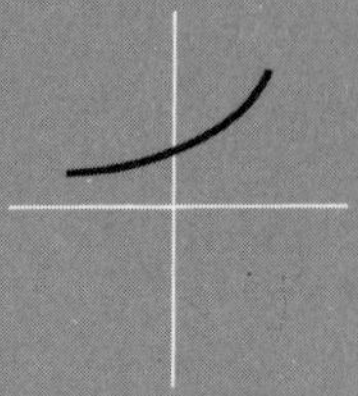
Polarizability derivative	$\neq 0$	$\neq 0$	$\neq 0$
Raman activity	Yes	Yes	Yes
Variation of dipole moment with normal coordinate (schematic)	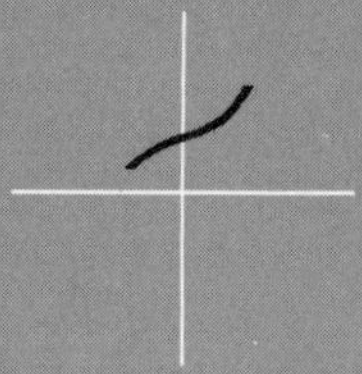	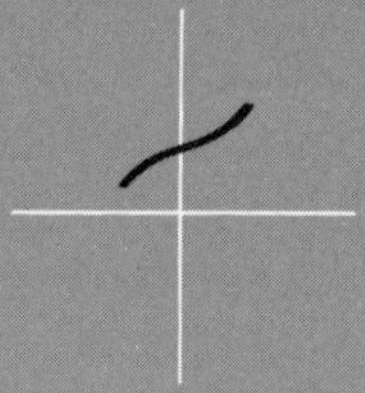	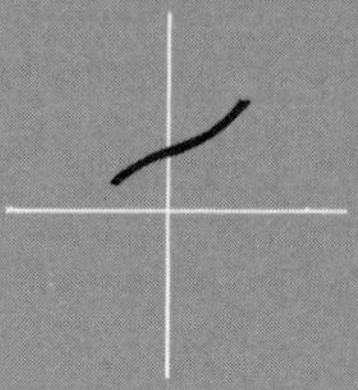
Dipole moment derivative	$\neq 0$	$\neq 0$	$\neq 0$
Infrared activity	Yes	Yes	Yes

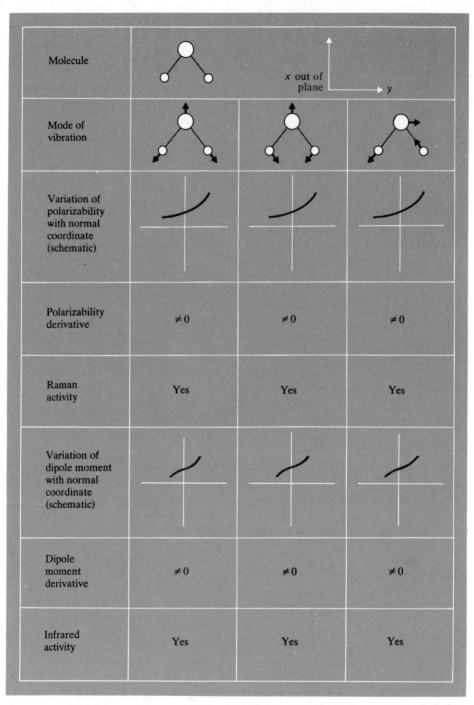

Fig. 3.9 Polarizability and dipole moment variations in the neighbourhood of the equilibrium position and vibrational Raman and infrared activities for a non-linear *A—B—A* molecule (axes chosen in accord with convention[4])

appear that only fundamental vibrations can occur. When anharmonicities are taken into account, overtone and combination bands are also permitted. Fortunately, in Raman spectra, unlike infrared spectra, such bands are almost invariably much weaker than the fundamentals. Vibrational Raman spectra are therefore usually much simpler than infrared spectra. It is, however, worthwhile examining briefly the factors controlling the activity of overtone and combination bands in the Raman effect.

If mechanical anharmonicity is taken into account, the time dependence of the normal coordinate Q_k will not depend solely on $\cos(\omega_k t+\delta_k)$, as in eq. (3.35), but will include terms involving $\cos(2\omega_k t+\delta_{2k})$, $\cos(3\omega_k t+\delta_{3k})$, etc., which relate to overtones, and also terms involving $\cos(\omega_k t+\delta_{kl})\cos(\omega_l t+\delta'_{kl})$, etc., which relate to combination tones. It is easy to see that, as a consequence of the presence of these terms, there will be additional induced dipoles with frequencies $\omega_0 \pm 2\omega_k$, etc., and $\omega_0 \pm (\omega_k \pm \omega_l)$, etc. Thus, mechanical anharmonicity can lead to the observation of overtones and combinations in the Raman effect. Since $\boldsymbol{\alpha}_k'$ is unaffected by the introduction of mechanical anharmonicity, the same selection rules apply to the overtones and combination tones as to the fundamentals. Thus, for example, in a linear symmetric molecule ABA overtones of $\tilde{\nu}_1$ (i.e., $2\tilde{\nu}_1$, $3\tilde{\nu}_1, \ldots,$ etc.) would be Raman active, but not overtones of $\tilde{\nu}_2$ or $\tilde{\nu}_3$. Similarly, some combinations of $\tilde{\nu}_1$ with $\tilde{\nu}_2$ or $\tilde{\nu}_3$ would be Raman active, but no combinations of $\tilde{\nu}_2$ with $\tilde{\nu}_3$.

The situation is different if electrical anharmonicity is taken into account. This means that the third (and possibly higher) terms in the Taylor series expansion of the polarizability in terms of the normal coordinates (eq. 3.31) must be considered. The consequence of this is that there will again be additional induced dipoles with frequencies $\omega_0 \pm 2\omega_k$, etc., and $\omega_0 \pm (\omega_k \pm \omega_l)$, etc., but in this case these dipoles involve a new derived tensor with components of the type

$$\left(\frac{\partial^2 \alpha_{ij}}{\partial Q_k^2}\right)_0, \left(\frac{\partial^2 \alpha_{ij}}{\partial Q_k \partial Q_l}\right)_0, \text{ and so on.}$$

Thus, overtones and combinations arising from electrical anharmonicity can be Raman active even if the fundamental vibration is not. For example, in a linear symmetric molecule ABA, although $\tilde{\nu}_2$ and $\tilde{\nu}_3$ are not Raman active, $2\tilde{\nu}_2$ and $2\tilde{\nu}_3$ are Raman active, since, as can be seen from Fig. 3.8, although the first derivative of the polarizability is zero, the second derivative is not for $\tilde{\nu}_2$ or $\tilde{\nu}_3$.

References

1. G. Placzek, Rayleigh-Streuung und Raman-Effekt, in E. Marx (Ed.), *Handbuch der Radiologie*, **VI**, 2, 205–374, Akademische Verlag, Leipzig, 1934.
2. T. C. Damen, S. P. S. Porto, and B. Tell, *Phys. Rev.*, **142**, 570, 1960; *Phys. Rev.*, **144**, 771, 1966.
3. R. P. Bell and D. A. Long, *Proc. Roy. Soc.*, *A*, **203**, 364, 1950.
4. Report on Notation for the Spectra of Polyatomic Molecules, *J. Chem. Phys.*, **23**, 1997, 1955.

4 A partial quantum mechanical treatment of Rayleigh and rotational and vibrational Raman scattering

"Next when I cast mine eyes and see
That brave vibration each way free;
O how that glittering taketh me!"

Robert Herrick

"I have been ever of the opinion that
revolutions are not to be evaded."

Benjamin Disraeli

4.1 Introduction

According to the quantum theory, radiation is emitted or absorbed as a result of a system making a downward or upward transition between two discrete energy levels; and the radiation itself is also quantized with the energy in discrete photons. A quantum theory of spectroscopic processes should, therefore, treat the radiation and molecule together as a complete system, and explore how energy may be transferred between the radiation and the molecule as a result of their interaction. Such a treatment is beyond the scope of this book, but essentially correct results for many spectroscopic processes, including Rayleigh and Raman scattering, can be obtained by a compromise procedure which is partly classical and partly quantum mechanical in nature. The radiation is treated classically and is regarded as the source of a perturbation of the molecular system which is treated quantum mechanically.

It transpires from such a treatment that transitions between energy levels of the molecular system take place with the emission or absorption of radiation, provided the *transition moment* associated with the initial and final molecular states is non-zero. Using the Dirac bracket notation[1] the transition moment is defined by

$$\boldsymbol{P}_{fi} = \langle \Psi_f | \boldsymbol{P} | \Psi_i \rangle \tag{4.1}$$

where Ψ_i and Ψ_f are the wave functions for the initial and final states, respectively, and $\boldsymbol{P}$ is the appropriate dipole moment operator. For direct absorption $\boldsymbol{P}$ corresponds to the permanent dipole moment operator of the system, and for light scattering $\boldsymbol{P}$ is the induced dipole moment operator of the system. In general, not only Ψ_i and Ψ_f but also $\boldsymbol{P}$ are functions of both the molecular coordinates and the time.

In the treatment to be followed in this chapter we shall concern ourselves only with the calculation of transition moment *amplitudes*. This avoids the complications of time-dependent perturbation theory, but denies us direct information regarding the frequency dependence of the transition moment. However, we shall find no difficulty in determining this frequency dependence by analogy with the classical results already obtained; and a knowledge of the transition moment amplitude suffices to determine the intensity of radiation of a given frequency for non-absorbing systems. It is simply a matter of using the transition moment amplitude in place of its analogue, the classical amplitude, in the formulae for the intensity of radiation from an oscillating dipole derived in chapter 2.

We shall consider time-dependent perturbation theory in the next chapter. This will enable us to extend the theory of scattering to absorbing systems.

4.2 Transition moment amplitudes for scattering of radiation

4.2.1 *General considerations*

As we saw in chapter 3, according to classical radiation theory, Rayleigh and Raman scattering is associated with an *induced* dipole moment whose amplitude is given, in the linear induced dipole approximation, by eq. (3.2)

$$\boldsymbol{P}_0^{(1)} = \boldsymbol{\alpha} \cdot \boldsymbol{E}_0 \tag{4.2}$$

Thus, in quantum mechanical terms, if a transition from an initial state i to a final state f is induced by incident radiation of circular frequency ω_0, the transition moment *amplitude* associated with this change is given by

$$[\boldsymbol{P}_0^{(1)}]_{fi} = \langle \psi_f | \boldsymbol{\alpha} | \psi_i \rangle \cdot \boldsymbol{E}_0 \tag{4.3}$$

where ψ_i and ψ_f are the time-independent wave functions of the initial and final states. In general, the wave functions ψ_i and ψ_f and the polarizability tensor $\boldsymbol{\alpha}$ are functions of all the coordinates of the system, and the integral is over all the coordinate space.

It will be noted that the electric field vector has not been included in the space integral in eq. (4.3). Provided the wavelength of the incident radiation is much larger than the size of the scattering molecule, the electric field vector will remain sensibly constant over the molecule and eq. (4.3) is adequate for the calculation of the transition moment amplitude. For molecular systems whose dimensions are of the order of Angstroms, and visible radiation whose wavelength lies in the region 4000–7000 Å (400·0–700·0 nm), this is usually a very good approximation. In higher order approximations and for large molecules, the variation of the electric field over the molecular system must be taken into account. Scattering which is of electric quadrupole and magnetic dipole origin arises from this higher order treatment. Scattering of this kind, although very much weaker than dipole scattering, is important in the theory of Raman scattering from chiral systems and will be considered in chapter 5.

4.2.2 *Matrix elements of polarizability components*

Because of the vector nature of $\boldsymbol{P}$ and $\boldsymbol{E}$ and the tensor nature of $\boldsymbol{\alpha}$, there is implicit in eq. (4.3) the following set of linear equations:

$$\left.\begin{aligned}[P^{(1)}_{x0}]_{fi} &= \{[\alpha_{xx}]_{fi}E_{x_0}+[\alpha_{xy}]_{fi}E_{y_0}+[\alpha_{xz}]_{fi}E_{z_0}\}\\ [P^{(1)}_{y0}]_{fi} &= \{[\alpha_{yx}]_{fi}E_{x_0}+[\alpha_{yy}]_{fi}E_{y_0}+[\alpha_{yz}]_{fi}E_{z_0}\}\\ [P^{(1)}_{z0}]_{fi} &= \{[\alpha_{zx}]_{fi}E_{x_0}+[\alpha_{zy}]_{fi}E_{y_0}+[\alpha_{zz}]_{fi}E_{z_0}\}\end{aligned}\right\} \tag{4.4}$$

where

$$\begin{aligned}[\alpha_{xx}]_{fi} &= \langle\psi_f|\alpha_{xx}|\psi_i\rangle\\ [\alpha_{xy}]_{fi} &= \langle\psi_f|\alpha_{xy}|\psi_i\rangle\end{aligned} \tag{4.5}$$

and so on. $[\alpha_{xx}]_{fi}$, $[\alpha_{xy}]_{fi}$, . . . are termed *matrix elements* of the polarizability tensor components α_{xx}, α_{yy}, . . . for the transition $f \leftrightarrow i$. Alternatively, $[\alpha_{xy}]_{fi}$ is termed a component of the *transition polarizability tensor* $[\boldsymbol{\alpha}]_{fi}$. It can be seen that the equations of (4.4) have the same general form as the equations of (3.49) and (3.50) obtained in the classical treatment, but the elements of the polarizability tensor and the derived polarizability tensor which appear in the equations of (3.49) and (3.50) have been replaced in the equations of (4.4) by components of the transition polarizability tensor. The equations of (4.4) correspond to Rayleigh scattering if the initial and final states are the same, and to Raman scattering if the initial and final states are different.

4.2.3 *Separation of vibration and rotation*

We shall now examine in more detail the nature of a typical matrix element of the polarizability tensor $[\alpha_{xy}]_{fi}$ for Raman scattering. We shall confine ourselves to consideration of vibrational and rotational transitions induced by incident radiation of circular frequency ω_0 in a molecule which remains throughout in its ground electronic state. Provided certain conditions are satisfied, we may then write

$$[\alpha_{xy}]_{fi} = \langle\phi_f\Theta_f|\alpha_{xy}|\phi_i\Theta_i\rangle \tag{4.6}$$

where ϕ_i and ϕ_f are the time-independent vibrational wave functions and Θ_i and Θ_f are the time-independent rotational wave functions, in the initial and final states, and α_{xy} is the xy component of the polarizability for the molecule in its ground electronic state. In eq. (4.6) not only the vibrational and rotational wave functions but also the polarizability are functions of the nuclear coordinates only. While it might seem that eq. (4.6) follows naturally from eq. (4.5) as a result of the separation of the total wave functions into their component parts and the classical properties of the polarizability discussed in chapter 3, the more complete treatment given in chapter 5 will reveal that eq. (4.6) is valid only if the following conditions[2] are satisfied: (a) the excitation frequency must be much larger than the frequency of any vibrational or rotational transition of the molecule; (b) the excitation frequency must be much less than any electronic transition frequency of the molecule; and (c) the ground electronic state must not, normally, be degenerate. Where the electronic ground state is degenerate, the expansion given by eq. (4.6) will not necessarily apply, but there are special cases when it does. We shall assume these conditions to be satisfied in what follows.

Returning now to the development of eq. (4.6), we recall from eq. (3.17) that, if x, y, z are space-fixed axes and x', y', z' are molecule-fixed axes, then

$$\alpha_{xy} = \sum_{x'y'} \alpha_{x'y'} \cos(xx') \cos(yy') \tag{4.7}$$

The components α_{xy} are functions of both the vibrational and rotational coordinates, but the components $\alpha_{x'y'}$ are functions of the vibrational coordinates only. Thus eq. (4.7) achieves a separation of vibrational and rotational coordinates for the polarizability.

Substituting eq. (4.7) into eq. (4.6), we obtain

$$[\alpha_{xy}]_{fi} = \sum_{x'y'} [\alpha_{x'y'}]_{v^f v^i} \langle \Theta_{R^f} | \cos(xx') \cos(yy') | \Theta_{R^i} \rangle \tag{4.8}$$

where

$$[\alpha_{x'y'}]_{v^f v^i} = \langle \phi_{v^f} | \alpha_{x'y'} | \phi_{v^i} \rangle \tag{4.9}$$

In these equations v^i and v^f are the *sets* of vibrational quantum numbers, and R^i and R^f are the *sets* of rotational quantum numbers in the initial and final states. We can see from eq. (4.8) that the vibrational selection rules will be determined by the properties of the first term (a vibrational transition polarizability component) and the rotational selection rules by the properties of the second term. We shall now consider these terms in more detail.

4.3 Vibrational transitions

4.3.1 *General consideration of the vibrational transition polarizability tensor*

In general the evaluation of eq. (4.8) involves the simultaneous consideration of the vibrational and rotational terms. However, for a system of freely orienting molecules, if we are not interested in the rotational fine structure, we can avoid the evaluation of the rotational term. It can be shown that a classical space averaging is equivalent to summing and averaging over all the permitted rotational transitions. Thus we can obtain formulae for the intensities and polarization properties of the scattered radiation arising from vibrational transitions in freely orienting molecules simply by using space averages of the squares of vibrational transition polarizability components. These are of the following types

$$\overline{[\alpha_{xx}]^2_{v^f v^i}}; \quad \overline{[\alpha_{xy}]^2_{v^f v^i}}; \quad \overline{[\alpha_{xx}]_{v^f v^i}[\alpha_{yy}]_{v^f v^i}}$$

and are related to the invariants of the transition polarizability tensor $[a]^2_{v^f v^i}$ and $[\gamma]^2_{v^f v^i}$ by equations of the same form as eqs. (3.23) to (3.25) but with transition polarizabilities replacing polarizabilities or polarizability derivatives.

Consequently the formulae for the polarization properties of Raman scattering will have the same form as eqs. (3.83) to (3.85), (3.107) and (3.108) with $[a]^2_{v^f v^i}$ and $[\gamma]^2_{v^f v^i}$ replacing $[a']^2_k$ and $(\gamma')^2_k$; and formulae for Raman intensities may be obtained by substituting $[a]^2_{v^f v^i}$ and $[\gamma]^2_{v^f v^i}$ for $(a')^2_k Q^2_{k_0}$ and $(\gamma')^2_k Q^2_{k_0}$ respectively in the classical formulae (e.g. eqs. (3.80) to (3.82) and (3.103) and (3.104)). The resulting formulae are summarized in Tables A, B, C and D in the central reference section. These formulae relate to a *general* vibrational transition $v^f \leftarrow v^i$; for Rayleigh scattering $v^f = v^i$.

4.3.2 *Expansion of vibrational transition polarizability tensor components in the normal coordinates*

The formulae in terms of the transition polarizabilities reveal nothing about the selection rules. To obtain this information, we must examine the dependence of the transition polarizabilities on the normal coordinates of vibration. Following the classical treatment in chapter 3, we shall expand each component of the polarizability tensor in terms of the normal coordinates as follows:

$$\alpha_{x'y'} = (\alpha_{x'y'})_0 + \sum_k \left(\frac{\partial \alpha_{x'y'}}{\partial Q_k}\right)_0 Q_k + \tfrac{1}{2}\sum_{k,l}\left(\frac{\partial^2 \alpha_{x'y'}}{\partial Q_k \partial Q_l}\right) Q_k Q_l + \ldots \tag{4.10}$$

In doing this, we must emphasize that we are making the specific assumption that the polarizability tensor components are functions of the normal coordinates only, and that this assumption is only valid if the conditions given on page 76 are satisfied.

Introducing the first two terms of the expansion of eq. (4.10) into eq. (4.9), (i.e. assuming electrical harmonicity) we obtain

$$[\alpha_{x'y'}]_{v^f v^i} = (\alpha_{x'y'})_0 \langle \phi_{v^f} | \phi_{v^i} \rangle + \sum_k \left(\frac{\partial \alpha_{x'y'}}{\partial Q_k}\right)_0 \langle \phi_{v^f} | Q_k | \phi_{v^i} \rangle \tag{4.11}$$

In the harmonic oscillator approximation, the total vibrational wave function is the product of harmonic oscillator wave functions for each of the normal modes of vibration. Thus, for ϕ_{v^i} and ϕ_{v^f}, we may write

$$\phi_{v^i} = \prod_k \phi_{v_k^i}(Q_k) \tag{4.12}$$

$$\phi_{v^f} = \prod_k \phi_{v_k^f}(Q_k) \tag{4.13}$$

where $\phi_{v_k^i}(Q_k)$ and $\phi_{v_k^f}(Q_k)$ are harmonic oscillator wave functions associated with the normal coordinate Q_k, and have vibrational quantum numbers v_k^i and v_k^f in the initial and final states, respectively. The product is taken over all normal coordinates. When these expansions are introduced into eq. (4.11), we obtain

$$\begin{aligned}[\alpha_{x'y'}]_{v^f v^i} = {} & (\alpha_{x'y'})_0 \left\langle \prod_k \phi_{v_k^f}(Q_k) \middle| \prod_k \phi_{v_k^i}(Q_k) \right\rangle \\ & + \sum_k \left(\frac{\partial \alpha_{x'y'}}{\partial Q_k}\right)_0 \left\langle \prod_k \phi_{v_k^f}(Q_k) \middle| Q_k \middle| \prod_k \phi_{v_k^i}(Q_k) \right\rangle \end{aligned} \tag{4.14}$$

To proceed further we shall need to invoke the following properties of harmonic oscillator functions:

$$\langle \phi_{v_k^f}(Q_k) | \phi_{v_k^i}(Q_k) \rangle = \begin{Bmatrix} 0 \text{ for } v_k^f \neq v_k^i \\ 1 \text{ for } v_k^f = v_k^i \end{Bmatrix} \tag{4.15}$$

and

$$\langle \phi_{v_k^f}(Q_k) | Q_k | \phi_{v_k^i}(Q_k) \rangle = \begin{Bmatrix} 0 \text{ for } v_k^f = v_k^i \\ (v_k^i + 1)^{\frac{1}{2}} b_{v_k} \text{ for } v_k^f = v_k^i + 1 \\ (v_k^i)^{\frac{1}{2}} b_{v_k} \text{ for } v_k^f = v_k^i - 1 \end{Bmatrix} \tag{4.16}$$

where

$$b_{v_k}^2 = \frac{h}{8\pi^2 \nu_k} = \frac{h}{8\pi^2 c\tilde{\nu}_k} = \frac{h}{4\pi\omega_k} \tag{4.17}$$

The quantity b_{v_k} is the quantum mechanical analogue of the amplitude Q_{k_0} in the classical treatment of an oscillator. The definition of b_{v_k} given by eq. (4.17) assumes that the normal coordinate Q_k is mass-adjusted.

4.3.3 *General selection rules*

We can now examine the conditions which have to be satisfied if the terms in eq. (4.14) are to be non-zero. Consider the first term on the right-hand side of eq. (4.14). This term, which involves the equilibrium value of the polarizability tensor component, clearly relates to Rayleigh scattering, and by analogy with the classical theory has the circular frequency dependence ω_0. Because of the orthogonality of the harmonic oscillator functions, this term will be non-zero if none of the vibrational quantum numbers change in going from the initial to the final state, i.e., if $v_k^i = v_k^f$ for all k. When this condition is satisfied, because of the normalization properties of the harmonic oscillator functions, the integral in this term is unity and the term reduces simply to $(\alpha_{x'y'})_0$, the equilibrium value of the $x'y'$ component of the polarizability tensor. Since not all of the components of the equilibrium polarizability tensor of a system can be zero, Rayleigh scattering will always occur. Thus, for Rayleigh scattering, the quantum mechanical treatment and the classical theory give the same results.

We now come to the second term on the right-hand side of eq. (4.14). This term, which involves derived polarizability tensor components, clearly relates to Raman scattering. Consider the k-th summand. This will vanish unless every term in the product is non-vanishing, and to achieve this the following conditions must be satisfied: for all modes except the k-th, the vibrational quantum numbers must be the same in the initial and final states (i.e., $v_j^i = v_j^f$, except for $j = k$) and, for the k-th mode, the vibrational quantum number must change by unity (i.e., $v_k^f = v_k^i \pm 1$). The transition moment is associated with Stokes Raman scattering for $\Delta v_k = 1$ (i.e., $v_k^f = v_k^i + 1$), and with anti-Stokes Raman scattering when $\Delta v_k = -1$ (i.e., $v_k^f = v_k^i - 1$).These conditions are a consequence of the properties of harmonic oscillator wave functions given above. These arguments apply to each normal mode and to each of the six components of the polarizability matrix elements. Thus, the k-th summand in eq. (4.14) takes the values

$$(v_k^i + 1)^{\frac{1}{2}} b_{v_k} \left(\frac{\partial \alpha_{x'y'}}{\partial Q_k} \right)_0 \quad \text{if } v_k^f = v_k^i + 1 \text{ and } v_j^f = v_j^i \text{ for all other } Q_j \tag{4.18}$$

i.e.,

$$[\alpha_{x'y'}]_{v_k^i+1, v_k^i} = (v_k^i + 1)^{\frac{1}{2}} b_{v_k} \left(\frac{\partial \alpha_{x'y'}}{\partial Q_k} \right)_0 = (v_k^i + 1)^{\frac{1}{2}} b_{v_k} (\alpha'_{x'y'})_k \tag{4.19}$$

and

$$(v_k^i)^{\frac{1}{2}} b_{v_k} \left(\frac{\partial \alpha_{x'y'}}{\partial Q_k} \right)_0 \quad \text{if } v_k^f = v_k^i - 1 \text{ and } v_j^f = v_j^i \text{ for all other } Q_j \tag{4.20}$$

i.e.,

$$[\alpha_{x'y'}]_{v_k^i-1,v_k^i} = (v_k^i)^{\frac{1}{2}} b_{v_k}\left(\frac{\partial \alpha_{x'y'}}{\partial Q_k}\right)_0 = (v_k^i)^{\frac{1}{2}} b_{v_k}(\alpha'_{x'y'})_k \tag{4.21}$$

The k-th summand is otherwise zero.

It follows from these arguments that, in the approximation of electrical and mechanical harmonicity, only vibrational fundamentals, i.e., transitions for which only one vibrational quantum number changes by unity, can be observed in the Raman effect. It must be noted, however, that the restriction $\Delta v_k = \pm 1$ is a *necessary*, but not a *sufficient*, condition for the occurrence of Raman scattering associated with the k-th vibrational mode. It is also necessary that at least one of the elements of the derived polarizability tensor be non-zero. This latter condition is exactly the same as that obtained from the classical treatment. Indeed, we can see that there is a close correspondence between the quantum mechanical transition moment amplitude and the classical induced moment amplitude associated with Raman scattering. In the expression for the transition moment, the quantum mechanical amplitude factor b_{v_k} replaces the classical amplitude Q_{k_0}, but in addition there appears the dependence on the quantum number v_k^i which, of course, has no classical analogue.

If we do not make the assumption of electrical harmonicity when considering the expansion of the polarizability in terms of the normal coordinates, we have to include the second (and, if necessary, higher) derivatives in eq. (4.10). This leads to integrals of the type

$$\left(\frac{\partial^2 \alpha_{x'y'}}{\partial Q_k{}^2}\right)_0 \langle \phi_{v_k^f}(Q_k)|Q_k^2|\phi_{v_k^i}(Q_k)\rangle \tag{4.22}$$

and

$$\left(\frac{\partial^2 \alpha_{x'y'}}{\partial Q_k \partial Q_l}\right)_0 \langle \phi_{v_k^f}(Q_k)|Q_k|\phi_{v_k^i}(Q_k)\rangle\langle \phi_{v_l^f}(Q_l)|Q_l|\phi_{v_l^i}(Q_l)\rangle \tag{4.23}$$

In consequence, the following vibrational transition polarizabilities can be non-zero: $[\alpha_{x'y'}]_{v_k^i\pm 2,v_k^i}$ and $[\alpha_{x'y'}]_{v_k^i\pm 1,v_l^i\pm 1;v_k^i,v_l^i}$. The former relates to overtone transitions ($\Delta v_k = \pm 2$) and the latter to binary combination transitions ($\Delta v_k = \pm 1$ and $\Delta v_l = \pm 1$) which are permitted if the appropriate second derivatives are non-zero. We shall not consider these further, but complete expressions for vibrational transition polarizabilities for fundamental, overtone, and combination transitions are presented in Table H in the central reference section.

4.3.4 *Scattering from a single space-fixed molecule*

For a space-fixed molecule so oriented that the molecule-fixed $x'y'z'$-axis system is coincident with the space-fixed xyz-axis system, then eq. (4.8) reduces to

$$[\alpha_{xy}]_{v^f v^i} = [\alpha_{x'y'}]_{v^f v^i} = \langle \phi_{v^f}|\alpha_{x'y'}|\phi_{v^i}\rangle = \langle \phi_{v^f}|\alpha_{xy}|\phi_{v^i}\rangle \tag{4.24}$$

since the integral over the rotational coordinates is unity under these conditions. Then, introducing this simplification, together with the results we have just obtained for the matrix elements of the polarizability, the set of linear equations (4.4) takes on the following form for Rayleigh scattering for which $f=i$

$$\left.\begin{aligned}
[P^{(1)}_{x_0}]_{ii} &= \{(\alpha_{xx})_0 E_{x_0} + (\alpha_{xy})_0 E_{y_0} + (\alpha_{xz})_0 E_{z_0}\} \\
[P^{(1)}_{y_0}]_{ii} &= \{(\alpha_{yx})_0 E_{x_0} + (\alpha_{yy})_0 E_{y_0} + (\alpha_{yz})_0 E_{z_0}\} \\
[P^{(1)}_{z_0}]_{ii} &= \{(\alpha_{zx})_0 E_{x_0} + (\alpha_{zy})_0 E_{y_0} + (\alpha_{zz})_0 E_{z_0}\}
\end{aligned}\right\} \tag{4.25}$$

For Stokes Raman scattering associated with the single vibrational transition $v_k^i+1 \leftarrow v_k^i$, the corresponding set of linear equations is

$$\left.\begin{aligned}
[P^{(1)}_{x_0}]_{v_k+1,v_k^i} &= (v_k^i+1)^{\frac{1}{2}} b_{v_k}\{(\alpha'_{xx})_k E_{x_0} + (\alpha'_{xy})_k E_{y_0} + (\alpha'_{xz})_k E_{z_0}\} \\
[P^{(1)}_{y_0}]_{v_k+1,v_k^i} &= (v_k^i+1)^{\frac{1}{2}} b_{v_k}\{(\alpha'_{yx})_k E_{x_0} + (\alpha'_{yy})_k E_{y_0} + (\alpha'_{yz})_k E_{z_0}\} \\
[P^{(1)}_{z_0}]_{v_k+1,v_k^i} &= (v_k^i+1)^{\frac{1}{2}} b_{v_k}\{(\alpha'_{zx})_k E_{x_0} + (\alpha'_{zy})_k E_{y_0} + (\alpha'_{zz})_k E_{z_0}\}
\end{aligned}\right\} \tag{4.26}$$

For anti-Stokes Raman scattering, $v_k^i-1 \leftarrow v_k^i$, and the factor $(v_k^i+1)^{\frac{1}{2}}$ in eq. (4.26) must be replaced by $(v_k^i)^{\frac{1}{2}}$ to yield $[P^{(1)}_{x_0}]_{v_k^i-1,v_k^i}$ etc.

It can be seen immediately that eq. (4.25) will give expressions for depolarization ratios for Rayleigh scattering from a single space-fixed molecule which are identical with the classical formulae of eqs. (3.74) to (3.76). Similarly, eq. (4.26) will give expressions for depolarization ratios for Raman scattering from a single space-fixed molecule which are identical with the classical formulae. In the case of Raman scattering, we note that equivalence of the quantum mechanical and classical formulae for the depolarization ratios results from the cancellation of the factors (v_k^i+1) (or v_k^i for anti-Stokes) and $b^2_{v_k}$, since *ratios* of scattered powers or radiant intensities are involved in the depolarization values.

4.3.5 *Scattering from a crystal*

It is also clear from a comparison of eqs. (3.50) and (4.26) that the procedures for determining relative magnitudes of derived polarizability components from orientation studies of Raman scattering from crystals are equally applicable when quantum considerations are taken into account.

4.3.6 *Scattering from an assembly of randomly oriented molecules: depolarization ratios, reversal coefficients and degrees of circularity*

We now proceed to the case of an assembly of molecules freely rotating. The relation between the scattering from $\mathcal{N}$ molecules and the scattering from one molecule has been discussed in chapter 3, and we shall assume that Rayleigh and Raman scattering from $\mathcal{N}$ molecules is $\mathcal{N}$ times that from a single molecule. As we are not interested in the fine structure associated with transitions between rotational states, we shall average over all permitted molecular orientations. However, as we have explained, averaging over all the discrete molecular orientations permitted by quantum theory leads to the same result as a classical average over all orientations. Thus, we may still use the classical space averages of the squares of the polarizability tensor components given by eqs. (3.23) to (3.25), and the analogous equations for the derived polarizability tensor components, to calculate the scattering from an assembly of $\mathcal{N}$ molecules freely rotating. It is clear that we shall obtain for the polarization properties of Rayleigh scattering, formulae which are identical with the classical results of eqs, (3.86) to (3.88), (3.105), and (3.106). Similarly, we obtain for the polarization properties of vibrational Raman scattering, ($\Delta v_k = \pm 1$), formulae which are the same as those given by the classical eqs. (3.83) to (3.85), (3.107), and (3.108). The restrictions on the ρ, $\mathscr{P}$, and $\mathscr{C}$ values for Rayleigh and Raman scattering given on page 62 will therefore also apply in quantum theory. We note again that in the calculation of ρ, $\mathscr{P}$ and $\mathscr{C}$ values *ratios* of scattered powers or radiant intensities are involved, and so the factors (v_k^i+1) (or v_k^i for anti-Stokes) and $b_{v_k}^2$ cancel.

4.3.7 *Intensities of scattering from an assembly of randomly oriented molecules*

When we proceed to calculate scattered powers or radiant intensities rather than ratios of these quantities, differences emerge between the classical and quantum mechanical treatments for Raman scattering. The space-averaged squares of the transition moments for Raman scattering formed from eq. (4.26) will involve the factor $(v_k^i+1)b_{v_k}^2$, whereas the classical eq. (3.50) leads to a factor $Q_{k_0}^2$. In the case of Rayleigh scattering, eq. (4.25) gives the same result as the classical eq. (3.49).

The appearance of the factor (v_k^i+1) has important consequences when we consider the scattering from an assembly of $\mathscr{N}$ molecules rather than from one molecule. Although at normal temperatures most of the molecules will be in the lowest vibrational state ($v_k^i=0$), there will also be some population of the higher vibrational states ($v_k^i=1, 2$, etc.). Thus, in an assembly of $\mathscr{N}$ molecules, the scattering associated with a given Stokes vibrational transition $v_k^i+1 \leftarrow v_k^i$ depends on $\mathscr{N}(v_k^i+1)f_{v_k^i}$, where $f_{v_k^i}$ is the fraction of molecules in the state v_k^i. Since, in the harmonic approximation, the wavenumbers of the vibrational transitions $1\leftarrow 0$, $2\leftarrow 1$, $3\leftarrow 2$ are identical, (see page 93), the calculation of the total observed identity involves summing $\mathscr{N}(v_k^i+1)f_{v_k^i}$ over all i for a given mode k. Now, from the Boltzmann distribution law,

$$f_{v_k^i} = \frac{\exp\{-(v_k^i+\frac{1}{2})hc\tilde{\nu}_k/kT\}}{\sum\limits_i \exp\{-(v_k^i+\frac{1}{2})hc\tilde{\nu}_k/kT\}} \tag{4.27}$$

where $\tilde{\nu}_k$ is the wavenumber of the k-th normal mode and T is the absolute temperature. Utilizing this expression, it is found that

$$\mathscr{N}\sum_i (v_k^i+1)f_{v_k^i} = \frac{\mathscr{N}}{1-\exp\{-hc\tilde{\nu}_k/kT\}} \tag{4.28}$$

For the associated anti-Stokes Raman wavenumber shift, it is found that the corresponding expression is given by

$$\mathscr{N}\sum_i (v_k^i)f_{v_k^i} = \frac{\mathscr{N}}{\exp\{hc\tilde{\nu}_k/kT\}-1} \tag{4.29}$$

We can now obtain expressions for the scattered power or radiant intensity in the Raman effect associated with vibrational transitions in an assembly of $\mathscr{N}$ freely rotating molecules. Illumination will be along the z-axis and the scatter plane will be the xz-plane. We shall investigate in some detail the case of illumination with plane polarized radiation for which $E_y \neq 0$ ($E_x = E_z = 0$), and then tabulate the corresponding formulae for other cases of illumination. The general procedure follows that already given in some detail in chapter 3.

Consider first a typical space-averaged square of a Stokes transition moment amplitude, e.g., $\overline{[P_{y_0}^{(1)}]^2_{v_k^i+1, v_k^i}}$. It follows immediately from eq. (4.26) that

$$\overline{[P_{y_0}^{(1)}]^2_{v_k^i+1, v_k^i}} = (v_k^i+1)b_{v_k}^2\{\overline{(\alpha'_{yy})_k^2}\}E_{y_0}^2 \tag{4.30}$$

Using the space-averaged square of the appropriate derived polarizability tensor component (eq. 3.23), this becomes

$$\overline{[P_{y_0}^{(1)}]^2_{v_k^i+1, v_k^i}} = (v_k^i+1)b_{v_k}^2\left\{\frac{45(a')_k^2+4(\gamma')_k^2}{45}\right\}E_{y_0}^2 \tag{4.31}$$

We now sum over $\mathcal{N}$ molecules distributed amongst the various levels v_k of the vibrational state k, using eq. (4.28). Hence, *in the harmonic approximation* we obtain for the radiant intensity ${}^{\perp}I_{\perp}(\pi/2)$ observed along the x direction

$$\begin{aligned} {}^{\perp}I_{\perp}(\pi/2) &= \frac{{}^{\perp}\mathrm{d}\Phi_{\perp}(\pi/2)}{\mathrm{d}\Omega} \\ &= \frac{k_{\tilde{\nu}}h\mathcal{N}(\tilde{\nu}_0-\tilde{\nu}_k)^4\{45(a')_k^2+4(\gamma')_k^2\}\mathcal{I}}{8\pi^2c\tilde{\nu}_k\{1-\exp[-hc\tilde{\nu}_k/kT]\}45} \end{aligned} \tag{4.32}$$

where eq. (4.17) has been used for $b_{v_k}^2$, $E_{y_0}^2$ has been replaced by the irradiance, and $k_{\tilde{\nu}}$ is defined by eq. (3.59).

Similarly, for ${}^{\perp}I_{\parallel}(\pi/2)$ we have

$${}^{\perp}I_{\parallel}(\pi/2) = \frac{{}^{\perp}\mathrm{d}\Phi_{\parallel}(\pi/2)}{\mathrm{d}\Omega} = \frac{k_{\tilde{\nu}}h\mathcal{N}(\tilde{\nu}_0-\tilde{\nu}_k)^4(\gamma')_k^2\mathcal{I}}{8\pi^2c\tilde{\nu}_k\{1-\exp[-hc\tilde{\nu}_k/kT]\}15} \tag{4.33}$$

Thus, the total intensity ${}^{\perp}I(\pi/2)$ observed along the x-direction is given by

$$\begin{aligned} {}^{\perp}I(\pi/2) &= {}^{\perp}I_{\perp}(\pi/2) + {}^{\perp}I_{\parallel}(\pi/2) \\ &= \frac{k_{\tilde{\nu}}h\mathcal{N}(\tilde{\nu}_0-\tilde{\nu}_k)^4\{45(a')_k^2+7(\gamma')_k^2\}\mathcal{I}}{8\pi^2c\tilde{\nu}_k\{1-\exp[-hc\tilde{\nu}_k/kT]\}45} \end{aligned} \tag{4.34}$$

Equations (4.32) to (4.34) may be compared with the classical results given by eqs. (3.80) to (3.82).

Finally, for observation along a direction Ow lying in the xz-plane and making an angle θ with the z-axis, the intensity ${}^{\perp}I_{\parallel}(\theta)$ of the radiation polarized in the scatter plane is given by

$$\begin{aligned} {}^{\perp}I_{\parallel}(\theta) = \frac{{}^{\perp}\mathrm{d}\Phi_{\parallel}(\theta)}{\mathrm{d}\Omega} &= {}^{\perp}I_{\parallel}(\pi/2)\sin^2\theta + {}^{\perp}I_{\parallel}(0)\cos^2\theta \\ &= {}^{\perp}I_{\parallel}(\pi/2) \end{aligned} \tag{4.35}$$

since ${}^{\perp}I_{\parallel}(0) = {}^{\perp}I_{\parallel}(\pi/2)$ in this case. ${}^{\perp}I_{\perp}(\theta)$ the intensity of the radiation polarized perpendicular to the scatter plane is equal to ${}^{\perp}I_{\perp}(\pi/2)$.

The differential cross-section $\mathrm{d}\sigma/\mathrm{d}\Omega$ per molecule is obtained simply by dividing the above formulae for radiant intensities by $\mathcal{N}\mathcal{I}$ (see page 36).

The corresponding equations for anti-Stokes Raman scattering are obtained by substituting $(\tilde{\nu}_0+\tilde{\nu}_k)^4$ for $(\tilde{\nu}_0-\tilde{\nu}_k)^4$ and $\exp\{hc\tilde{\nu}_k/kT\}-1$ for $1-\exp\{-hc\tilde{\nu}_k/kT\}$.

The ratio of Stokes to anti-Stokes radiant intensities or differential scattering cross-sections is given by

$$\frac{(\tilde{\nu}_0-\tilde{\nu}_k)^4}{(\tilde{\nu}_0+\tilde{\nu}_k)^4}\exp\{hc\tilde{\nu}_k/kT\} \tag{4.36}$$

It should be noted that in the case of vibrational degeneracy, if there are d_{v_k} vibrational modes each with wavenumber $\tilde{\nu}_k$, the vibrational Raman intensity observed at $\tilde{\nu}_0 \pm \tilde{\nu}_k$ will be d_{v_k} times that given by the formulae (4.32) to (4.35).

It can be seen from eqs. (4.32) to (4.34) that the intensity of Raman scattering is proportional to the concentration of the scattering species; thus it can be used to measure concentration, provided there is no departure from a linear law as a result of intermolecular interactions. Further, since the intensity of vibrational Raman scattering is temperature-dependent, it can be used to measure temperatures. Appropriate measurements of vibrational Raman intensities and depolarization ratios can also be used to determine the magnitudes but not the signs of $(a')_k$ and $(\gamma')_k$, since I and $\rho_\perp$ involve different combinations of these two invariants.

The formulae corresponding to other types of illumination may be obtained in a similar way. The central reference section explains how such formulae may be obtained from the general formulae in Tables A–D.

4.3.8 *Stokes parameters for scattering from an assembly of randomly oriented molecules*

We now consider how vibrational Raman scattering may be characterized by Stokes parameters. As we have already indicated, the virtue of these parameters is that they enable all states of polarization of the incident radiation to be included in a general treatment which can be subsequently particularized to any chosen state of polarization.

We shall consider the exciting radiation of frequency ω_0 to be incident along the positive z-axis and to be characterized by its irradiance $\mathcal{I}$, a degree of polarization P, an azimuth ψ defined with respect to the x-axis, and an ellipticity χ. It is readily seen from Fig. 4.1 that the complex amplitudes of the electric field intensity associated with the polarized part of the radiation are given by

$$\tilde{E}_{x_0} = E_0\{\cos\chi\cos\psi + i\sin\chi\sin\psi\} \tag{4.37}$$

$$\tilde{E}_{y_0} = E_0\{\cos\chi\sin\psi - i\sin\chi\cos\psi\} \tag{4.38}$$

where E_0 is the amplitude of the electric vector of the polarized part of the radiation.

It follows from the equations of (4.4) that for one space-fixed molecule the components of the amplitude of the transition moment induced by $\tilde{E}_{x_0}$ and $\tilde{E}_{y_0}$ are given by

$$[\tilde{P}_{x_0}]_{v^f v^i} = [\alpha_{xx}]_{v^f v^i}\tilde{E}_{x_0} + [\alpha_{xy}]_{v^f v^i}\tilde{E}_{y_0} \tag{4.39}$$

$$[\tilde{P}_{y_0}]_{v^f v^i} = [\alpha_{yx}]_{v^f v^i}\tilde{E}_{x_0} + [\alpha_{yy}]_{v^f v^i}\tilde{E}_{y_0} \tag{4.40}$$

$$[\tilde{P}_{z_0}]_{v^f v^i} = [\alpha_{zx}]_{v^f v^i}\tilde{E}_{x_0} + [\alpha_{zy}]_{v^f v^i}\tilde{E}_{y_0} \tag{4.41}$$

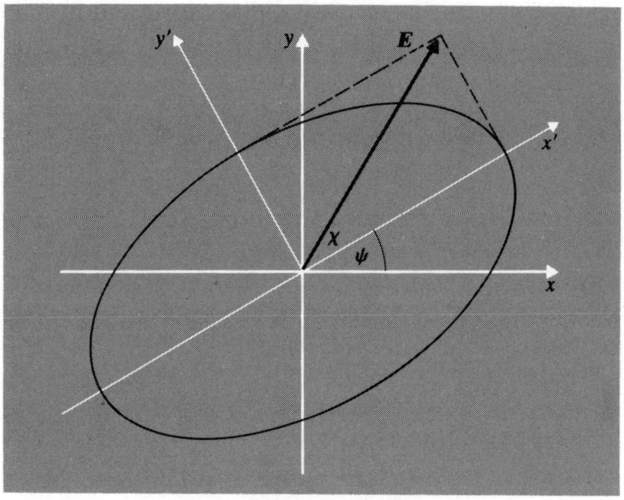

Fig. 4.1 Polarization ellipse

Equations (4.37) to (4.41) enable us to relate the space-averaged squares of the transition moment amplitudes to the space-averaged transition polarizabilities and the polarization characteristics of the polarized part of the incident radiation; and we can proceed similarly for the unpolarized part. However, the Stokes parameters of the scattered radiation are expressed in terms of the electric field amplitudes of the scattered radiation, and we must note from eq. (2.95) that for a dipole along the z-direction the square of the electric field amplitude at a distance r along the x-direction is given by

$$\tilde{E}^s_{z_0}\tilde{E}^{s*}_{z_0} = K\tilde{P}_{z_0}\tilde{P}^*_{z_0} \tag{4.42}$$

where the superscript s refers to scattered radiation and

$$K = \frac{(\omega')^4}{16\pi^2\varepsilon_0^2 c^4 r^2} = \frac{(\tilde{\nu}')^4\pi^2}{\varepsilon_0^2 r^2} \tag{4.43}$$

where ω' is the circular frequency and $\tilde{\nu}'$ the wavenumber of the *scattered* radiation. Similarly,

$$\tilde{E}^s_{y_0}\tilde{E}^{s*}_{y_0} = K\tilde{P}_{y_0}\tilde{P}^*_{y_0}; \quad \tilde{E}^s_{y_0}\tilde{E}^{s*}_{z_0} = K\tilde{P}_{y_0}\tilde{P}^*_{z_0} \quad \text{etc.} \tag{4.44}$$

After some algebraic and trigonometric manipulation, we find, using appropriate forms of eqs. (2.60) to (2.63), and the space-averages of eqs. (3.23) to (3.25) that the Stokes parameters for the scattered radiation along the x-axis are given for *a single scattering molecule* by

$$S_0(\pi/2) = \tfrac{1}{2}KE_0^2\left\{\left(\frac{45a^2+13\gamma^2}{45}\right) - \left(\frac{45a^2+\gamma^2}{45}\right)P\cos 2\chi\cos 2\psi\right\} \tag{4.45}$$

$$S_1(\pi/2) = \tfrac{1}{2}KE_0^2\left\{\left(\frac{45a^2+\gamma^2}{45}\right)(1-P\cos 2\chi\cos 2\psi)\right\} \tag{4.46}$$

$$S_2(\pi/2) = 0 \tag{4.47}$$

$$S_3(\pi/2) = 0 \tag{4.48}$$

where the total irradiance of the incident radiation is given by $\mathcal{I} = \frac{1}{2}c\varepsilon_0 E_0^2$, and to simplify the notation we have used a^2 for $[a]^2_{v^f v^i}$ and γ^2 for $[\gamma]^2_{v^f v^i}$. This simplified notation will be used throughout this section. Similarly, for scattering along the positive z-axis ($\theta = 0°$) and along the negative z-axis ($\theta = 180°$), the Stokes parameters of the scattered radiation are found to be

$$S_0(0) = S_0(\pi) = KE_0^2\left(\frac{45a^2 + 7\gamma^2}{45}\right) \tag{4.49}$$

$$S_1(0) = S_1(\pi) = KE_0^2\left(\frac{45a^2 + \gamma^2}{45}\right)P \cos 2\chi \cos 2\psi \tag{4.50}$$

$$S_2(0) = -S_2(\pi) = KE_0^2\left(\frac{45a^2 + \gamma^2}{45}\right)P \cos 2\chi \sin 2\psi \tag{4.51}$$

$$S_3(0) = -S_3(\pi) = KE_0^2\left(\frac{45a^2 - 5\gamma^2}{45}\right)P \sin 2\chi \tag{4.52}$$

We now illustrate how these Stokes parameters may be used to generate the intensities, depolarization ratios, reversal coefficients, and degrees of circularity of the scattered radiation for incident radiation of various states of polarization. We recall that for natural radiation $P = 0$; for linearly polarized $P = 1$, $\chi = 0$ with $\psi = 0$ if the radiation has $E_x \neq 0$ (in this case, parallel to the scatter plane) and $\psi = \pi/2$ if $E_y \neq 0$ (in this case, perpendicular to the scatter plane); and for circularly polarized radiation $P = 1$, $\psi = 0$ with $\chi = \pi/4$ for right circularly polarized radiation and $\chi = -\pi/4$ for left circularly polarized radiation.

Using eq. (2.90), we obtain the general expression,

$$\frac{{}^{p_i}I_{\|}(\pi/2)}{{}^{p_i}I_{\perp}(\pi/2)} = \frac{6\gamma^2}{(45a^2 + 7\gamma^2) - (45a^2 + \gamma^2)P \cos 2\chi \cos 2\psi} \tag{4.53}$$

from which it is readily found that

$$\rho_{\perp}(\pi/2) = \frac{3\gamma^2}{45a^2 + 4\gamma^2} \tag{4.54}$$

$$\rho_{\|}(\pi/2) = 1 \tag{4.55}$$

and

$$\rho_n(\pi/2) = \frac{6\gamma^2}{45a^2 + 7\gamma^2} \tag{4.56}$$

Using eq. (2.91), we find that for either right or left circularly polarized incident radiation the reversal coefficients are given by

$$\mathcal{P}(0) = \frac{6\gamma^2}{45a^2 + \gamma^2} = \mathcal{P}^{-1}(\pi) \tag{4.57}$$

Using eq. (2.92), the degrees of circularity are given by

$${}^{p_i}\mathcal{C}(0) = \frac{45a^2 - 5\gamma^2}{45a^2 + 7\gamma^2}P \sin 2\chi = -{}^{p_i}\mathcal{C}(\pi) \tag{4.58}$$

It follows from the definitions of Stokes parameters on p. 27 that for observation along x

$$^{p_i}I_{\perp}(\pi/2)=\tfrac{1}{2}K'\{S_0(\pi/2)+S_1(\pi/2)\} \tag{4.59}$$

and

$$^{p_i}I_{\parallel}(\pi/2)=\tfrac{1}{2}K'\{S_0(\pi/2)-S_1(\pi/2)\} \tag{4.60}$$

To relate the scattered intensities to the irradiance of the incident radiation we define

$$K'=\frac{r^2\mathcal{I}}{E_0^2} \tag{4.61}$$

Hence using eqs. (4.45) and (4.46) with (4.59) and (4.60)

$$^{p_i}I_{\perp}(\pi/2)=\tfrac{1}{90}k_{\tilde{\nu}}(\tilde{\nu}')^4[(45a^2+7\gamma^2)-(45a^2+\gamma^2)P\cos 2\chi\cos 2\psi]\mathcal{I} \tag{4.62}$$

and

$$^{p_i}I_{\parallel}(\pi/2)=k_{\tilde{\nu}}(\tilde{\nu}')^4\left(\frac{\gamma^2}{15}\right)\mathcal{I} \tag{4.63}$$

where $k_{\tilde{\nu}}$ is defined by eq. (3.59). Similarly, for observation at 0° along z, we have

$$^{p_i}I_{\circledR}(0)=\tfrac{1}{2}K'\{S_0(0)+S_3(0)\} \tag{4.64}$$

and

$$^{p_i}I_{\copyright}(0)=\tfrac{1}{2}K'\{S_0(0)-S_3(0)\} \tag{4.65}$$

Hence using eqs. (4.52) and (4.49) with (4.64) and (4.65)

$$^{p_i}I_{\circledR}(0)=\tfrac{1}{90}k_{\tilde{\nu}}(\tilde{\nu}')^4[(45a^2+7\gamma^2)+(45a^2-5\gamma^2)P\sin 2\chi]\mathcal{I} \tag{4.66}$$

$$^{p_i}I_{\copyright}(0)=\tfrac{1}{90}k_{\tilde{\nu}}(\tilde{\nu}')^4[(45a^2+7\gamma^2)-(45a^2-5\gamma^2)P\sin 2\chi]\mathcal{I} \tag{4.67}$$

The Stokes parameters for various situations are summarized in Table E in the central reference section. General formulae for the intensities and polarization properties of the scattered radiation derived from the Stokes parameters are given in Table F in the central reference section.

4.3.9 *Vibrational selection rules*

We have already seen (page 80) that the quantum mechanical treatment presented here leads to the requirement that, in the approximation of electrical and mechanical harmonicity, at least one of the components of the derived polarizability tensor taken at the equilibrium position must be non-zero for a fundamental vibrational transition to be Raman active. In this form, the selection rule corresponds to that obtained from the classical treatment. The disadvantages of this form of the selection rule were discussed in chapter 3: it depends on the assumption of harmonicity, and it is not easy to assess whether components of the derived polarizability tensor will be non-zero, for a given vibration, particularly in larger molecules.

Fortunately, we can establish a much better basis for determining selection rules for vibrational transitions in the Raman effect, if we consider the properties of the vibrational transition polarizability components, rather than the derived polarizability tensor components. For a particular vibrational transition to be Raman active, at least one of the six tensor components of the type $[\alpha_{xy}]_{v^f v^i}$ must be non-zero. Now, for a *fundamental* vibrational transition, where in the initial state all vibrational quantum numbers are zero and in the final state only the k-th vibrational quantum number has changed to unity, we may write

$$[\alpha_{xy}]_{v^f v^i} = \langle \phi_1(Q_k)|\alpha_{xy}|\phi_0(Q_k)\rangle \tag{4.68}$$

It is possible to ascertain whether the integral in eq. (4.68) is zero or non-zero without actually evaluating it, by considering its symmetry properties. It can be rigorously established by group theory arguments that *this integral will be non-zero, only if* α_{xy} *and* $\phi_1(Q_k)$ *belong to the same symmetry species*, which implies that, under each symmetry operation of the molecule in question, α_{xy} and $\phi_1(Q_k)$ transform in the same way. This constitutes a *general* selection rule for the Raman activity of a *fundamental* vibrational transition. In its most general form, covering all types of transitions, the selection rule is as follows: a transition between two states characterized by the wave functions ψ_i and ψ_f is Raman forbidden unless at least one of the triple products of the type $\psi_f\alpha_{xy}\psi_i$ belongs to a representation whose structure contains the totally symmetric species. We shall restrict ourselves here to the case of fundamentals. Since overtones and combinations are relatively rarely observed in Raman scattering, this restriction is not of great practical consequence.

We may gain an insight into the nature of the above rule if we consider in more detail the case of *non-degenerate* fundamental vibrations. Then, the condition that the integral in eqn. (4.68) is non-zero amounts to requiring that it is totally symmetric, i.e., that it goes into itself under all the symmetry operations of the molecule in question. Let us examine in turn the symmetry properties of each of the terms in the integral. A simple harmonic wave function for the ground state ($v_k = 0$) is always totally symmetric; and a simple harmonic wave function with $v_k = 1$ has the same symmetry species as the normal coordinate Q_k. Thus, the integral in eq. (4.68) is totally symmetric if the product $Q_k\alpha_{xy}$ is totally symmetric. For non-degenerate vibrations, this condition is satisfied if for each of the symmetry operations of the molecule Q_k and α_{xy} either both change sign or both remain unchanged; i.e., Q_k and α_{xy} belong to the same symmetry species. Similar arguments based on symmetry can, of course, be applied to Rayleigh scattering for which, in eq. (4.68), $v^f = v^i$.

It should be noted that the above conclusions were based on the properties of simple harmonic oscillator wave functions. However, it can be established that they also apply for anharmonic oscillator functions. Mathematically, this is a consequence of the fact that anharmonic oscillator functions can always be expressed as linear combinations of harmonic oscillator functions. Physically, we can recognize that this is because the presence of anharmonicity, while modifying the classical form of vibration and its energy, cannot change the symmetry of the potential energy so that the symmetry of the anharmonic mode is still that of the harmonic mode. Now, the harmonic oscillator functions have known mathematical forms whereas, in general, the anharmonic oscillator functions do not, so it is a most fortunate circumstance that we can make deductions about symmetry on the basis of the harmonic functions and apply them with equal validity to the anharmonic case.

It should also be noted that the above arguments did not involve any assumptions about electrical harmonicity. Thus, the selection rule stated above is a truly general one for fundamental vibrations and is completely independent of any assumptions about mechanical or electrical harmonicity.

The application of this general rule requires a knowledge of the symmetry species of the polarizability tensor components and the normal coordinates for a particular molecule of a given symmetry. The symmetry species of the polarizability tensor components depend only on the point group to which the molecule belongs, and so can be determined once and for all for each point group. However, the symmetry species of the normal coordinates depend on the molecule in question, as well as on the point group to which it belongs, and so have to be established for each particular case.

We shall have to confine ourselves to illustrating the application of the general selection rule to a particular molecule, in this case the non-linear $A—B—A$ molecule which belongs to the point group C_{2v}. The molecular geometry and the axis system is shown in Fig. 4.2(a). There are four symmetry operations which send the molecule into itself: the identity E, $C_2(z)$, the rotation through 180° about the z-axis, and σ_{yz} and σ_{xz}, reflections in the yz and xz planes. The permitted symmetry species for this point group and their conventional labels are tabulated in Fig. 4.2(b). For the present purposes, we may regard each of the four rows of Fig. 4.2(b) as representing one of the permitted transformation patterns for a normal coordinate, or a polarizability tensor component, or a dipole moment component under the operations of the group. The labelling of the axes and the symmetry species in Fig. 4.2 follows the recommended practice.[3]

We now consider the symmetry species of the normal coordinates. In this treatment, we shall have to assume the forms of the three normal modes of vibration for this molecule. The forms of these vibrations and their behaviour under the symmetry operations of the C_{2v} point group are shown in Fig. 4.2(c), and enable us to identify Q_1 and Q_2 as belonging to the a_1 symmetry species and Q_3 to the b_2 symmetry species. A more detailed acquaintance with group theory procedures than we are assuming here would enable the symmetry species of the fundamental vibrations to be deduced without assuming the form of the vibrations, provided the *equilibrium* symmetry of the molecule is known.

We must now consider the symmetry properties of the polarizability tensor components. Suppose a static electric field is applied to the molecule along the y-direction ($E_y \neq 0$, $E_x = E_z = 0$). Then P_x, the induced dipole along the x-direction, is given by

$$P_x = \alpha_{xy} E_y \tag{4.69}$$

Now P_x and E_y can only remain unchanged or change sign for a symmetry operation in the point group C_{2v}, and thus α_{xy} will remain unchanged or change sign, depending on whether P_x and E_y behave in the same or in the opposite way for a given symmetry operation. The behaviour of P_x and E_y under each of the symmetry operations for the point group C_{2v} and the behaviour of α_{xy} that follows from this are given below:

	E	$C_2(z)$	σ_{xz}	σ_{yz}
$E_y \rightarrow$	E_y	$-E_y$	$-E_y$	E_y
$P_x \rightarrow$	P_x	$-P_x$	P_x	$-P_x$
Hence $\alpha_{xy} \rightarrow$	α_{xy}	α_{xy}	$-\alpha_{xy}$	$-\alpha_{xy}$

(a)

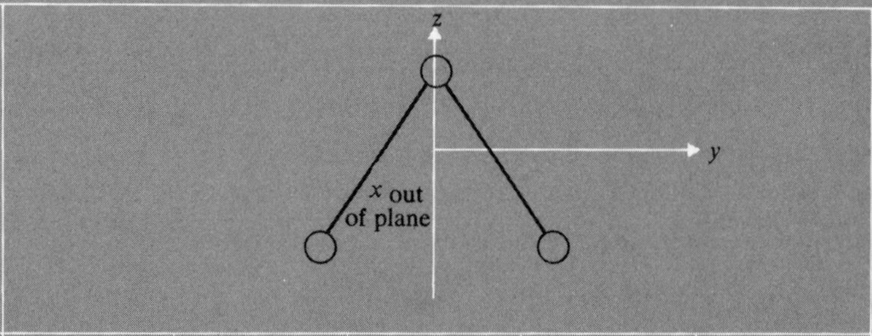

(b)

	E	$C_2(z)$	σ_{xz}	σ_{yz}
a_1	1	1	1	1
a_2	1	1	−1	−1
b_1	1	−1	1	−1
b_2	1	−1	−1	1

(c)

(d)

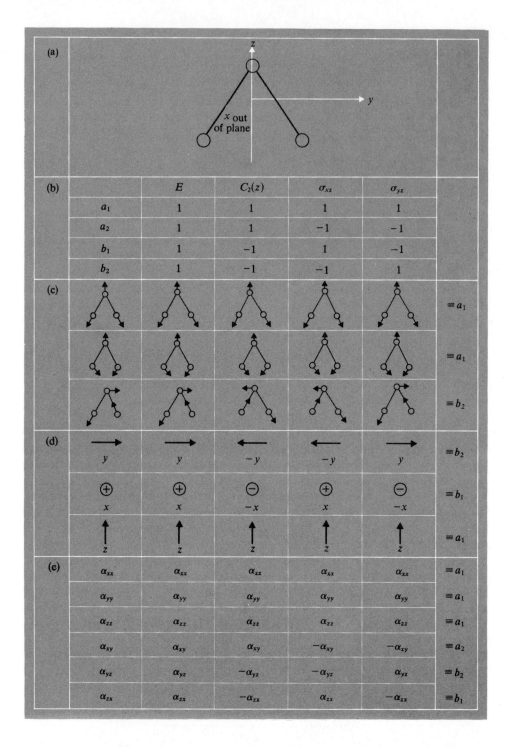

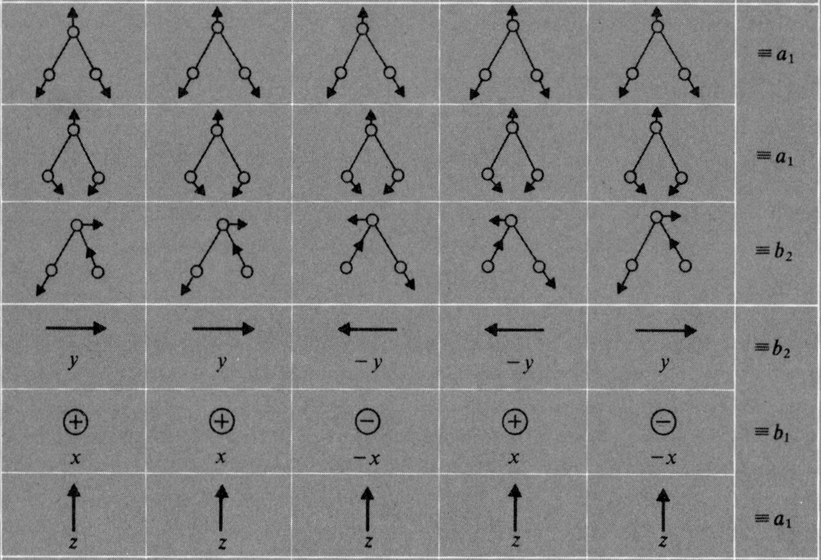

(e)

α_{xx}	α_{xx}	α_{xx}	α_{xx}	α_{xx}	≡ a_1
α_{yy}	α_{yy}	α_{yy}	α_{yy}	α_{yy}	≡ a_1
α_{zz}	α_{zz}	α_{zz}	α_{zz}	α_{zz}	≡ a_1
α_{xy}	α_{xy}	α_{xy}	$-\alpha_{xy}$	$-\alpha_{xy}$	≡ a_2
α_{yz}	α_{yz}	$-\alpha_{yz}$	$-\alpha_{yz}$	α_{yz}	≡ b_2
α_{zx}	α_{zx}	$-\alpha_{zx}$	α_{zx}	$-\alpha_{zx}$	≡ b_1

Fig. 4.2 A non-linear $A—B—A$ molecule, point group C_{2v}: (a) definition of molecular geometry and axis system; (b) symmetry species; (c) symmetry of fundamental vibrations; (d) symmetry of translational displacements; and (e) symmetry of polarizability components

It can be seen immediately that E_y belongs to the b_2 symmetry species, P_x to the b_1 species, and α_{xy} to the a_2 species. Similarly, it may be established that α_{xx}, α_{yy}, and α_{zz} belong to the a_1 representation, α_{yz} to the b_2 representation, and α_{xz} to the b_1 representation (see Fig. 4.2(e)). It is not difficult to see that the components of the vector quantities $\boldsymbol{P}$ and $\boldsymbol{E}$ transform in the same way as unit displacements x, y, z along the corresponding axes, and the components of the polarizability tensor in the same way as the products of these displacements, i.e., α_{xy} as xy, and so on. The transformation properties of x, y, z are shown in Fig. 4.2(d) for the point group C_{2v}. It can also be seen from these considerations that the symmetry properties of the translational displacements, the dipole moment and electric field components, and the polarizability tensor components depend only on the point group concerned.

With the symmetry species of the normal coordinates and the polarizability components established, application of the general selection rule (see page 88) shows that all three fundamental vibrations of this molecule are Raman active. The totally symmetric vibrations Q_1 and Q_2 of the a_1 symmetry species have α_{xx}, α_{yy}, and α_{zz} non-zero, and the vibration Q_3 of the b_2 symmetry species has α_{yz} non-zero. We can also see that, in general, for molecules belonging to the C_{2v} point group, fundamental vibrations of all four symmetry classes will be Raman active.

For the sake of completeness, we shall consider briefly the infrared activity of the fundamental vibrations of this molecule. It is not difficult to see that, since the transition moment for infrared absorption involves the permanent dipole moment operator instead of the induced dipole moment operator, infrared activity of a non-degenerate fundamental vibration Q_k requires products of the type Q_kx, Q_ky, Q_kz to be totally symmetric since x transforms as the x component of the dipole moment, and so on. Thus, Q_k and at least one of the translational displacements x, y, z must belong to the same symmetry species. Thus, it can be seen (see Fig. 4.2d) that all three fundamental vibrations of this molecule are infrared active. We also note that, in general, for molecules belonging to the C_{2v} point group, fundamental vibrations of a_1, b_1, and b_2 symmetry will be infrared active, but those of a_2 symmetry will be infrared inactive.

Although the above considerations have been restricted to fundamental transitions, other types of vibrational transitions are possible in principle: e.g., overtones for which $\Delta v_k = \pm 2, \pm 3$, etc., and binary combination bands for which $\Delta v_k = \pm 1$ and $\Delta v_k = \pm 1$ simultaneously. Such transitions may be infrared and/or Raman active, and symmetry arguments can also be used to relate the activity to the symmetry of the vibrations. Such transitions generally have a much lower intensity in Raman scattering than in infrared absorption.

Symmetry arguments can also be used to make predictions about depolarization ratios. By analogy with eq. (3.83), the depolarization ratio $\rho_\perp(\pi/2)$ may be written in terms of invariants of the transition polarizability tensor associated with the vibrational transition $v^f \leftarrow v^i$:

$$\rho_\perp(\pi/2) = \frac{3[\gamma]^2_{v^f v^i}}{45[a]^2_{v^f v^i} + 4[\gamma]^2_{v^f v^i}} \tag{4.70}$$

where $[a]_{v^f v^i}$ and $[\gamma]_{v^f v^i}$ are, respectively, the mean value and the anisotropy invariants of a tensor whose components are of the type $[\alpha_{xx}]_{v^f v^i}$, $[\alpha_{xy}]_{v^f v^i}$, and so on, as defined by eq. (4.9). It follows that

$$[a]_{v^f v^i} = \tfrac{1}{3}\langle \phi_{v^f} | (\alpha_{xx} + \alpha_{yy} + \alpha_{zz}) | \phi_{v^i} \rangle \tag{4.71}$$

Now, for the k-th fundamental transition, symmetry arguments require this integral to be zero unless $(\alpha_{xx}+\alpha_{yy}+\alpha_{zz})$ belongs to the same symmetry species as the Q_k. But the symmetry of $(\alpha_{xx}+\alpha_{yy}+\alpha_{zz})$ is the same as that of $(x^2+y^2+z^2)$. Since $(x^2+y^2+z^2)$ is spherically symmetrical, it will remain unchanged under all symmetry operations and so can only belong to the totally symmetric species. Thus for all fundamental transitions which are not totally symmetric, $[a]_{v^f v^i}=0$, but of course $[\gamma]^2_{v^f v^i}$ must be non-zero if the vibration is Raman active. This leads us to a general rule concerning depolarization ratios for fundamental vibrations in Raman scattering; only vibrational fundamentals belonging to the totally symmetric class can have $\rho_\perp(\pi/2)<\frac{3}{4}$; for all other fundamentals, $\rho_\perp(\pi/2)=\frac{3}{4}$. In the special case of spherically symmetric molecules as, for example, molecules of tetrahedral symmetry like CH_4 or octahedral symmetry like SF_6, $[\gamma]^2_{v^f v^i}$ is zero for totally symmetric vibrations and so $\rho_\perp(\pi/2)=0$.

It can be seen that the experimental determination of depolarization ratios enables totally symmetric vibrations to be distinguished from the rest. Of course, if the measured $\rho_\perp(\pi/2)$ value is close to $\frac{3}{4}$, the certainty of this distinction is reduced. Unfortunately, $\rho_\perp(\pi/2)$ values do not enable distinction to be made between non-totally symmetric vibrations of different symmetry classes. Such information can often be extracted from measurements on oriented single crystals when in favourable cases the relative magnitudes of the individual tensor components can be determined (see chapter 3, page 63). In liquids and gases, this information is lost as a result of the need to average over all configurations.

We shall not pursue further, considerations of symmetry and Raman and infrared activity. However, in Appendix II will be found the symmetry properties of the translational displacements x, y, z and the polarizability components for all the point groups. It is clear from the example considered above that this information enables the infrared and Raman activity of vibrations in a molecule to be immediately ascertained, provided the point group is known and the distribution of the vibrations among the symmetry classes has been calculated. The procedures for assigning normal vibrations to symmetry classes must be sought in standard texts on group theory and molecular vibrations. Appendix II also includes the symmetry properties of the hyperpolarizability components which, we shall see later (chapter 8), control the activity of vibrations in the hyper Raman effect.

Appendix III outlines how the transformation between tensor components referred to molecular axes and tensor components referred to crystal axes may be effected, and illustrates how two non-totally symmetric vibrations of different symmetries may be distinguished by measurements on oriented single crystals.

4.3.10 *Vibrational wavenumber patterns*

Now that the selection rules have been established, it is a straightforward matter to obtain expressions for the magnitudes of the wavenumber *shifts* $\Delta\tilde{\nu}$ observed in the Raman effect as a result of permitted vibrational transitions. The absolute wavenumbers of the Stokes and anti-Stokes Raman lines are, of course, given by $\tilde{\nu}_0-|\Delta\tilde{\nu}|$ and $\tilde{\nu}_0+|\Delta\tilde{\nu}|$ respectively.

The energy of a harmonic oscillator E_{vib} is given by

$$E_{\text{vib}}=(v+\tfrac{1}{2})hc\tilde{\nu} \tag{4.72}$$

where $\tilde{\nu}$ is the oscillator frequency in wavenumbers. For spectroscopic purposes, the vibrational energy is expressed more conveniently in wavenumbers. It is then described as a vibrational term and denoted by $G(v)$. Clearly, $G(v) = E_{\text{vib}}/hc$ and thus, in the harmonic approximation,

$$G(v) = (v + \tfrac{1}{2})\tilde{\nu} \tag{4.73}$$

The energy levels of a harmonic oscillator are therefore equally spaced (see Fig. 4.3(a)). The vibrational selection rule $\Delta v = \pm 1$ will thus lead to the appearance of a Raman line associated with the vibrational transition $v_k + 1 \leftrightarrow v_k$ with a wavenumber shift equal to $\tilde{\nu}_k$, the wavenumber of the k-th normal mode, *provided* the symmetry

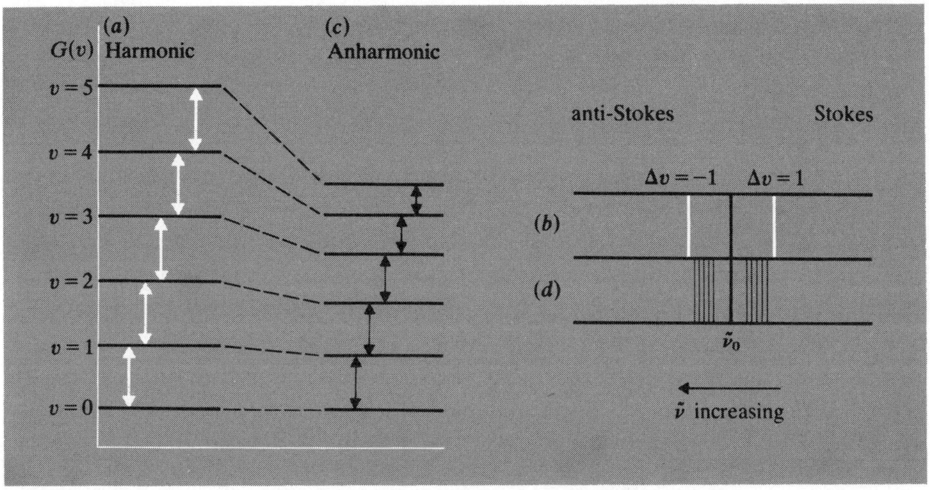

Fig. 4.3 Energy levels (a) and Stokes and anti-Stokes vibration Raman spectra (b) for a harmonic oscillator. Energy levels (c) and Stokes and anti-Stokes vibration Raman spectra (d) for an anharmonic oscillator (x_e positive)

properties of the k-th normal coordinate Q_k and at least one of the polarizability tensor components satisfy the necessary conditions. Thus, in the harmonic oscillator approximation, the Stokes and anti-Stokes Raman lines associated with all transitions of the k-th oscillator of the type $v_k + 1 \leftrightarrow v_k$ will have the same wavenumber shift irrespective of the value of v (see Fig. 4.3(b)).

When anharmonicity is taken into account, the vibrational term for a diatomic molecule has the form

$$G(v) = (v + \tfrac{1}{2})\tilde{\nu}_e - (v + \tfrac{1}{2})^2\tilde{\nu}_e x_e \ldots \tag{4.74}$$

where higher order terms are neglected. Here $\tilde{\nu}_e$ is the frequency in wavenumbers of the oscillator at the equilibrium internuclear separation and x_e is the anharmonicity. The term $\tilde{\nu}_e x_e$ is very much smaller than $\tilde{\nu}_e$, and, for the choice of signs in eq. (4.74), x_e is practically always positive. In the anharmonic case, the energy levels are not equally spaced (see Fig. 4.3(c)). Applying the selection rule $\Delta v = \pm 1$, the wavenumber shift of a Raman line associated with the transition $v + 1 \leftrightarrow v$ will be given by

$$|\Delta\tilde{\nu}| = \tilde{\nu}_e - 2(v + 1)\tilde{\nu}_e x_e \tag{4.75}$$

Thus, in the anharmonic oscillator approximation, the wavenumber shift of a Stokes or anti-Stokes vibrational line associated with the normal mode now depends on the vibrational quantum number v. Consequently, Raman wavenumber shifts arising from different transitions of the type $v+1 \leftrightarrow v$ will not coincide exactly (see Fig. 4.3(d)). For the special case of the transitions $1 \leftrightarrow 0$, eq. (4.75) becomes

$$|\Delta\tilde{\nu}| = \tilde{\nu}_e - 2\tilde{\nu}_e x_e \tag{4.76}$$

The above formulae may be expressed in an alternative form. Since the lower energy level of the anharmonic oscillator corresponds to $v = 0$, we have

$$G(0) = \tfrac{1}{2}\tilde{\nu}_e - \tfrac{1}{4}\tilde{\nu}_e x_e \tag{4.77}$$

Then, if the vibrational energy levels are related to $G(0)$, we have

$$G_0(v) = v\tilde{\nu}_0 - v^2\tilde{\nu}_0 x_0 \tag{4.78}$$

where

$$\tilde{\nu}_0 = \tilde{\nu}_e - \tilde{\nu}_e x_e \tag{4.79}$$

and

$$\tilde{\nu}_0 x_0 = \tilde{\nu}_e x_e \ldots \tag{4.80}$$

In terms of $\tilde{\nu}_0$ and x_0, eq. (4.76) becomes

$$|\Delta\tilde{\nu}| = \tilde{\nu}_0 - \tilde{\nu}_0 x_0 \tag{4.81}$$

If the wavenumber shifts of, say, a fundamental transition ($v = 1 \leftrightarrow v = 0$) and a first overtone ($v = 2 \leftrightarrow v = 0$) are observed, then it is possible to determine $\tilde{\nu}_e$ and x_e (or $\tilde{\nu}_0$ and x_0).

For a diatomic molecule, the above considerations suffice to determine the pattern of the vibrational Raman spectrum, at least qualitatively, but, of course, the actual wavenumber shift(s) and their intensities depend on the mechanical and electrical properties of the molecule; i.e., upon the masses and force constants and the invariants $(a')_k$ and $(\gamma')_k$ of the derived polarizability tensor. In the case of polyatomic molecules, similar considerations determine only the wavenumber pattern associated with a particular normal mode which is Raman active. The number of such Raman active modes depends on the number of atoms in the molecule and the symmetry of their spatial arrangement; and their relative intensities depend on the relative values of $(a')_k$ and $(\gamma')_k$. Thus, we can make no useful general predictions about the pattern of a vibrational Raman spectrum. However, in specific cases the number of vibrational Raman lines can be predicted and experience enables estimates to be made of the likely wavenumbers and intensities of particular types of vibrations, as we shall see in chapter 7.

4.4 Rotation and vibration–rotation Raman transitions

4.4.1 *General consideration of the rotation and vibration–rotation transition polarizability*

The selection rules, states of polarization, and intensities of rotational Raman scattering may be obtained by considering the properties of matrix elements of the type:

$$[\alpha_{xy}]_{fi} = \sum_{x'y'} [\alpha_{x'y'}]_{v^f v^i} \langle \Theta_{R^f} | \cos(x'x)\cos(y'y) | \Theta_{R^i} \rangle \tag{4.82}$$

where x, y, z are space-fixed Cartesian axes and x', y', z' are molecule-fixed Cartesian axes. The vibrational matrix elements $[\alpha_{x'y'}]_{v^f v^i}$ have already been evaluated for the case of electrical and mechanical harmonicity (pages 78–80). Using these results, if there is no change of vibrational quantum number, $(v^f = v^i = v)$, eq. (4.82) yields the rotational transition polarizability component

$$[\alpha_{xy}]_{v,R^f;v,R^i} = \sum_{x'y'} (\alpha_{x'y'})_0 \langle \Theta_{R^f} | \cos(x'x)\cos(y'y) | \Theta_{R^i} \rangle \tag{4.83}$$

where $(\alpha_{x'y'})_0$ is a component of the equilibrium polarizability tensor. Likewise, if the k-th vibrational quantum number changes from v_k^i to v_k^i+1 and all the others remain unchanged $(v^f = [v], v_k^i+1; v^i = [v], v_k^i)$ then eq. (4.82) yields the Stokes vibration–rotation transition polarizability component

$$[\alpha_{xy}]_{[v],v_k^i+1,R^f;[v],v_k^i,R^i} = (v_k^i+1)^{1/2} b_{v_k} \sum_{x'y'} (\alpha'_{x'y'})_k \langle \Theta_{R^f} | \cos(x'x)\cos(y'y) | \Theta_{R^i} \rangle \tag{4.84}$$

where $(\alpha'_{x'y'})_k$ is a component of the derived polarizability tensor associated with the k-th normal coordinate. The anti-Stokes vibration–rotation transition polarizability component is obtained from eq. (4.84) by replacing $(v_k^i+1)^{1/2}$ by $(v_k^i)^{1/2}$.

It must be noted that, whereas for all molecules the vibrational wave functions have the same mathematical form, the rotational wave functions depend on the symmetry of the molecule, and in the absence of symmetry cannot even be expressed in an explicit closed form. To treat all the possible cases in detail is beyond the scope of this book and we shall content ourselves with an analysis of the simplest case which will illustrate the principles involved. Some results for other, common cases will then be tabulated.

4.4.2 *Rotation and vibration–rotation transition polarizabilities for diatomic molecules*

We shall examine the case of a diatomic molecule for which $\alpha_{X'X'} = \alpha_{Y'Y'} \neq \alpha_{Z'Z'}$, where X', Y', Z', are molecule-fixed Cartesian axes chosen to coincide with the principal axes of the polarizability ellipsoid. The molecule will be assumed to be in a non-degenerate ground electronic state. Consider illumination of this molecule with plane polarized radiation such that $E_z \neq 0$ and $E_x = E_y = 0$, and let us proceed to evaluate $[P_z^{(1)}]_0$, the transition moment amplitude associated with pure rotational transitions. It follows from eq. (4.4) that

$$[P_{z0}^{(1)}] = [\alpha_{zz}]_{v,R^f;v,R^i} E_{z0} \tag{4.85}$$

On expansion of the summation in eq. (4.83) we obtain the following expression for $[\alpha_{zz}]_{v,R^f;v,R^i}$:

$$[\alpha_{zz}]_{v,R^f;v,R^i} = \langle \Theta_{R^f} | \alpha_{X'X'} \cos^2(X'z) + \alpha_{Y'Y'} \cos^2(Y'z) + \alpha_{Z'Z'} \cos^2(Z'z) | \Theta_{R^i} \rangle \tag{4.86}$$

The rotational wave function for a diatomic molecule is given by

$$\Theta_R = N_R P_J^{|M|} (\cos\theta)\, e^{iM\phi} \tag{4.87}$$

and

$$d\tau_R = \sin\theta \, d\theta \, d\phi \tag{4.88}$$

It can be seen that for this system the *set* of rotational quantum numbers R involves two quantum numbers J and M. J can take the values 0, 1, 2, . . . and defines the rotational energy. M can take the values 0, ±1, ±2, . . . , $\pm J$ and relates to the spatial degeneracy of the rotational state J which is $2J+1$ degenerate. The value of M represents, in units of $h/2\pi$, the component of the angular momentum associated with J in the direction of the space-fixed axis, z. The function $P_J^{|M|}(\cos\theta)$ is an associated Legendre polynomial and N_R is a normalization constant. The angles θ and ϕ are defined in Fig. 4.4: θ is the angle between the bond axis, Z', and the space-fixed axis, z, and ϕ is the azimuth of the bond axis taken about this space-fixed axis. The explicit forms of the associated Legendre polynomials need not concern us, but we shall be involved with certain of their properties which we shall introduce as required.

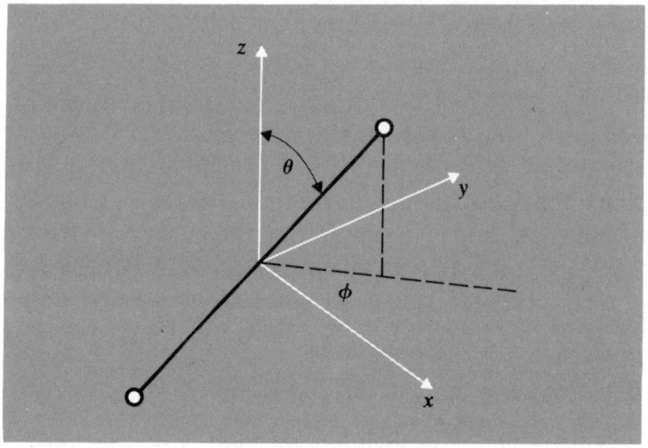

Fig. 4.4 Coordinate system for a rotating diatomic molecule

Equation (4.86) may be simplified by utilizing the property of direction cosines that

$$\cos^2(X'z) + \cos^2(Y'z) + \cos^2(Z'z) = 1 \tag{4.89}$$

putting $\alpha_{X'X'} = \alpha_{Y'Y'}$ and introducing $\cos(Z'z) = \cos\theta$. We then obtain

$$[\alpha_{zz}]_{v,R^f;v,R^i} = \alpha_{X'X'}\langle\Theta_{R^f}|\Theta_{R^i}\rangle + (\alpha_{Z'Z'} - \alpha_{X'X'})\langle\Theta_{R^f}|\cos^2\theta|\Theta_{R^i}\rangle \tag{4.90}$$

Since, as shown in chapter 3,

$$(a)_0 = \tfrac{1}{3}\{(\alpha_{X'X'}) + (\alpha_{Y'Y'}) + (\alpha_{Z'Z'})\} \tag{4.91}$$

and

$$(\gamma)_0 = \{(\alpha_{Z'Z'}) - (\alpha_{X'X'})\} \tag{4.92}$$

eq. (4.90) may be rewritten in terms of $(a)_0$ and $(\gamma)_0$, the invariants of the equilibrium polarizability tensor, as

$$[\alpha_{zz}]_{v,R^f;v,R^i} = (a)_0\langle\Theta_{R^f}|\Theta_{R^i}\rangle + (\gamma)_0\langle\Theta_{R^f}|(\cos^2\theta - \tfrac{1}{3})|\Theta_{R^i}\rangle \tag{4.93}$$

The properties of the associated Legendre polynomials are such that

$$\langle\Theta_{R^f}|\Theta_{R^i}\rangle = 0 \tag{4.94}$$

unless $R^f = R^i$, i.e., $\Delta J = 0$, $\Delta M = 0$, and

$$\langle\Theta_{R^f}|\cos^2\theta|\Theta_{R^i}\rangle = 0 \tag{4.95}$$

unless $\Delta J = 0, \pm 2$ and $\Delta M = 0$.

These properties immediately establish the selection rules for the J and M quantum numbers. As a result of these selection rules, only the following matrix elements of the polarizability tensor component $(\alpha_{zz})_0$ can be non-zero:

$$[\alpha_{zz}]_{v,J,M;v,J,M} \text{ and } [\alpha_{zz}]_{v,J\pm 2,M;v,J,M} \tag{4.96}$$

In a similar manner, it is possible to deduce which pure rotational matrix elements of the other polarizability tensor components are also non-zero. It is found that for all such components the same selection rules for J operate, namely, $\Delta J = 0, \pm 2$. The selection rules for M are either $\Delta M = 0$ or $\Delta M = \pm 1$ depending on the tensor component.

A similar analysis of harmonic vibration–rotation matrix elements for a diatomic molecule gives the following selection rules for v and J: $\Delta v = \pm 1$, $\Delta J = 0, \pm 2$. The M selection rules can be $\Delta M = 0$ or ± 1 depending on the tensor component.

4.4.3 *Rotation and vibration–rotation wavenumber patterns for diatomic molecules*

The wavenumber patterns in rotational and vibration–rotation Raman scattering can be deduced without further consideration of the M selection rules (since the M levels are degenerate in the absence of external fields) and without actual evaluation of the non-zero matrix elements. Of course, when scattered powers, scattering cross-sections, and polarization properties are required, all permitted transitions between M levels must be taken into account and the matrix elements evaluated. We shall return to these problems subsequently; meanwhile, we shall consider the wavenumber patterns in the pure rotation and vibration–rotation Raman spectra of diatomic molecules resulting from the operation of the v and J selection rules.

First, some general considerations. When $\Delta v = 0$, $\Delta J = 0$ relates simply to Rayleigh scattering. The selection rule $\Delta v = 0$, $\Delta J = +2$, relates to both Stokes *and* anti-Stokes rotational Raman scattering. The selection rule $\Delta J = -2$, which might be thought to relate to anti-Stokes rotational Raman scattering, does not arise for pure rotational Raman scattering. This is because ΔJ is defined as $J' - J''$, where J' refers to the state with the higher energy and J'' to the state with the lower energy. In consequence, if the system stays in the ground vibrational state, J' must always be greater than J'' and $\Delta J = -2$ cannot arise.

The selection rule $\Delta v = +1$ refers to Stokes Raman scattering and $\Delta v = -1$ to anti-Stokes Raman scattering. Now, when $\Delta v \neq 0$, there is a change in the vibrational energy. Thus, if there is also a change in the rotational energy, $\Delta J = -2$ $(J' < J'')$ can arise as well as $\Delta J = +2$, $(J' > J'')$, since then a rotational level in the higher vibrational state will necessarily have a higher energy than a rotational level in the lower vibrational state, irrespective of the relative J values.

The rotational energy E_R of a diatomic molecule is most conveniently expressed in wavenumbers, using the rotational term $F(J)$, where $F(J)=E_R/hc$. In the rigid rotor approximation, assuming the moment of inertia about the bond axis is zero, $F(J)$ is given by

$$F(J)=BJ(J+1) \tag{4.97}$$

where

$$B=\frac{h}{8\pi^2 Ic} \tag{4.98}$$

and I is the moment of inertia about an axis through the centre of gravity and perpendicular to the bond axis. If the masses of the two atoms are m_1 and m_2, and their distance apart is r, then it is easily shown that

$$I=\mu r^2 \tag{4.99}$$

where μ is termed the reduced mass and is given by

$$\mu=\frac{m_1 m_2}{m_1+m_2} \tag{4.100}$$

It should be noted that in real molecules the rigid rotor approximation does not apply, and centrifugal stretching arises because of the non-rigidity of the bond. Also the distance r and hence I and B become functions of the vibrational quantum number because of anharmonicity. These complications will be considered subsequently. Proceeding with the rigid rotor approximation, and combining the selection rule $\Delta J=+2$ with $F(J)$ as defined in eq. (4.97), we may write for the *magnitude* of the wavenumber shift associated with such a pure rotation transition

$$\begin{aligned}|\Delta\tilde{\nu}_S| &= F(J+2)-F(J)\\ &= B(J+2)(J+3)-BJ(J+1)\\ &= 4B(J+\tfrac{3}{2})\end{aligned} \tag{4.101}$$

where $J=0, 1, 2, \ldots$. The series of lines arising from transitions with $\Delta J=+2$ is termed an S branch, hence the designation of the wavenumber shift as $|\Delta\tilde{\nu}_S|$. The spectroscopic nomenclature for rotational branches is as follows: $\Delta J=0$ is termed a Q branch, $\Delta J=+1$, an R branch, $\Delta J=+2$, an S branch, and so on; $\Delta J=-1$, a P branch, $\Delta J=-2$, an O branch, and so on. Thus, provided the rotational levels are reasonably populated, the pure rotational Raman spectrum will consist of two series of lines or S branches, one on each side of the exciting line of wavenumber $\tilde{\nu}_0$. One series will be the Stokes lines (involving transitions of the type $J+2\leftarrow J$, for which $\Delta J=+2$), with absolute wavenumbers $\tilde{\nu}_0-|\Delta\tilde{\nu}_S|$, where the $|\Delta\tilde{\nu}_S|$ values are given by eq. (4.101). The other series will be the anti-Stokes lines (involving transitions of the type $J+2\rightarrow J$, for which ΔJ is also $+2$), with absolute wavenumbers given by $\tilde{\nu}_0+|\Delta\tilde{\nu}_S|$. The pattern of spacings is now easily deduced. When $J=0$, $|\Delta\tilde{\nu}_S|=6B$, when $J=1$, $|\Delta\tilde{\nu}_S|=10B$, when $J=2$, $|\Delta\tilde{\nu}_S|=14B$, and so on. Thus, in each S branch the first rotational line has a wavenumber shift of $6B$ from the exciting line and successive rotational lines have equal interline spacings of $4B$. These considerations apply to all heteronuclear diatomic molecules. In some homonuclear diatomic molecules, nuclear spin statistics can lead to the absence of odd or even J levels with a consequential effect on the pure rotational Raman spectrum. We shall consider nuclear spin statistics and their effects later. The pattern of rotational energy levels and the associated pure rotation Raman spectrum for a rigid heteronuclear diatomic molecule are shown in Fig. 4.5(a) and (b).

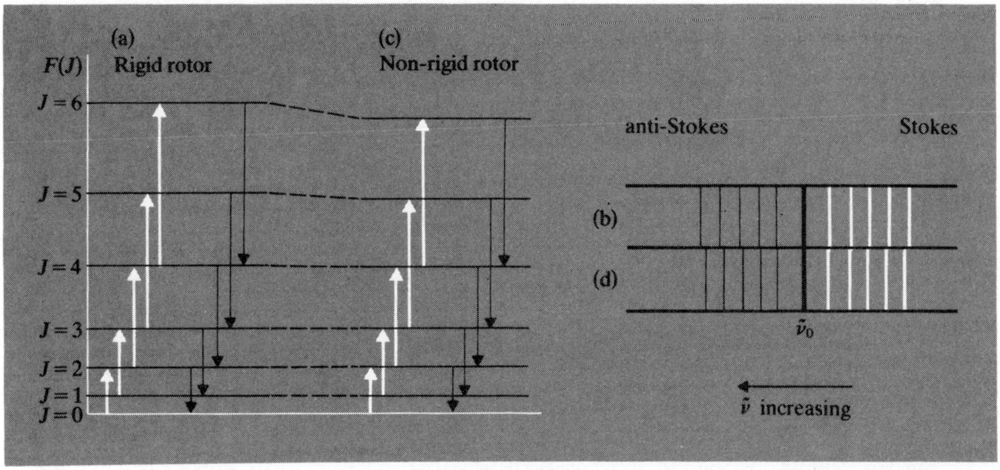

Fig. 4.5 Rigid *AB* molecule rotational energy levels (a) and Stokes and anti-Stokes rotational Raman spectra (b); non-rigid *AB* molecule rotational energy levels (c) and Stokes and anti-Stokes rotational Raman spectra (d)

Formulae for the wavenumber shifts associated with vibration–rotation transitions may be found in a similar way. We shall confine ourselves to Stokes Raman scattering. Then, using the rigid rotor term $F(J)$ (eq. 4.97), the harmonic oscillator vibration term $G(v)$ (eq. 4.73), and the selection rules $\Delta v=+1$, $\Delta J=+2$, we may write for the wavenumber shift of the lines in the Stokes S branch

$$\begin{aligned}|\Delta\tilde{\nu}_S| &= \{G(v+1)+F(J+2)\}-\{G(v)+F(J)\}\\ &= \tilde{\nu}+4B(J+\tfrac{3}{2})\end{aligned} \tag{4.102}$$

where $J=0, 1, 2, \ldots$.

Similarly, for the lines in the Stokes O branch ($\Delta v=+1$, $\Delta J=-2$), we have

$$\begin{aligned}|\Delta\tilde{\nu}_O| &= \{G(v+1)+F(J-2)\}-\{G(v)+F(J)\}\\ &= \tilde{\nu}-4B(J-\tfrac{1}{2})\end{aligned} \tag{4.103}$$

where $J=2, 3, 4, \ldots$, and, for the Stokes Q branch ($\Delta v=+1$, $\Delta J=0$),

$$|\Delta\tilde{\nu}_Q|=\tilde{\nu} \tag{4.104}$$

Thus for heteronuclear diatomic molecules, the Stokes Q branch will be a single line with wavenumber shift $\tilde{\nu}$. The S branch lines will have wavenumber shifts $\tilde{\nu}+6B$, $\tilde{\nu}+10B$, $\tilde{\nu}+14B$ etc., and the O branch lines $\tilde{\nu}-6B$, $\tilde{\nu}-10B$, $\tilde{\nu}-14B$ etc. Nuclear spin statistics can modify this pattern for certain homonuclear diatomic molecules.

We now consider some refinements to these formulae. When centrifugal stretching is taken into account, the rotational term is given by

$$F(J) = BJ(J+1) - DJ^2(J+1)^2 \tag{4.105}$$

where D is the centrifugal stretching constant which is always positive for the choice of sign in eq. (4.105). The ratio D/B is less than 10^{-4} so that centrifugal stretching effects are small. Using eq. (4.105), the magnitude of the wavenumber shift associated with pure rotation transitions ($\Delta J = +2$, S branch) is now given by

$$|\Delta\tilde{\nu}_S| = (4B - 6D)(J + \tfrac{3}{2}) - 8D(J + \tfrac{3}{2})^3 \tag{4.106}$$

where $J = 0, 1, 2 \ldots$. Thus, the rotational lines no longer have exactly the same spacing as in the rigid rotor approximation; the spacing decreases slightly as J increases (see Fig. 4.5(c) and (d)).

Equations (4.102) to (4.104) for vibration–rotation transitions must also be modified. The moment of inertia of a real molecule, and hence B, depends on its vibrational state. Thus, we must write for the vibrational state with quantum number v,

$$B_v = \frac{h}{8\pi^2 c\mu}\overline{\left[\frac{1}{r^2}\right]}_v \tag{4.107}$$

where $\overline{[1/r^2]}_v$ is the mean value of $[1/r^2]$ during the vibration. The use of a mean value for B can be seen to be reasonable since the period of vibration is small

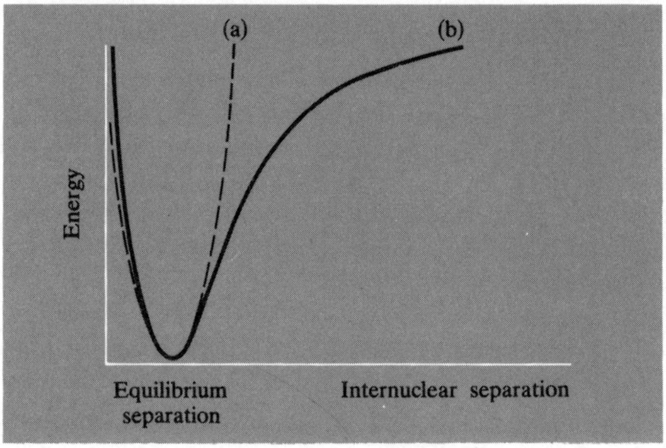

Fig. 4.6 Comparison of (a) harmonic and (b) anharmonic potentials for a diatomic molecule

compared to the period of rotation. Because of vibrational anharmonicity (see Fig. 4.6), it is to be expected that

$$B_v < B_e \tag{4.108}$$

where the subscript e refers to the equilibrium configuration. We may give expression to this inequality by writing

$$B_v = B_e - \alpha_e(v + \tfrac{1}{2}) \tag{4.109}$$

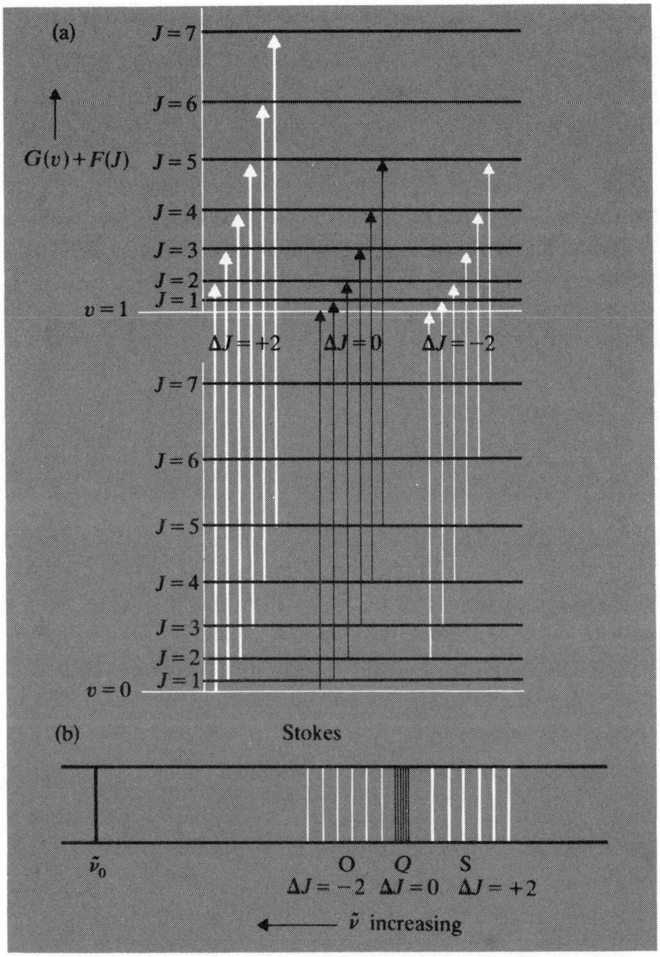

Fig. 4.7 Non-rigid and anharmonic AB molecule vibration and rotation energy levels (a) and Stokes vibration–rotation Raman spectra (b)

where α_e is positive. The rotational term for a diatomic molecule in the vibrational state with quantum number v is thus

$$F_v(J) = B_v J(J+1) \tag{4.110}$$

This formula neglects centrifugal stretching, but in vibration–rotation transitions the effect of change of B with vibrational quantum number is much more important than the effect of centrifugal stretching.

Thus, using eq. (4.110) for the rotational term $F_v(J)$, we may write for the wavenumber shift of a line in the Stokes S branch of a vibration–rotation Raman band ($\Delta J = +2$)

$$\begin{aligned} |\Delta\tilde{\nu}_S| &= \{G(v') + F_{v'}(J+2)\} - \{G\,(v'') + F_{v''}(J)\} \\ &= \{G(v') - G(v'')\} + \{B_{v'}(J+3)(J+2) - B_{v''}J(J+1)\} \end{aligned} \tag{4.111}$$

where $G(v)$ is the anharmonic vibrational term given by eq. (4.74), v' refers to the upper vibrational level and v'' to the lower vibrational level. Equation (4.111) reduces to

$$|\Delta\tilde{\nu}_S| = \{G(v')-G(v'')\}+6B_{v'}+(5B_{v'}-B_{v''})J+(B_{v'}-B_{v''})J^2 \qquad (4.112)$$

where $J=0, 1, 2\ldots$. Vibration–rotation bands in the Raman effect are almost invariably associated with the fundamental vibration transition $v=1 \leftarrow v=0$, and then $G(v')-G(v'')$ is given by eq. (4.76) or eq. (4.81). The corresponding formula for the Stokes O branch ($\Delta J=-2$) is given by

$$|\Delta\tilde{\nu}_O| = \{G(v')-G(v'')\}+2B_{v'}-(3B_{v'}+B_{v''})J+(B_{v'}-B_{v''})J^2 \qquad (4.113)$$

where $J=2, 3, \ldots$, and for the Q branch, ($\Delta J=0$) by

$$|\Delta\tilde{\nu}_Q| = \{G(v')-G(v'')\}+(B_{v'}-B_{v''})J+(B_{v'}-B_{v''})J^2 \qquad (4.114)$$

where $J=0, 1, 2, \ldots$. The general pattern of a vibration–rotation Raman band may be deduced from these equations if we remember that $B_{v'}<B_{v''}$ and thus $(B_{v'}-B_{v''})$ is negative. Then, from eq. (4.112), we see that the S branch consists of a series of lines, with higher wavenumber shifts than the pure vibration line, the first line having a shift of about $6B$ from the pure vibration line and successive lines having spacings of about $4B$, but with this spacing *decreasing* as J increases. From eq. (4.113) we see that the O branch consists of a series of lines with lower wavenumber shifts than the pure vibration line, the first line having a shift of about $6B$ from the pure vibration line and successive lines having spacings of about $4B$, but with this spacing *increasing* as J increases. From eq. (4.114), the Q branch will consist of a series of closely spaced lines with lower wavenumber shifts than the pure vibration line. Fig. 4.7 shows typical energy levels and the resultant Stokes Raman spectrum. Often the difference between $B_{v'}$ and $B_{v''}$ is so small that the lines of the Q branch will overlap and so the Q branch will appear much more intense than the O and S branches.

4.4.4 *Intensities and polarization characteristics for rotation and vibration Raman scattering from diatomic molecules*

To obtain expressions for scattered intensities it is necessary to evaluate the squares of the transition moment amplitudes for the appropriate transitions; and this involves calculating the squares of the matrix elements of the components of the polarizability tensor associated with the transition in question. In the absence of a magnetic field, the $2J+1$ states ($M=0, \pm 1, \pm 2, \ldots, \pm J$) associated with a given J value are energetically identical (i.e., degenerate) and so the ΔM transitions have the same frequency for a given ΔJ transition. Therefore, we need to calculate the averages over all M values of the squares of the matrix elements of the components of the polarizability tensor. We achieve this by squaring the matrix elements, summing over all M values, and dividing by $2J+1$. Thus, for the xy component, for which $\Delta M=0$, we have

$$\frac{1}{(2J+1)}\sum_{M''}[\alpha_{xy}]^2_{v',J',M'';v''J'',M''} = \overline{[\alpha_{xy}]^2_{v',J';v'',J''}} \qquad (4.115)$$

Some rather tedious algebra leads to the results given in Table I (central reference section) which involve J-dependent anisotropic coefficients $b_{J',J''}$.

For pure rotational transitions as might have been anticipated, these averages are expressed in terms of the invariants of the equilibrium polarizability tensor. By comparing eqs. (4.83) and (4.84), it is not difficult to see that the space-averaged squares of the Stokes vibration–rotation matrix elements associated with the k-th normal mode Q_k are obtained from the space-averaged squares of the pure rotational matrix elements by replacing $(a)_0^2$ and $(\gamma)_0^2$ by the corresponding invariants of the derived polarizability tensor $(a')_k^2$ and $(\gamma')_k^2$ and including the factors (v_k^i+1) and $b_{v_k}^2$.

Some important generalizations emerge from Table I. The mean polarizability $(a)_0$ can make a contribution only when $\Delta J=0$, whereas the anisotropy $(\gamma)_0$ can make a contribution when $\Delta J=0$ or ± 2. Thus, Rayleigh scattering can have an isotropic and an anisotropic contribution, whereas pure rotational Raman scattering depends only on $(\gamma)_0$ and is entirely anisotropic. Likewise in vibration–rotation Raman scattering the Q branch depends on $(a')_k$ and $(\gamma')_k$, and so can have isotropic and anisotropic contributions, whereas the O and S branches depend only on $(\gamma')_k$, and so are entirely anisotropic.

We are now in a position to calculate scattered intensities (and so polarization properties) for rotation and vibration–rotation Raman scattering from an assembly of $\mathcal{N}$ molecules randomly oriented. We recall that we are dealing with the case of a diatomic molecule with a non-degenerate ground electronic state. Consider first the case of a Stokes pure rotational transition $0, J' \leftarrow 0, J''$. To calculate the scattered intensity, we use a form of eq. (2.110) in which the dipole moment is replaced by the appropriate transition moment matrix element and multiply by the number of molecules in the initial state $0, J''$. Then, for illumination along the space-fixed z-axis with radiation of wavenumber $\tilde{\nu}_0$, plane polarized so that $E_y \neq 0$ (see Fig. 3.4a), the intensity observed along the x-axis is given by

$$^{\perp}I_{\parallel}(\pi/2)=\frac{^{\perp}\mathrm{d}\Phi_{\parallel}(\pi/2)}{\mathrm{d}\Omega}=k'_{\tilde{\nu}}(\tilde{\nu}_0-\tilde{\nu}_{0,J';0,J''})^4\mathcal{N}_{0,J''}\overline{[\alpha_{zy}]^2_{0,J';0,J''}}E^2_{y0} \qquad (4.116)$$

for scattered radiation plane polarized with the z electric vector non-zero, and

$$^{\perp}I_{\perp}(\pi/2)=\frac{^{\perp}\mathrm{d}\Phi_{\perp}(\pi/2)}{\mathrm{d}\Omega}=k'_{\tilde{\nu}}(\tilde{\nu}_0-\tilde{\nu}_{0,J';0,J''})^4\mathcal{N}_{0,J''}\overline{[\alpha_{yy}]^2_{0,J';0,J''}}E^2_{y0} \qquad (4.117)$$

for the scattered radiation plane polarized with the y electric vector non-zero. Here $k'_{\tilde{\nu}}$ is defined by eq. (3.56), $\tilde{\nu}_{0,J';0,J''}$ is the wavenumber associated with the transition $0, J' \leftarrow 0, J''$ and $\mathcal{N}_{0,J''}$ is the number of molecules in the initial state $0, J''$. The corresponding forms of eqs. (4.116) and (4.117) for anti-Stokes scattering are obtained by replacing $(\tilde{\nu}_0-\tilde{\nu}_{0,J';0,J''})^4$ by $(\tilde{\nu}_0+\tilde{\nu}_{0,J';0,J''})^4$ and $\mathcal{N}_{0,J''}$ by $\mathcal{N}_{0,J'}$.

We shall return subsequently to a consideration of the factors determining $\mathcal{N}_{0,J''}$ but we can proceed immediately to obtain expressions for depolarization ratios since ratios of scattered powers are involved and $\mathcal{N}_{0,J''}$ cancels.

Substitution of the appropriate values of the space-averaged squares of the matrix elements of the polarizability tensor components (Table I) into eqs. (4.116) and (4.117) enables us to obtain expressions for $\rho_{\perp}(\pi/2)$. Thus, for transitions of the type $0, J \leftarrow 0, J$ (i.e., $J'=J''=J$, $\Delta J=0$, Rayleigh scattering) we obtain

$$\rho_{\perp}(\pi/2)=\frac{3b_{J,J}(\gamma)_0^2}{45(a)_0^2+4b_{J,J}(\gamma)_0^2} \qquad (4.118)$$

and for transitions of the type $0, J+2, \leftrightarrow 0, J$ (i.e., $J'=J+2$, $J''=J$, $\Delta J=+2$, Stokes or anti-Stokes rotational Raman scattering) we obtain

$$\rho_{\perp}(\pi/2)=\tfrac{3}{4} \tag{4.119}$$

Similarly, we find that

$$\rho_{\parallel}(\pi/2)=1 \text{ (provided } (\gamma)_0 \neq 0) \tag{4.120}$$

for both $\Delta J=0$ (Rayleigh scattering) and $\Delta J=+2$ (Stokes or anti-Stokes rotational Raman scattering). $\mathscr{P}(0)$ and $\mathscr{P}(\pi)$ may be calculated similarly.

These calculations may be easily extended to vibration–rotation transitions. For example, for the Stokes vibration–rotation transition $1, J', \leftarrow 0, J''$ we have, corresponding to eq. (4.116),

$$^{\perp}I_{\parallel}(\pi/2)=\frac{^{\perp}\mathrm{d}\Phi_{\parallel}(\pi/2)}{\mathrm{d}\Omega}=k'_{\tilde{\nu}}(\tilde{\nu}_0-\tilde{\nu}_{1,J';0,J''})^4\mathcal{N}_{0,J''}\overline{[\alpha_{zy}]^2_{1,J';0,J''}}E^2_{y_0} \tag{4.121}$$

and so on. Introduction of the appropriate vibration–rotation matrix elements of the polarizability tensor components (Table I) and formation of the appropriate ratios of scattered intensities enables $\rho_{\perp}(\pi/2)$ and $\rho_{\parallel}(\pi/2)$ to be evaluated for vibration–rotation transitions. Formulae for intensities and polarization properties of Rayleigh, and rotation and vibration–rotation Raman scattering from diatomic molecules are given in Table J.

We recall that whereas $(\gamma)_0$ may be zero or non-zero, $(a)_0$ may never be zero. Thus, Rayleigh scattering will always be observed and will have $0\leqslant\rho_{\perp}(\pi/2)<\frac{3}{4}$. For pure rotational Raman scattering to exist, $(\gamma)_0$ must be non-zero and then $\rho_{\perp}(\pi/2)=\frac{3}{4}$ for each rotational line in the S branch ($\Delta v=0$, $\Delta J=+2$). For O and S branches to exist in vibration–rotation Raman scattering ($\Delta v_k=\pm1$, $\Delta J=\pm2$), $(\gamma')_k$ must be non-zero and then $\rho_{\perp}(\pi/2)=\frac{3}{4}$ for each line in the O and S branches. The minimum requirement for a Q branch ($\Delta v_k=\pm1$, $\Delta J=0$) to exist in vibration–rotation Raman scattering is that either $(a')_k$ or $(\gamma')_k$ is non-zero. If only $(a')_k$ is non-zero, then the Q branch exists without O and S branches and has $\rho_{\perp}(\pi/2)=0$, since only the isotropic part of the scattering is non-zero. If only $(\gamma')_k$ is non-zero, then the Q branch exists together with the O and S branches and all three branches have $\rho_{\perp}(\pi/2)=\frac{3}{4}$. If both $(a')_k$ and $(\gamma')_k$ are non-zero, then all three branches exist but the Q branch has $0<\rho_{\perp}(\pi/2)<\frac{3}{4}$, whereas the O and S branches have $\rho_{\perp}(\pi/2)=\frac{3}{4}$.

We now consider the population factors. For a system initially in the ground vibrational state and the rotational state J we may write

$$\mathcal{N}_{0,J}=\frac{\mathcal{N}g_J(2J+1)\exp\{-F(J)hc/kT\}}{Q_R} \tag{4.122}$$

where

$$J\equiv J'' \text{ for Stokes transitions} \quad J\equiv J' \text{ for anti-Stokes transitions} \tag{4.123}$$

and Q_R is the rotational partition function which for a diatomic molecule is given by

$$Q_R=\sum_J g_J(2J+1)\exp\{-F(J)hc/kT\} \tag{4.124}$$

The term $(2J+1)$, as already mentioned, appears because of the M degeneracy of the rotational state J; g_J is a further degeneracy term arising from nuclear spin effects. g_J is the same for all rotational levels in heteronuclear diatomic molecules, but for homonuclear diatomics it takes different values for odd and even J values, these values depending on the nuclear spin of the atoms involved and on the statistics obeyed by the nuclei. We cannot consider here the theory underlying nuclear statistics, but we give in Table 4.1 values of g_J and related nuclear statistical information for some typical diatomic molecules.

Table 4.1 Nuclear spin statistics and relative intensities of rotational Raman lines in some diatomic molecules

Species	$^{16}O_2$, $^{18}O_2$	H_2, T_2, $^{19}F_2$	D_2, $^{14}N_2$	$^{35}Cl_2$
Nuclear spin, I	0	$\frac{1}{2}$	1	$\frac{3}{2}$
Nuclear spin degeneracy, $T=2I+1$	1	2	3	4
Statistics	Bose	Fermi	Bose	Fermi
Ground electronic state	$^3\Sigma_g^-$	Σ_g^+	Σ_g^+	Σ_g^+
Nuclear spin statistical weight g_J for				
J odd	1	3	3	5
J even	—[a]	1	6	3
Relative intensities of rotational Raman lines				
J odd	1	3	1	5
J even	0	1	2	3

[a] Even J levels unpopulated.

The factor $(2J+1)\exp\{-F(J)hc/kT\}$ is shown plotted against J for $T=300$ K and $F(J)=10.44\,J(J+1)$ cm^{-1} in Fig. 4.8. We can see that since $(2J+1)$ increases linearly with J, the factor $(2J+1)\exp(-F(J)(J+1)hc/kT)$ first increases with increasing J, despite the steady decrease of the exponential term with J; the whole

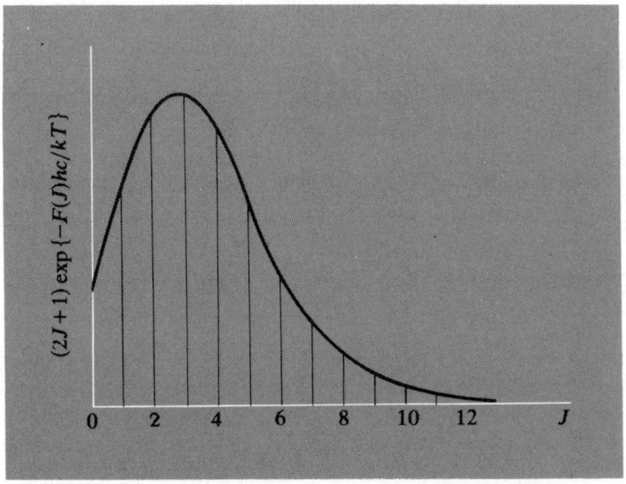

Fig. 4.8 A plot of $(2J+1)\exp\{-F(J)hc/kT\}$ against J; $T=300$ K, and $F(J)/\text{cm}^{-1}=10{\cdot}44J(J+1)$

factor then reaches a maximum value and subsequently decreases as J increases. The maximum in the factor occurs at

$$J_{\max}=\sqrt{\frac{kT}{2Bhc}}-\frac{1}{2} \tag{4.125}$$

Thus, for heteronuclear diatomic molecules where g_J is the same for all J values, the intensity distribution in the Stokes pure rotational Raman spectrum will follow a pattern very similar to that of Fig. 4.8. The pattern will be slightly modified because of the dependence of the factors $b_{J+2,J}$ and $(\tilde{\nu}-\tilde{\nu}_{J+2,J})^4$ on J.

In the case of homonuclear diatomic molecules where g_J takes on different values according to whether J is even or odd, there will be a distinct *alternation* in the intensities of successive lines. For molecules where g_J is zero for odd or even J values, every other rotational line will be missing. Specific examples of the effect of nuclear spin statistics will be considered in chapter 7.

4.4.5 *Selection rules and wavenumber shift formulae for rotational Raman scattering from polyatomic molecules*

In this section we shall summarize, without proof, the selection rules and wavenumber shift formulae for pure rotational Raman lines in other types of molecule; discussion of other factors will be in qualitative terms only.

For a linear molecule in a Σ^+ ground electronic state, the selection rules for Raman active rotation transitions are exactly the same as for a diatomic molecule in a non-degenerate electronic ground state, namely, $\Delta J=0, +2$. Thus, eqs. (4.101) and (4.106) also give the magnitudes of the rotational wavenumber shifts for linear molecules in the rigid and non-rigid rotor approximations, respectively; and the corresponding rotation Raman spectra have the forms shown in Fig. 4.5(b) and (d). However, the moment of inertia I about an axis perpendicular to the internuclear axis and passing through the centre of mass is now given by

$$I=\sum_i m_i r_i^2 \tag{4.126}$$

where r_i is the distance of the i-th nucleus of mass m_i from the centre of mass.

The rotational Raman scattering will again be entirely anisotropic and hence depends only on $(\gamma)_0$. The population factor $\mathcal{N}_{0,J}$ which largely determines the intensity distribution in the pure rotational Raman spectrum will again be given by eq. (4.122). In centrosymmetric molecules, nuclear spin statistics will again lead to different weightings for alternate levels or to the absence of alternate levels. For example, in $^{32}S^{16}C^{32}S$ the levels with odd J values are absent and in $^{1}H—^{12}C\equiv^{12}C—^{1}H$ the nuclear spin weighting of the levels with odd J values is 3 and of the levels with even J values is 1. These considerations apply to the ground vibrational state. In CS_2 the degenerate bending vibration has $\tilde{\nu}_M=397\ \text{cm}^{-1}$, and 20 per cent of the molecules will be in the first vibrational level at 300 K. Rotation–vibration interaction splits each J level into two components one of which is antisymmetric in the nuclei. A family of J levels with the same nuclear spin weighting is thus produced. Thus, the rotational Raman spectrum of CS_2 will consist of the superposition of a strong S branch, arising from the ground vibrational state, with interline spacings of approximately $8B$ (alternate bands missing), and a weaker S

branch, arising from an excited bending vibrational state, with interline spacings of approximately $4B$. Alternate lines of the weaker spectrum will coincide with lines in the stronger spectrum producing an overall intensity alternation. A very weak R branch is not observed.

We now consider non-linear molecules. In general, such molecules have three moments of inertia, I_A, I_B, and I_C. The corresponding rotational constants, A, B, and C, are defined by

$$A = \frac{h}{8\pi^2 c I_A}, \qquad B = \frac{h}{8\pi^2 c I_B}, \qquad \text{and} \qquad C = \frac{h}{8\pi^2 c I_C} \tag{4.127}$$

By convention the labels are assigned so that $I_A < I_B < I_C$, i.e., so that $A > B > C$. If two of the moments of inertia are equal, the molecule is said to be a symmetric top; if $I_A \neq I_B = I_C$, as, for example, in CH_3Cl, then the molecule is said to be a prolate symmetric top; if $I_A = I_B \neq I_C$, as, for example, in BCl_3, then the molecule is said to be an oblate symmetric top. If all three moments of inertia are equal, i.e., $I_A = I_B = I_C$, the molecule is said to be a spherical top. If all three moments of inertia are unequal, the molecule is described as an asymmetric top. In what follows we shall have to confine ourselves to the most important case, the pure rotation Raman spectrum of a symmetric top molecule.

The rotational term for a rigid prolate symmetric top is given by

$$F(J, K) = BJ(J+1) + (A-B)K^2 \tag{4.128}$$

and the rotational term for an oblate symmetric top is obtained by replacing A in eq. (4.128) by C. We shall therefore only need to discuss in detail the prolate symmetric top in what follows. In eq. (4.128), J is the rotational quantum number and K is the quantum number associated with rotation about the molecular symmetry axis (the figure axis for the prolate symmetric top). K cannot be greater than J, i.e.,

$$J = K, K+1, K+2, \ldots \tag{4.129}$$

The internal angular momentum about the molecular symmetry axis is given by $Kh/2\pi$. K is restricted to integer values in the range J, $(J-1)$, $(J-2)$, ...0... $-(J-1)$, $-J$, and since the energy depends on K^2, all levels except those with $K = 0$ are twofold degenerate. This degeneracy is essentially associated with the clockwise or anticlockwise rotations about the symmetry axis and is additional to the $(2J+1)$-fold orientation degeneracy which arises because the rotational energy is independent of the orientation quantum number M in the absence of fields. The rotational levels also possess symmetry properties associated with the symmetry of the wave functions and statistical weights arising from nuclear spins, but we shall not consider these here.

The energy level diagram for a rigid prolate symmetric top is shown in Fig. 4.9(a) for $K = 0, 1, 2, 3$. It should be noted that for $K = 1$, the lowest value of J is 1, for $K = 2$, the lowest J value is 2, and so on. It can also be seen that for each value of K the rotational levels have exactly the same *spacing*.

The selection rules for pure rotational Raman scattering associated with a symmetric top may be established by procedures similar to those used for a diatomic molecule, and are found to be

$$\Delta J = 0, \pm 1 \text{ (except for } K = 0\text{)}, \pm 2 \quad \text{and } \Delta K = 0 \tag{4.130}$$

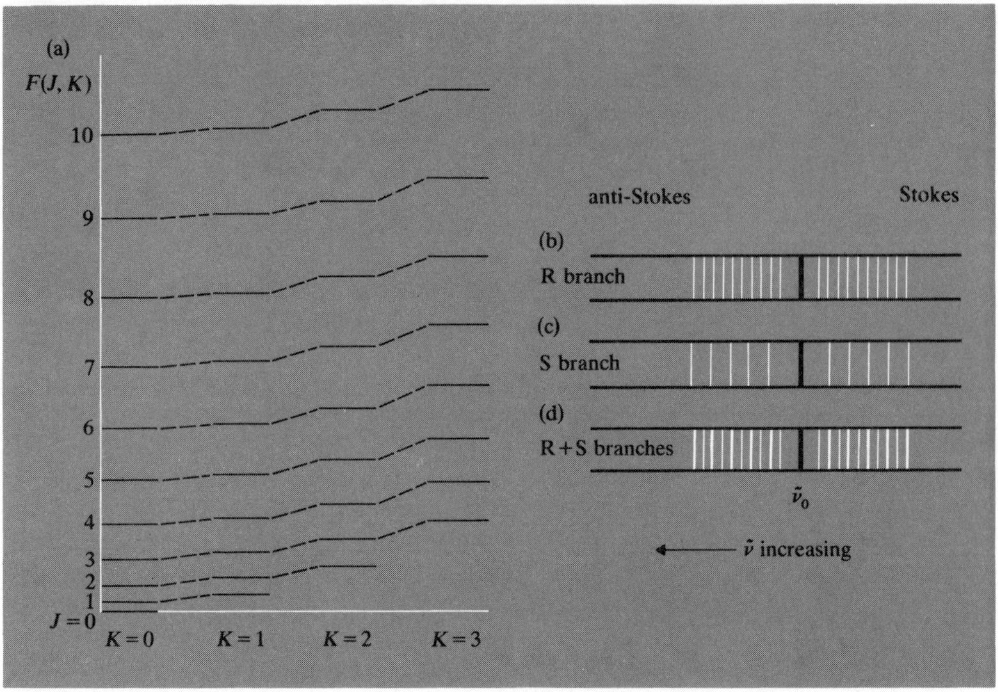

Fig. 4.9 (a) Rigid prolate symmetric top molecule energy levels; and (b) *R* branch, (c) *S* branch, (d) *R* and *S* branches combined, in the Stokes and anti-Stokes rotation Raman spectra

As already explained, the cases $\Delta J = -1$ and $\Delta J = -2$ do not arise in practice in pure rotational scattering.

Using these rules, the magnitudes of the wavenumber shifts for the S branch ($\Delta J = +2$, $\Delta K = 0$) lines are given in terms of the rotational terms by

$$|\Delta\tilde{\nu}_S| = F(J+2, K) - F(J, K) \tag{4.131}$$

Using eq. (4.128) this is readily shown to give

$$|\Delta\tilde{\nu}_S| = 4B(J+\tfrac{3}{2}) \tag{4.132}$$

where $J = 0, 1, 2, \ldots$. Similarly, for the R branch ($\Delta J = +1$, $\Delta K = 0$) lines, we have

$$|\Delta\tilde{\nu}_R| = F(J+1, K) - F(J, K) \tag{4.133}$$

which leads to

$$|\Delta\tilde{\nu}_R| = 2B(J+1) \tag{4.134}$$

where $J = 1, 2, 3, \ldots$. The $J = 0$ case is excluded since $\Delta J = 1$ does not operate for $K = 0$ and for $K = 1$ the lowest J value is 1.

The R and S branches are shown diagrammatically in Fig. 4.9(b) to (d). Figure 4.9(b) shows the R branch, Fig. 4.9(c) the S branch, and Fig. 4.9(d) the R and S branches combined. It can be seen from these diagrams, or from comparison of eqs. (4.132) and (4.134), that the lines of the R branches with even J coincide with the lines of the S branches. As a result, there will be an *apparent* intensity alternation.

This is quite distinct from any intensity alternations arising from nuclear spin effects which we shall not consider here. Also, since the S branch extends to larger wavenumber shifts than does the R branch, this *apparent* intensity alternation does not extend over the whole rotational spectrum.

If centrifugal stretching is taken into account, the rotational term for a symmetric top has the form

$$F(J, K) = BJ(J+1) + (A-B)K^2 - D_J J^2(J+1)^2 - D_{JK}J(J+1)K^2 - D_K K^4 \tag{4.135}$$

and as a result the formulae for the magnitudes of the wavenumber shifts of the R and S branches become

$$|\Delta\tilde{\nu}_S| = (4B - 6D_J)(J+\tfrac{3}{2}) - 4D_{JK}K^2(J+\tfrac{3}{2}) - 8D_J(J+\tfrac{3}{2})^3 \tag{4.136}$$

where $J = 0, 1, 2, \ldots,$ and

$$|\Delta\tilde{\nu}_R| = 2B(J+1) - 2D_{JK}K^2(J+1) - 4D_J(J+1)^3 \tag{4.137}$$

where $J = 1, 2, \ldots$. These formulae contain two centrifugal stretching terms, D_J and D_{JK}. The term D_{JK} is K-dependent and should therefore lead to a very small splitting of each rotational line.

The formulae for intensities, depolarization ratios and reversal coefficients of scattering arising from rotational transitions in symmetric top molecules have the same forms as those for diatomic molecules; they differ only in the anisotropic coefficients $b_{J',K';J'',K''}$ which are now functions of J and K and are non-zero for $\Delta J = 0, +1, +2$ and $\Delta K = 0$. The formulae for intensities, $\rho_\perp(\pi/2)$ and $\mathscr{P}(0)$ and the definitions of $b_{J',K';J'',K''}$ are given in the central reference section (see Table K and the introductory notes (d)). From comparison of Tables J and K we see that the generalisations relating to rotational scattering from a diatomic molecule apply also to a symmetric top. Thus pure rotation Raman scattering from a symmetric top is always anisotropic with $\rho_\perp(\pi/2) = \frac{3}{4}$ and $\mathscr{P}(0) = 6$ and requires $(\gamma)_0$ to be non-zero, whereas pure rotational Rayleigh scattering can have $0 \leq \rho_\perp(\pi/2) < \frac{3}{4}$ or $0 \leq \mathscr{P}(0) < 6$.

A spherical top molecule can be regarded as a degenerate symmetric top molecule with $A = B = C$, and hence the pure rotation term is given in the rigid rotor approximation by eq. (4.97) and for the non-rigid rotor by eq. (4.105). However a spherical top molecule necessarily has an isotropic polarizability (i.e. $(\gamma)_0 = 0$) and hence can have no pure rotation Raman spectrum.

4.4.6 *Vibration–rotation Raman spectra of polyatomic molecules*

In polyatomic molecules additional factors like Coriolis interaction and vibrational degeneracy can operate to create a much more complicated pattern of energy levels than we have considered hitherto. An account of these energy levels and the vibration–rotation Raman spectra that can result from transitions between them is felt to be outside the scope of this book.

However the formulae for intensities, depolarization ratios and reversal coefficients for scattering arising from vibration–rotation scattering have the same forms as those for pure rotation transitions, except that $(a')_k^2$ and $(\gamma')_k^2$ replace $(a)_0^2$ and $(\gamma_0)^2$, and non-zero anisotropic coefficients $b_{J',K';J''K''}$ now arise when $\Delta J = 0, \pm 1, \pm 2$

and $\Delta K = 0, \pm 1, \pm 2$. The general forms of these anisotropic coefficients are given in Table M in the central reference section, and if used with the general scattering formulae in Table L will yield expressions for intensities and polarization properties for scattering arising from rotation and vibration–rotation transitions for linear, symmetric top and spherical top molecules, the selection rules for which are summarized in Table 4.2. The use of these tables is also discussed in paragraph (d) of the notes in the central reference section. It can be seen that Tables I, J and K are special cases derivable from Tables L, M and 4.2.

In those cases where it is not possible to resolve individual rotation lines useful structural information can be obtained from an analysis of the contour of the band.[4]

Table 4.2 Selection rules for rotation and vibration–rotation transitions

Rotor	Vibrational species[a]	Selection rules
Linear	Totally symmetric	$\Delta J = 0, \pm 2$
	Degenerate	$\Delta J = 0, \pm 1, \pm 2$ $(J' + J'' \geqslant 0)$
Symmetric top	Totally symmetric	$\Delta J = 0, \pm 1, \pm 2, \Delta K = 0$
	Non-totally symmetric non-degenerate[e]	$\Delta J = 0, \pm 1, \pm 2, \Delta K = \pm 2$
	Degenerate	$\Delta J = 0, \pm 1, \pm 2, \Delta K = \pm 2^{b}$ $\Delta J = 0, \pm 1, \pm 2, \Delta K = \pm 1^{c}$ $(J' + J'' \geqslant 2)^{d}$
Spherical top	Totally symmetric	$\Delta J = 0$
	Doubly degenerate	$\Delta J = 0, \pm 1, \pm 2$
	Triply degenerate	$\Delta J = 0, \pm 1, \pm 2$ $(J' + J'' \geqslant 2)^{d}$

[a] Pure rotation transitions have the same selection rules as totally symmetric vibration–rotation transitions.
[b] If $\alpha_{xx} - \alpha_{yy}$ or α_{xy} or both non-zero.
[c] If α_{yz} or α_{zx} or both non-zero.
[d] For pure rotation transitions $J' + J'' \geqslant 0$.
[e] Only in molecules with fourfold axes.

References

1. P. W. Atkins, *Molecular Quantum Mechanics*, Clarendon Press, Oxford, 1970, for example, provides an account of this elegant but condensed nomenclature.
2. G. Placzek, Rayleigh-Streuung und Raman-Effekt, in E. Marx (Ed.), *Handbuch der Radiologie*, vol. **VI**, 2, pp. 205–374, Akademische Verlag, Leipzig, 1934.
3. Report on Notation for the Spectra of Polyatomic Molecules, *J. Chem. Phys.*, **23**, 1997, 1955.
4. F. N. Masri and W. H. Fletcher, *J. Chem. Phys.*, **52**, 5759, 1970.
   F. N. Masri and W. H. Fletcher, *J. Raman Spectroscopy*, **1**, 221, 1973.

5 Time-dependent perturbation theory and Rayleigh and Raman scattering

"Oh polished perturbation! golden care!
That keepst the ports of slumber open wide
To many a watchful night!"

William Shakespeare

5.1 Introduction

In this chapter we shall consider a more general quantum mechanical treatment of the interaction of a scattering system with radiation. The radiation, which we shall continue to treat classically, will now be regarded as a source of perturbation of the energy levels of the scattering system; and we shall use the techniques of quantum mechanics to investigate transitions between the energy levels of the perturbed system and the scattering that ensues. This is a much more fundamental procedure than that used in chapter 4, and will enable us to relate the scattering tensor to the wave functions and energy levels of the scattering system. Fortunately, the classical treatment of the radiation does not restrict the general validity of the results for normal Rayleigh and Raman scattering, but some other phenomena like the inverse Raman effect cannot be explained without quantization of the field.

5.2 Time-dependent perturbation treatment: general considerations

We now seek expressions for the electric dipole transition moment associated with a transition from an initial state i of the system to a final state f, when the system is subject to the perturbation of electromagnetic radiation of frequency ω_0. The Hamiltonian for such a perturbation can be closely approximated to by an electric dipole term only, and we shall restrict ourselves to this approximation here. In a subsequent section, we shall consider the small additional contributions to the electric dipole transition moment when the perturbation Hamiltonian includes magnetic dipole and electric quadrupole terms and also the magnetic dipole and electric quadrupole transition moments.

For the unperturbed system in the state i the time-dependent wave function $\Psi_i^{(0)}$ is given by

$$\Psi_i^{(0)} = \psi_i \exp(-i\omega_i t) \tag{5.1}$$

where ψ_i is the corresponding time-independent wave function and $\hbar\omega_i$ is the energy of the state i. We note that ψ_i is associated with an arbitrary phase factor $\exp(i\delta_i)$.

When the system is perturbed the time-dependent wave function Ψ_i' may be expressed as

$$\Psi_i' = \Psi_i^{(0)} + \Psi_i^{(1)} + \Psi_i^{(2)} + \ldots \tag{5.2}$$

where $\Psi_i^{(1)}$ is the first-order perturbation term, $\Psi_i^{(2)}$ the second-order perturbation term, and so on.

Hence $[\boldsymbol{P}]_{fi}$, the electric dipole transition moment for the transition $f \leftarrow i$, when the system is perturbed, is given by

$$[\boldsymbol{P}]_{fi} = \langle \Psi_f' | \boldsymbol{P} | \Psi_i' \rangle \tag{5.3}$$

$\boldsymbol{P}$ is the electric dipole moment operator for the system defined by

$$\boldsymbol{P} = \sum_j e_j \boldsymbol{r}_j \tag{5.4}$$

where e_j is the charge, $\boldsymbol{r}_j$ the position vector of the j-th particle, and the summation is over all particles.

On introducing into eq. (5.3) the expansions of the perturbed wave functions Ψ_f' and Ψ_i' of the form given by eq. (5.2), we find that

$$[\boldsymbol{P}]_{fi} = [\boldsymbol{P}^{(0)}]_{fi} + [\boldsymbol{P}^{(1)}]_{fi} + [\boldsymbol{P}^{(2)}]_{fi} + \ldots \tag{5.5}$$

where

$$[\boldsymbol{P}^{(0)}]_{fi} = \langle \Psi_f^{(0)} | \boldsymbol{P} | \Psi_i^{(0)} \rangle \tag{5.6}$$

$$[\boldsymbol{P}^{(1)}]_{fi} = \langle \Psi_f^{(1)} | \boldsymbol{P} | \Psi_i^{(0)} \rangle + \langle \Psi_f^{(0)} | \boldsymbol{P} | \Psi_i^{(1)} \rangle \tag{5.7}$$

$$[\boldsymbol{P}^{(2)}]_{fi} = \langle \Psi_f^{(1)} | \boldsymbol{P} | \Psi_i^{(1)} \rangle + \langle \Psi_f^{(2)} | \boldsymbol{P} | \Psi_i^{(0)} \rangle + \langle \Psi_f^{(0)} | \boldsymbol{P} | \Psi_i^{(2)} \rangle \tag{5.8}$$

The transition moment $[\boldsymbol{P}^{(0)}]_{fi}$ relates to a direct transition between the unperturbed states f and i, and will not be considered further here. We shall find that the first-order term $[\boldsymbol{P}^{(1)}]_{fi}$ includes terms that relate to normal Rayleigh and Raman scattering, and that the second-order term contains terms relating to hyper Rayleigh and Raman scattering, and so on. We see, therefore, that eq. (5.5), with the direct transition moment $[\boldsymbol{P}^{(0)}]_{fi}$ omitted, is closely analogous to eq. (3.1).

We shall now consider in detail the first-order transition moment $[\boldsymbol{P}^{(1)}]_{fi}$. The second-order transition moment $[\boldsymbol{P}^{(2)}]_{fi}$ will be treated briefly in chapter 8.

5.3 Time-dependent perturbation treatment: first-order perturbation and dipole approximation

Time-dependent perturbation theory enables us to write

$$\Psi_i^{(1)} = \sum_r a_{ir} \Psi_r^{(0)} \quad \text{and} \quad \Psi_f^{(1)} = \sum_r a_{fr} \Psi_r^{(0)} \tag{5.9}$$

where the summation is over all states r of the system except i (or f). If the system is initially in the state $\Psi_i^{(0)}$ then the coefficients a_{ir} are obtained by integrating

$$\dot{a}_{ir} = -\frac{i}{\hbar} \langle \Psi_r^{(0)} | H' | \Psi_i^{(0)} \rangle \tag{5.10}$$

where H' is the perturbation Hamiltonian; and similarly for the a_{fr}. In the problem we are considering, the perturbation arises from electromagnetic radiation of frequency ω_0. We shall write the time dependence of the electric field intensity of the radiation in its most general form (see eq. 2.54) as

$$\boldsymbol{E} = \tilde{\boldsymbol{E}}_0 \exp(-i\omega_0 t) + \tilde{\boldsymbol{E}}_0^* \exp(i\omega_0 t) \tag{5.11}$$

If we make the approximation that the electric field intensity does not vary over the molecule (see page 75), then the perturbation Hamiltonian contains only an electric dipole term H'_p, and we have

$$H' = H'_p = -\boldsymbol{P} \cdot \boldsymbol{E} \tag{5.12}$$

where $\boldsymbol{P}$ is the dipole moment operator defined by eq. (5.4). If the variation of the electric field intensity over the molecule is not neglected, then the perturbation Hamiltonian has additional terms, H'_m, a magnetic dipole term, and H'_Θ, an electric quadrupole term, whose contributions we shall consider later.

When the perturbation Hamiltonian is given by eq. (5.12), then

$$a_{ir} = \frac{1}{\hbar}[\boldsymbol{P}]_{ri} \cdot \left[\frac{\tilde{\boldsymbol{E}}_0^* \exp\{i(\omega_{ri}+\omega_0)t\}}{\omega_{ri}+\omega_0} + \frac{\tilde{\boldsymbol{E}}_0 \exp\{i(\omega_{ri}-\omega_0)t\}}{\omega_{ri}-\omega_0}\right] \tag{5.13}$$

and

$$a_{fr}^* = \frac{1}{\hbar}[\boldsymbol{P}]_{fr} \cdot \left[\frac{\tilde{\boldsymbol{E}}_0 \exp\{-i(\omega_{rf}+\omega_0)t\}}{\omega_{rf}+\omega_0} + \frac{\tilde{\boldsymbol{E}}_0^* \exp\{-i(\omega_{rf}-\omega_0)t\}}{\omega_{rf}-\omega_0}\right] \tag{5.14}$$

where

$$\omega_{ri} = \omega_r - \omega_i \quad \text{etc.,} \tag{5.15}$$

$$[\boldsymbol{P}]_{ri} = \langle \psi_r | \boldsymbol{P} | \psi_i \rangle \quad \text{etc.} \tag{5.16}$$

The first-order transition moment $[\boldsymbol{P}^{(1)}]_{fi}$ can now be evaluated by introducing eqs. (5.9), (5.13), and (5.14) into eq. (5.7). We find that

$$\begin{aligned}[\boldsymbol{P}^{(1)}]_{fi} = {} & \frac{1}{\hbar}\sum_r \left[\frac{([\boldsymbol{P}]_{fr}\cdot\tilde{\boldsymbol{E}}_0)[\boldsymbol{P}]_{ri}}{(\omega_{rf}+\omega_0)} + \frac{[\boldsymbol{P}]_{fr}([\boldsymbol{P}]_{ri}\cdot\tilde{\boldsymbol{E}}_0)}{(\omega_{ri}-\omega_0)}\right] \exp\{-i(\omega_0-\omega_{fi})t\} \\ & + \frac{1}{\hbar}\sum_r \left[\frac{([\boldsymbol{P}]_{fr}\cdot\tilde{\boldsymbol{E}}_0^*)[\boldsymbol{P}]_{ri}}{(\omega_{rf}-\omega_0)} + \frac{[\boldsymbol{P}]_{fr}([\boldsymbol{P}]_{ri}\cdot\tilde{\boldsymbol{E}}_0^*)}{(\omega_{ri}+\omega_0)}\right] \exp\{i(\omega_0+\omega_{fi})t\}\end{aligned} \tag{5.17}$$

The two time-dependent transition moments in eq. (5.17) are complex. However, the radiation associated with a complex transition moment of the form $\tilde{\boldsymbol{P}}_{ml} \exp\{-i\omega_{lm}t\}$ corresponds to that from the real transition moment

$$\tilde{\boldsymbol{P}}_{ml} \exp\{-i\omega_{lm}t\} + \tilde{\boldsymbol{P}}_{ml}^* \exp\{i\omega_{lm}t\} \tag{5.18}$$

provided the condition

$$\omega_{lm} > 0 \tag{5.19}$$

is satisfied. Otherwise there is no radiation.

Thus, the first term in eq. (5.17) will have a corresponding real transition moment if

$$\omega_0 - \omega_{fi} > 0 \tag{5.20}$$

If ω_{fi} is negative, i.e., the final state is lower in energy than the initial state, this condition is always satisfied. Similarly, if ω_{fi} is zero, i.e., the initial and final states have the same energy, this condition is always satisfied. If ω_{fi} is positive, i.e., the final state is higher in energy than the initial state, then it is necessary for the energy of the incident quantum to be more than sufficient to reach the final state f. For rotational and vibrational transitions which involve no change of electronic state, this condition is always satisfied for excitation frequencies in the visible and ultraviolet regions of the spectrum. The real transition moment corresponding to the first term in eq. (5.17) is therefore

$$[\boldsymbol{P}^{(1)}]_{fi} = [\tilde{\boldsymbol{P}}_0^{(1)}]_{fi} \exp\{-i(\omega_0 - \omega_{fi})t\} + [\tilde{\boldsymbol{P}}_0^{(1)}]^*_{fi} \exp\{i(\omega_0 - \omega_{fi})t\} \tag{5.21}$$

where

$$[\tilde{\boldsymbol{P}}_0^{(1)}]_{fi} = \frac{1}{\hbar}\sum_r \left\{ \frac{([\tilde{\boldsymbol{P}}]_{fr}\cdot\tilde{\boldsymbol{E}}_0)[\tilde{\boldsymbol{P}}]_{ri}}{\omega_{rf}+\omega_0} + \frac{[\tilde{\boldsymbol{P}}]_{fr}([\tilde{\boldsymbol{P}}]_{ri}\cdot\tilde{\boldsymbol{E}}_0)}{\omega_{ri}-\omega_0} \right\} \tag{5.22}$$

$[\tilde{\boldsymbol{P}}_0^{(1)}]^*_{fi}$ is the conjugate complex, and to maintain complete generality we have now written $[\tilde{\boldsymbol{P}}]_{ri}$ instead of $[\boldsymbol{P}]_{ri}$ since such transition moments can be complex in those special cases where the only good wavefunctions are complex. If ω_{fi} is positive, this transition moment is associated with Stokes Raman scattering at $\omega_0 - |\omega_{fi}|$. If ω_{fi} is negative, this transition moment is associated with anti-Stokes Raman scattering at $\omega_0 + |\omega_{fi}|$. If ω_{fi} is zero, this transition moment is associated with Rayleigh scattering at ω_0.

The second term in eq. (5.17) will have a corresponding real transition moment if

$$-\omega_0 - \omega_{fi} > 0 \tag{5.23}$$

i.e.,

$$\omega_i - \omega_f > \omega_0 \tag{5.24}$$

This condition means that the energy of the initial state must exceed that of the final state by an amount which is greater than ω_0. Thus, if ω_0 lies in the visible region of the spectrum, the initial state must be an excited electronic state. We shall not consider this term further.

The transition moment amplitude defined by eq. (5.22) may be rearranged so that the amplitude of the electric field of the exciting radiation is disentangled from the other factors. We may note that

$$[\tilde{\boldsymbol{P}}]_{ri}\cdot\tilde{\boldsymbol{E}}_0 = [\tilde{P}_x]_{ri}\tilde{E}_{x_0} + [\tilde{P}_y]_{ri}\tilde{E}_{y_0} + [\tilde{P}_z]_{ri}\tilde{E}_{z_0} \tag{5.25}$$

where $[\tilde{P}_x]_{ri}\ldots$ and $\tilde{E}_x\ldots$ are components of the transition moment and the electric field amplitude respectively. When these expansions are introduced into eq. (5.22) and the terms collected, we find that the x components of the transition moment amplitudes can be related to the *complex electric field amplitudes* as follows

$$[\tilde{P}_{x_0}^{(1)}]_{fi} = \sum_y [\tilde{\alpha}_{xy}]_{fi}\tilde{E}_{y_0} \quad \text{and} \quad [\tilde{P}_{x_0}^{(1)*}]_{fi} = \sum_y [\tilde{\alpha}^*_{xy}]_{fi}\tilde{E}^*_{y_0} \tag{5.26}$$

where

$$[\tilde{\alpha}_{xy}]_{fi} = \frac{1}{\hbar}\sum_r \left\{ \frac{[\tilde{P}_y]_{fr}[\tilde{P}_x]_{ri}}{\omega_{rf}+\omega_0} + \frac{[\tilde{P}_x]_{fr}[\tilde{P}_y]_{ri}}{\omega_{ri}-\omega_0} \right\} \tag{5.27}$$

and

$$[\tilde{\alpha}^*_{xy}]_{fi} = \frac{1}{\hbar}\sum_r \left\{ \frac{[\tilde{P}_x]_{ir}[\tilde{P}_y]_{rf}}{\omega_{rf}+\omega_0} + \frac{[\tilde{P}_y]_{ir}[\tilde{P}_x]_{rf}}{\omega_{ri}-\omega_0} \right\} \tag{5.28}$$

Central reference section

Intensities, polarization properties and Stokes parameters for vibration, rotation and vibration–rotation Rayleigh and Raman scattering

(a) Introduction

This central section contains a series of tables, A–M, which give the functions of the invariants of the several scattering tensors necessary for calculating scattered intensities, depolarization ratios, reversal and circularity coefficients, and Stokes parameters for vibration, rotation, and vibration–rotation transitions in an assembly of randomly oriented molecules for various illumination geometries and states of polarization of the incident radiation. For vibration transitions, the treatment is fairly comprehensive: for non-chiral molecules the general case of an unsymmetric scattering tensor is considered, but for vibration transitions in chiral molecules the treatment is restricted to cases where the polarizability tensor is symmetric. For rotation and vibration–rotation transitions, the treatment is less comprehensive and is confined to non-chiral molecules and the case of a symmetric scattering tensor; further, only some illumination geometries, are considered in detail; and the asymmetric top is excluded.

(b) Illumination–observation geometry

For all the tables the illumination and observation geometry is defined as follows:

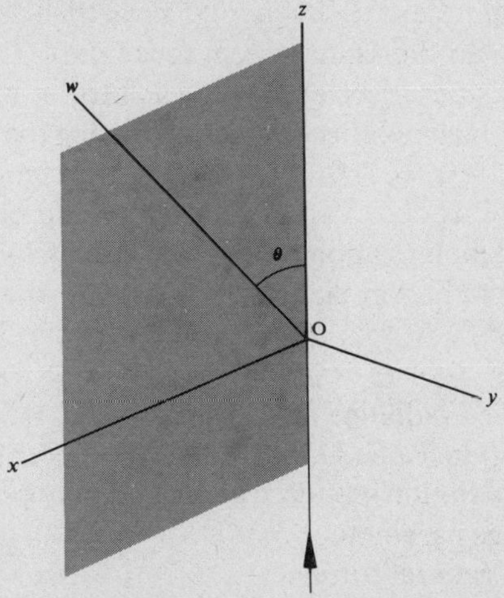

The scattering sample is located at the origin O and illumination is always along the positive z-axis. The scatter plane is the xz-plane, and a general direction of observation in this plane is Ow, which makes an angle θ with the z-axis; important special cases of the observation direction are $\theta = 0$ (along the positive z-axis), $\theta = \pi/2$ (along the positive x-axis), and $\theta = \pi$ (along the negative z-axis).

(c) Vibrational Raman (and Rayleigh) scattering

Consider an assembly of $\mathcal{N}$ randomly oriented molecules irradiated with monochromatic radiation of wavenumber $\tilde{\nu}_0$, irradiance $\mathcal{I}$, and polarization state p_i. Then ${}^{p_i}I_{p_s}(\theta)$, the intensity of the radiation of polarization state p_s scattered along Ow which arises from transitions from an initial state defined by the *set* of vibrational quantum numbers v^i to a final state defined by the *set* of vibrational quantum numbers v^f (or the power ${}^{p_i}\mathrm{d}\Phi_{p_s}(\theta)$, scattered into the solid angle dΩ about Ow) may be written in the general form

$$
{}^{p_i}I_{p_s}(\theta) = \frac{{}^{p_i}\mathrm{d}\Phi_{p_s}(\theta)}{\mathrm{d}\Omega} = k_{\tilde{\nu}}(\tilde{\nu}_0 \pm |\tilde{\nu}_{v^f v^i}|)^4 \mathcal{N}_{v^i} f(a^2, \gamma^2, \delta^2, \theta)\mathcal{I}
$$

The corresponding differential scattering cross-section per molecule is given by

$$
{}^{p_i}\sigma_{p_s} = \frac{{}^{p_i}I_{p_s}(\theta)}{\mathcal{N}\mathcal{I}}
$$

In these formulae

$$
k_{\tilde{\nu}} = \frac{\pi^2}{\varepsilon_0^2}
$$

$|\tilde{\nu}_{v^f v^i}|$ is the magnitude of the wavenumber associated with the vibration transition $v^f \leftrightarrow v^i$; the + and − signs refer to anti-Stokes and Stokes Raman scattering respectively; $\mathcal{N}_{v^i}$ is the number of molecules in the *initial* vibrational state v^i; and $f(a^2, \gamma^2, \delta^2, \theta)$ is a function of the invariants a^2, γ^2, and δ^2 of the transition polarizability tensor $[\boldsymbol{a}]_{v^f v^i}$ and the angle θ, provided the molecules are non-chiral. The intensity of vibrational Rayleigh scattering may be obtained by putting $|\tilde{\nu}_{v^f v^i}| = 0$ and using the invariants of the equilibrium polarizability tensor $\boldsymbol{a}_0$ in place of the transition polarizability tensor, since the initial and final states are the same and $v^f = v^i$. Where vibrational states are degenerate more than one transition of the same type may be associated with the same wavenumber shift. The observed intensity is then the sum of the intensities associated with each transition.

For illumination with plane polarized radiation having the electric vector ⊥ to the scatter plane ($p_i = \perp$, $E_y \neq 0$, $E_x = 0$, $\mathcal{I} = \frac{1}{2}c\varepsilon_0 E_{y0}^2$), the function $f(a^2, \gamma^2, \delta^2, \theta)$ is given in Table A for various observation geometries and polarization states of the scattered radiation; the depolarization ratios ρ are also included. The corresponding information for incident radiation plane polarized with the electric vector ‖ to the scatter plane ($p_i = \parallel$, $E_x \neq 0$, $E_y = 0$, $\mathcal{I} = \frac{1}{2}c\varepsilon_0 E_{x0}^2$) is given in Table B; for incident natural radiation ($p_i = n$, $E_x = E_y \neq 0$, $\mathcal{I} = \frac{1}{2}c\varepsilon_0[(E_{x0}^2 + E_{y0}^2)]$) in Table C; and for incident circularly polarized radiation (p_i = Ⓡ, $E_y = -iE_x \neq 0$ or p_i = Ⓛ, $E_y = iE_x \neq 0$ and $\mathcal{I} = \frac{1}{2}c\varepsilon_0[(E_{x0}^2 + E_{y0}^2)]$) in Table D, but with $\mathscr{P}$ and $\mathscr{C}$ in place of ρ.

The properties of the radiation scattered along Ow may be expressed alternatively in terms of the four Stokes parameters, $S_0(\theta)$, $S_1(\theta)$, $S_2(\theta)$, and $S_3(\theta)$, each of which may be expressed in the general form

$$
S(\theta) = K_{\tilde{\nu}}(\tilde{\nu}_0 \pm |\tilde{\nu}_{v^f v^i}|)^4 \mathcal{N}_{v^i} F(a^2, \gamma^2, \delta^2, \theta, P, \chi, \psi) E_0^2
$$

where

$$
K_{\tilde{\nu}} = \frac{\pi^2}{\varepsilon_0^2 r^2}
$$

and r is the distance along the observation direction at which the observation is made;

$$E_0^2 = \frac{2\mathcal{I}}{c\varepsilon_0}$$

and $F(a^2, \gamma^2, \delta^2, \theta, P, \chi, \psi)$ is a function of the invariants a^2, γ^2, and δ^2, the angle θ and the polarization characteristics P, χ, and ψ of the incident radiation, provided the molecules are non-chiral.

The function $F(a^2, \gamma^2, \delta^2, \theta, P, \chi, \psi)$ is given in Table E for the observation angles $\theta = 0$, $\pi/2$ and π, when the incident radiation has the general polarization state P, χ, ψ, is natural ($P = 0$) or has some of the following special polarization states:

p_i	P	χ	ψ
$\parallel$	1	0	0
$\perp$	1	0	$\pi/2$
Ⓡ	1	$+\pi/4$	0
Ⓛ	1	$-\pi/4$	0

The Stokes parameters of the scattered radiation may be used to express any desired property of the scattered radiation, as a function of a^2, γ^2, δ^2, θ, P, χ, and ψ. The appropriate forms of the function f for ${}^{p_i}I_{\parallel}(\theta)$, ${}^{p_i}I_{\perp}(\theta)$, ${}^{p_i}I_{Ⓡ}(\theta)$ and ${}^{p_i}I_{Ⓛ}(\theta)$, together with expressions for $\rho(\theta)$, ${}^{p_i}\mathcal{P}(\theta)$ and ${}^{p_i}\mathcal{C}(\theta)$ for the special cases of $\theta = 0$, $\pi/2$, and π, are given in Table F, for a general polarization state of the incident radiation, P, χ, ψ. When the values of P, χ, and ψ appropriate to the special polarization states $p_i = \parallel$, $\perp$, n, Ⓡ, or Ⓛ are introduced, the functions f and the expressions for $\rho(\theta)$, ${}^{p_i}\mathcal{P}(\theta)$ and ${}^{p_i}\mathcal{C}(\theta)$ reduce to those in Tables A, B, C, and D.

In the case of non-absorbing chiral molecules, the scattering involves a^2 and γ^2, invariants of the symmetric transition polarizability tensor, and aG', γ_G^2 and γ_A^2 invariants of the transition polarizability optical activity tensors; δ^2 is zero. Table G lists the functions $F(a^2, \gamma^2, aG', \gamma_G^2, \gamma_A^2, \theta, P, \chi, \psi)$ involved in the Stokes parameters and the functions $f(a^2, \gamma^2, aG', \gamma_G^2, \gamma_A^2, \theta, P, \chi, \psi)$ involved in the intensities ${}^{p_i}I_{\parallel}(\theta)$ and ${}^{p_i}I_{\perp}(\theta)$ for $\theta = \pi/2$ and a general polarization state P, χ, ψ of the incident radiation. Table G also includes the functions $f(a^2, \gamma^2, aG', \gamma_G^2, \gamma_A^2, \theta, P, \chi, \psi)$ for ${}^{Ⓡ}I_{\parallel}(\pi/2)$, ${}^{Ⓛ}I_{\parallel}(\pi/2)$, ${}^{Ⓡ}I_{\perp}(\pi/2)$, ${}^{Ⓛ}I_{\perp}(\pi/2)$ and expressions for $\Delta_{\parallel}(\pi/2)$, $\Delta_{\perp}(\pi/2)$, $\Delta(\pi/2)$, ${}^{\parallel}\mathcal{C}(\pi/2)$ and ${}^{\perp}\mathcal{C}(\pi/2)$.

When the polarizability theory is applicable (see pages 76 and 120 for conditions to be satisfied), the transition polarizability tensor is always symmetric and individual components may be related to derivatives of the polarizability tensor with respect to normal coordinates of vibration, the order of the derivatives and the number of normal coordinates involved depending on the vibrational transitions involved. The relations between the transition polarizability tensor and the derivatives of the polarizability tensor are summarized in Table H for fundamental, first overtone, and binary combination and difference transitions.

The general tables may be used together with Table H to obtain intensity formulae in the polarizability theory approximation. For example, for the single fundamental transition $\Delta v_k = +1$, intensity formulae may be obtained from the general formulae, if δ^2 is put equal to zero and a^2 and γ^2 are regarded as invariants of the symmetric tensor $(\boldsymbol{\alpha}')_k$ with elements $(\partial\alpha_{xy}/\partial Q_k)_0$ and are multiplied by $b_{v_k}^2(v_k+1)$. Similar procedures apply to other transitions.

In the simple harmonic approximation, all transitions of the type $\Delta v_k = +1$ have the same wavenumber and thus the transition $v_k = 1 \leftarrow v_k = 0$ cannot be separated from $v_k = 2 \leftarrow v_k = 1$, and so on. The observed intensity is thus obtained by summing the intensities associated with each transition of the type $\Delta v_k = +1$, taking into account the population and the factor $b_{v_k}^2(v_k+1)$ for each initial state. Again the required intensity formulae can be obtained from the general formulae if δ^2 is put equal to zero, a^2 and γ^2 are regarded as invariants of the symmetric tensor $(\boldsymbol{\alpha}')_k$ and are multiplied by $b_{v_k}^2\{1-\exp(-hc\tilde{\nu}_k/kT)\}^{-1}$, and $\mathcal{N}_{v_k}$ is replaced by $\mathcal{N}$. For the transitions $\Delta v_k = -1$ the multiplying factor is $b_{v_k}^2\{\exp(hc\tilde{\nu}_k/kT)-1\}^{-1}$. If the kth mode has a degeneracy of d_{v_k} the *observed* intensity will be d_{v_k} times greater. In formulae for quantities like ρ, $\mathscr{P}$, and $\mathscr{C}$ which involve intensity ratios, the multiplying factors like $b_{v_k}^2(v_k+1)$ cancel and the only changes needed in the general formulae are to put δ^2 equal to zero and to redefine a^2 and γ^2 as invariants of $(\boldsymbol{\alpha}')_k$.

(d) Rotation and vibration–rotation scattering

We shall consider only the following illumination geometries: illumination with plane polarized radiation such that $E_y \neq 0$, $E_x = 0$, $\mathscr{I} = \frac{1}{2}c\varepsilon_0 E_{y_0}^2$ and observation of ${}^{\perp}I_{\perp}(\pi/2)$, ${}^{\perp}I_{\|}(\pi/2)$ and $\rho_{\perp}\pi/2$; and illumination with circularly polarized radiation $\mathscr{I} = \frac{1}{2}c\varepsilon_0[(E_{x_0}^2 + E_{y_0}^2)]$ and observation of ${}^{\circledR}I_{\circledR}(0)\ (= {}^{\copyright}I_{\copyright}(0))$, ${}^{\circledR}I_{\copyright}(0)\ (= {}^{\copyright}I_{\circledR}(0))$, and $\mathscr{P}(0)\ (= \mathscr{P}^{-1}(\pi))$. The scattering system is again an assembly of $\mathcal{N}$ randomly oriented molecules, which are non-absorbing and non-chiral.

For a general *Stokes* transition from a lower state defined by the *set* of vibration and rotation quantum numbers $v''R''$ to an upper state defined by the *set* of vibration and rotation quantum numbers $v'R'$, we may write

$$ {}^{p_i}I_{p_s}(\theta) = k_{\tilde{\nu}}(\tilde{\nu}_0 - \tilde{\nu}_{v',R':v'',R''})^4 g''\mathcal{N}''\Phi(a^2, \gamma^2, \theta)\mathscr{I} $$

and for the corresponding differential scattering cross-section per molecule

$$ {}^{p_i}\sigma_{p_s}(\theta) = \frac{{}^{p_i}I_{p_s}(\theta)}{\mathcal{N}\mathscr{I}} $$

For randomly oriented molecules, transitions between M levels (degenerate in the absence of fields) are taken care of by space averaging; thus the quantum numbers M can be excluded from the set of rotation quantum numbers R. In these formulae $\tilde{\nu}_{v',R':v'',R''}$ is the wavenumber associated with the transition $v'R' \leftarrow v''R''$, $\mathcal{N}''$ is the number of molecules in the initial (lower) state, and g'' is the statistical weight of the initial (lower) state. (Here, following accepted convention, the upper level is indicated by $'$ and the lower level by $''$ and hence $\tilde{\nu}_{v',R':v'',R''}$ is always positive.) For pure rotational transitions g'' is the product of the nuclear spin degeneracy g_n'' and the rotational degeneracy g_R''; for vibration–rotation transitions vibrational degeneracy must also be taken into account.

For *anti-Stokes* transitions it is necessary to replace $(\tilde{\nu}_0 - \tilde{\nu}_{v',R':v'',R''})^4$ by $(\tilde{\nu}_0 + \tilde{\nu}_{v',R':v'',R''})^4$, g'' by g', and $\mathcal{N}''$ by $\mathcal{N}'$ since the initial state is now the upper state.

For a diatomic molecule in the ground electronic state, assuming zero angular momentum of the electrons about the bond axis ($\Lambda = 0$) and zero resultant electron spin ($S=0$) only one vibrational quantum number v and one rotational quantum number J are needed to define the energy for a system of randomly oriented molecules. The space-averaged values of the squares of the transition polarizability tensor components, i.e.,

$$\frac{1}{(2J+1)}\sum_{M''}[\alpha_{xy}]^2_{v',J',M':v''J'',M''}=\overline{[\alpha_{xy}]^2_{v',J':v'',J''}}$$

are given in Table I for permitted rotation and fundamental vibration–rotation transitions of a diatomic molecule in the approximation of the polarizability theory. These space averages are functions of $(a)_0^2$ and $(\gamma)_0^2$ or $(a')^2$ and $(\gamma')^2$, and J-dependent coefficients $b_{J'J''}$ which determine the anisotropic contribution.

Values of the function $\Phi(a^2, \gamma^2, \theta)$ for calculating scattered intensities, depolarization ratios, and reversal coefficients for permitted rotation and vibration–rotation transitions in a diatomic molecule are given in Table J. The functions $\Phi(a^2, \gamma^2, \theta)$ were obtained by using the appropriate space-averaged squares of the transition polarizability tensor components from Table I. For example, ${}^{\perp}I_{\|}(\pi/2)$ for the pure rotation transition $v, J+2 \leftarrow v, J$ was obtained using $\overline{[\alpha_{zy}]^2_{v,J+2:v,J}}$ from Table I. The functions Φ in Table J also apply to the pure rotation transitions of linear molecules and to the vibration–rotation transitions of linear molecules in which the vibration is totally symmetric. For the vibration–rotation transitions the factors $(v+1)b_v^2$ ($\Delta v = +1$) and vb_v^2 ($\Delta v = -1$) have been omitted.

For other rotors and vibrotors, the functions $\Phi(a^2, \gamma^2, \theta)$ and the formulae for $\rho_{\perp}(\pi/2)$ and $\mathscr{P}(0)=\mathscr{P}^{-1}(\pi)$, follow the same general pattern, differing essentially only in the anisotropic coefficients which of course depend on the rotation quantum numbers and the permitted transitions. This is illustrated by Table K which gives the function $\Phi(a^2, \gamma^2, \theta)$ for intensities and formulae for depolarization ratios, and reversal coefficients for pure rotational scattering from a symmetric top molecule. The rotational energy of a symmetric top molecule is determined by two quantum numbers J and K, and the anisotropic contribution to the scattering is determined by coefficients of the type $b_{J',K'':J'',K''}$ since the selection rule $\Delta K=0$ operates for pure rotation transitions. Table K also applies to vibration–rotation transitions of symmetric top molecules in which the vibration is totally symmetric ($\Delta K=0$ also applies), if $(a)_0^2$ is replaced by $(a')^2$ and $(\gamma)_0^2$ by $(\gamma')^2$, and $(v+1)b_v^2$ or vb_v^2 included.

Clearly generalisation to cover all rotation and vibration–rotation transitions is possible. Thus Table L gives the general forms of the function $\Phi(a^2, \gamma^2, \theta)$ and the formulae for $\rho_{\perp}(\pi/2)$ and $\mathscr{P}(0)=\mathscr{P}^{-1}(\pi)$; and Table M gives the general forms of the anisotropic coefficients $b_{R^fR^i}$. Using the selection rules summarized in Table 4.2 (page 110) formulae for intensities and polarization properties may be obtained for all permitted rotation and vibration–rotation transitions in linear, symmetric top and spherical top molecules.

Finally, the nuclear spin and rotation degeneracy factors: for heteronuclear diatomic molecules $g_N=1$; but for homonuclear diatomic molecules g_N depends on the nuclear spin and can be different according as J is odd or even. Some specific cases are considered in Table 4.1 (page 105). For polyatomic molecules the nuclear spin degeneracy cannot be generalized and reference must be made to other texts (see Appendix I). For linear molecules g_R is $(2J+1)$; for symmetric top molecules g_R is $2J+1$ for $K=0$ and $2(2J+1)$ for $K\neq 0$; and for spherical top molecules g_R is $(2J+1)^2$. g_N and g_R refer always to the *initial* state.

Table A $f(a^2, \gamma^2, \delta^2, \theta)$ for ${}^{\perp}I_{\perp}(\theta)$, ${}^{\perp}I_{\parallel}(\theta)$, ${}^{\perp}I(\theta)$; and $\rho_{\perp}(\theta)$; $\theta=\theta$, $\theta=\pi/2$, $\theta=0$, $\theta=\pi$.

Observation direction	Quantity observed	Contributing transition moment amplitudes	$f(a^2,\theta)$ or ρ (scalar part)	$f(\gamma^2,\theta)$ or ρ (anisotropic part)	$f(\delta^2,\theta)$ or ρ (antisymmetric part)	$f(a^2,\gamma^2,\theta)$ or ρ (scalar + anisotropic parts)	$f(a^2,\gamma^2,\delta^2,\theta)$ or ρ (total)
$Ow(\theta=\theta)$	${}^{\perp}I_{\perp}(\theta)$	$\overline{P^2_{yo}}$	a^2	$\frac{4\gamma^2}{45}$	0	$\frac{45a^2+4\gamma^2}{45}$	$\frac{45a^2+4\gamma^2}{45}$
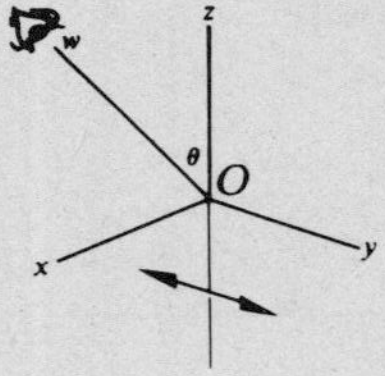	${}^{\perp}I_{\parallel}(\theta)$	$\overline{P^2_{xo}}c^2+\overline{P^2_{zo}}s^2$	0	$\frac{\gamma^2}{15}$	$\frac{\delta^2}{9}$	$\frac{\gamma^2}{15}$	$\frac{3\gamma^2+5\delta^2}{45}$
	${}^{\perp}I(\theta)$	$\overline{P^2_{xo}}c^2+\overline{P^2_{zo}}s^2+\overline{P^2_{yo}}$	a^2	$\frac{7\gamma^2}{45}$	$\frac{\delta^2}{9}$	$\frac{45a^2+7\gamma^2}{45}$	$\frac{45a^2+7\gamma^2+5\delta^2}{45}$
	$\rho_{\perp}(\theta)$	$\frac{\overline{P^2_{xo}}c^2+\overline{P^2_{zo}}s^2}{\overline{P^2_{yo}}}$	0	$\frac{3}{4}$	∞	$\frac{3\gamma^2}{45a^2+4\gamma^2}$	$\frac{3\gamma^2+5\delta^2}{45a^2+4\gamma^2}$
$Ox(\theta=\pi/2)$	${}^{\perp}I_{\perp}\left(\frac{\pi}{2}\right)$	$\overline{P^2_{yo}}$	a^2	$\frac{4\gamma^2}{45}$	0	$\frac{45a^2+4\gamma^2}{45}$	$\frac{45a^2+4\gamma^2}{45}$
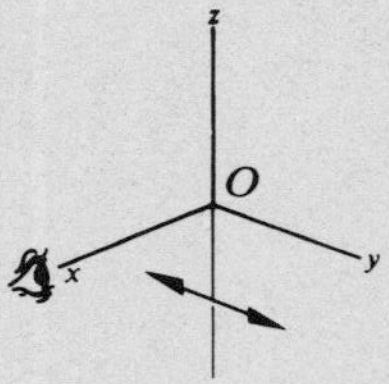	${}^{\perp}I_{\parallel}\left(\frac{\pi}{2}\right)$	$\overline{P^2_{zo}}$	0	$\frac{\gamma^2}{15}$	$\frac{\delta^2}{9}$	$\frac{\gamma^2}{15}$	$\frac{3\gamma^2+5\delta^2}{45}$
	${}^{\perp}I\left(\frac{\pi}{2}\right)$	$\overline{P^2_{yo}}+\overline{P^2_{zo}}$	a^2	$\frac{7\gamma^2}{45}$	$\frac{\delta^2}{9}$	$\frac{45a^2+7\gamma^2}{45}$	$\frac{45a^2+7\gamma^2+5\delta^2}{45}$
	$\rho_{\perp}(\pi/2)$	$\frac{\overline{P^2_{zo}}}{\overline{P^2_{yo}}}$	0	$\frac{3}{4}$	∞	$\frac{3\gamma^2}{45a^2+4\gamma^2}$	$\frac{3\gamma^2+5\delta^2}{45a^2+4\gamma^2}$
$Oz(\theta=0)$ and $-Oz(\theta=\pi)$	${}^{\perp}I_{\perp}(0)$	$\overline{P^2_{yo}}$	a^2	$\frac{4\gamma^2}{45}$	0	$\frac{45a^2+4\gamma^2}{45}$	$\frac{45a^2+4\gamma^2}{45}$
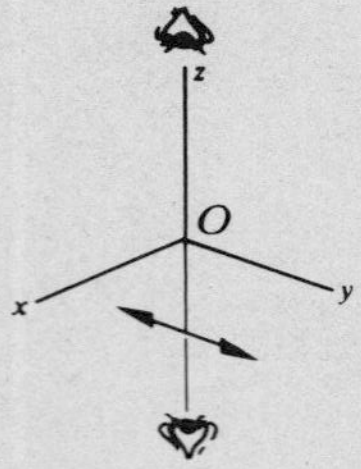	${}^{\perp}I_{\parallel}(0)$	$\overline{P^2_{xo}}$	0	$\frac{\gamma^2}{15}$	$\frac{\delta^2}{9}$	$\frac{\gamma^2}{15}$	$\frac{3\gamma^2+5\delta^2}{45}$
	${}^{\perp}I(0)$	$\overline{P^2_{xo}}+\overline{P^2_{yo}}$	a^2	$\frac{7\gamma^2}{45}$	$\frac{\delta^2}{9}$	$\frac{45a^2+7\gamma^2}{45}$	$\frac{45a^2+7\gamma^2+5\delta^2}{45}$
	$\rho_{\perp}(0)$	$\frac{\overline{P^2_{xo}}}{\overline{P^2_{yo}}}$	0	$\frac{3}{4}$	∞	$\frac{3\gamma^2}{45a^2+4\gamma^2}$	$\frac{3\gamma^2+5\delta^2}{45a^2+4\gamma^2}$

$c=\cos\theta$, $s=\sin\theta$.

Table B $f(a^2, \gamma^2, \delta^2, \theta)$ for ${}^{\|}I_{\perp}(\theta)$, ${}^{\|}I_{\|}(\theta)$, ${}^{\|}I(\theta)$; and $\rho_{\|}(\theta)$; $\theta = \theta$, $\theta = \pi/2$, $\theta = 0$, $\theta = \pi$.

Observation direction	Quantity observed	Contributing transition moment amplitudes	$f(a^2, \theta)$ or ρ (scalar part)	$f(\gamma^2, \theta)$ or ρ (anisotropic part)	$f(\delta^2, \theta)$ or ρ (antisymmetric part)	$f(a^2, \gamma^2, \theta)$ or ρ (scalar + anisotropic parts)	$f(a^2, \gamma^2, \delta^2, \theta)$ or ρ (total)
$Ow(\theta = \theta)$	${}^{\|}I_{\perp}(\theta)$	$\overline{P^2_{yo}}$	0	$\frac{\gamma^2}{15}$	$\frac{\delta^2}{9}$	$\frac{\gamma^2}{15}$	$\frac{3\gamma^2+5\delta^2}{45}$
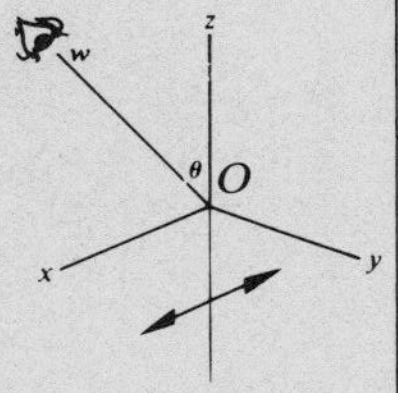	${}^{\|}I_{\|}(\theta)$	$\overline{P^2_{xo}}c^2+\overline{P^2_{zo}}s^2$	a^2c^2	$\frac{\gamma^2(3+c^2)}{45}$	$\frac{\delta^2 s^2}{9}$	$\frac{45a^2c^2+\gamma^2(3+c^2)}{45}$	$\frac{45a^2c^2+\gamma^2(3+c^2)+5\delta^2s^2}{45}$
	${}^{\|}I(\theta)$	$\overline{P^2_{xo}}c^2+\overline{P^2_{zo}}s^2+\overline{P^2_{yo}}$	a^2c^2	$\frac{\gamma^2(6+c^2)}{45}$	$\frac{\delta^2(1+s^2)}{9}$	$\frac{45a^2c^2+\gamma^2(6+c^2)}{45}$	$\frac{45a^2c^2+\gamma^2(6+c^2)+5\delta^2(1+s^2)}{45}$
	$\rho_{\|}(\theta)$	$\frac{\overline{P^2_{yo}}}{\overline{P^2_{xo}}c^2+\overline{P^2_{zo}}s^2}$	0	$\frac{3}{(3+c^2)}$	$\frac{1}{s^2}$	$\frac{3\gamma^2}{45a^2c^2+\gamma^2(3+c^2)}$	$\frac{3\gamma^2+5\delta^2}{45a^2c^2+\gamma^2(3+c^2)+5\delta^2s^2}$
$Ox(\theta = \pi/2)$	${}^{\|}I_{\perp}\left(\frac{\pi}{2}\right)$	$\overline{P^2_{yo}}$	0	$\frac{\gamma^2}{15}$	$\frac{\delta^2}{9}$	$\frac{\gamma^2}{15}$	$\frac{3\gamma^2+5\delta^2}{45}$
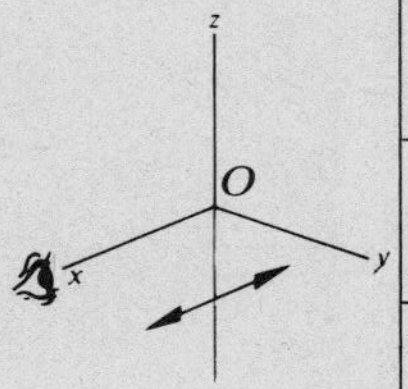	${}^{\|}I_{\|}\left(\frac{\pi}{2}\right)$	$\overline{P^2_{zo}}$	0	$\frac{\gamma^2}{15}$	$\frac{\delta^2}{9}$	$\frac{\gamma^2}{15}$	$\frac{3\gamma^2+5\delta^2}{45}$
	${}^{\|}I\left(\frac{\pi}{2}\right)$	$\overline{P^2_{yo}}+\overline{P^2_{zo}}$	0	$\frac{2\gamma^2}{15}$	$\frac{2\delta^2}{9}$	$\frac{2\gamma^2}{15}$	$\frac{6\gamma^2+10\delta^2}{45}$
	$\rho_{\|}(\pi/2)$	$\frac{\overline{P^2_{yo}}}{\overline{P^2_{zo}}}$	—	1	1	1	1
$Oz(\theta = 0)$ and $-Oz(\theta = \pi)$	${}^{\|}I_{\perp}(0)$	$\overline{P^2_{yo}}$	0	$\frac{\gamma^2}{15}$	$\frac{\delta^2}{9}$	$\frac{\gamma^2}{15}$	$\frac{3\gamma^2+5\delta^2}{45}$
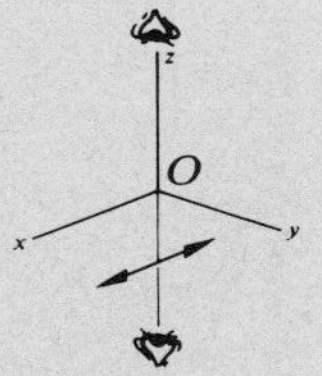	${}^{\|}I_{\|}(0)$	$\overline{P^2_{xo}}$	a^2	$\frac{4\gamma^2}{45}$	0	$\frac{45a^2+4\gamma^2}{45}$	$\frac{45a^2+4\gamma^2}{45}$
	${}^{\|}I(0)$	$\overline{P^2_{xo}}+\overline{P^2_{yo}}$	a^2	$\frac{7\gamma^2}{45}$	$\frac{\delta^2}{9}$	$\frac{45a^2+7\gamma^2}{45}$	$\frac{45a^2+7\gamma^2+5\delta^2}{45}$
	$\rho_{\|}(0)$	$\frac{\overline{P^2_{yo}}}{\overline{P^2_{xo}}}$	0	$\frac{3}{4}$	∞	$\frac{3\gamma^2}{45a^2+4\gamma^2}$	$\frac{3\gamma^2+5\delta^2}{45a^2+4\gamma^2}$

$c = \cos\theta$, $s = \sin\theta$.

Table C $f(a^2, \gamma^2, \delta^2, \theta)$ for ${}^nI_\perp(\theta)$, ${}^nI_\parallel(\theta)$, ${}^nI(\theta)$; and $\rho_n(\theta)$; $\theta=\theta$, $\theta=\pi/2$, $\theta=0$, $\theta=\pi$.

Observation direction	Quantity observed	Contributing transition moment amplitudes	$f(a^2, \theta)$ or ρ (scalar part)	$f(\gamma^2, \theta)$ or ρ (anisotropic part)	$f(\delta^2, \theta)$ or ρ (antisymmetric part)	$f(a^2, \gamma^2, \theta)$ or ρ (scalar + anisotropic parts)	$f(a^2, \gamma^2, \delta^2, \theta)$ or ρ (total)
$Ow(\theta=\theta)$	${}^nI_\perp(\theta)$	$\overline{P_{yo}^2}$	$\frac{a^2}{2}$	$\frac{7\gamma^2}{90}$	$\frac{\delta^2}{18}$	$\frac{45a^2+7\gamma^2}{90}$	$\frac{45a^2+7\gamma^2+5\delta^2}{90}$
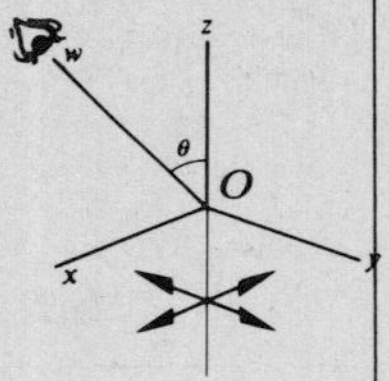	${}^nI_\perp(\theta)$	$\overline{P_{xo}^2}c^2+\overline{P_{zo}^2}s^2$	$\frac{a^2c^2}{2}$	$\frac{\gamma^2(6+c^2)}{90}$	$\frac{\delta^2(1+s^2)}{18}$	$\frac{45a^2c^2+\gamma^2(6+c^2)}{90}$	$\frac{45a^2c^2+\gamma^2(6+c^2)+5\delta^2(1+s^2)}{90}$
	${}^nI(\theta)$	$\overline{P_{xo}^2}c^2+\overline{P_{zo}^2}s^2+\overline{P_{yo}^2}$	$\frac{a^2(1+c^2)}{2}$	$\frac{\gamma^2(13+c^2)}{90}$	$\frac{\delta^2(2+s^2)}{18}$	$\frac{45a^2(1+c^2)+\gamma^2(13+c^2)}{90}$	$\frac{45a^2(1+c^2)+\gamma^2(13+c^2)+5\delta^2(2+s^2)}{90}$
	$\rho_n(\theta)$	$\frac{\overline{P_{xo}^2}c^2+\overline{P_{zo}^2}s^2}{\overline{P_{yo}^2}}$	c^2	$\frac{6+c^2}{7}$	$1+s^2$	$\frac{45a^2c^2+\gamma^2(6+c^2)}{45a^2+7\gamma^2}$	$\frac{45a^2c^2+\gamma^2(6+c^2)+5\delta^2(1+s^2)}{45a^2+7\gamma^2+5\delta^2}$
$Ox(\theta=\pi/2)$	${}^nI_\perp\left(\frac{\pi}{2}\right)$	$\overline{P_{yo}^2}$	$\frac{a^2}{2}$	$\frac{7\gamma^2}{90}$	$\frac{\delta^2}{18}$	$\frac{45a^2+7\gamma^2}{90}$	$\frac{45a^2+7\gamma^2+5\delta^2}{90}$
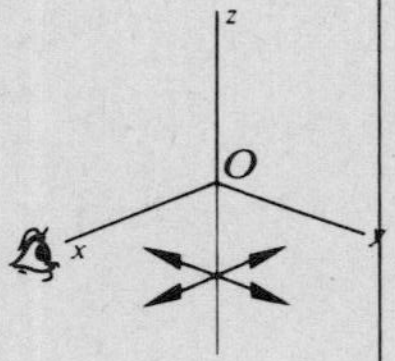	${}^nI_\parallel\left(\frac{\pi}{2}\right)$	$\overline{P_{zo}^2}$	0	$\frac{\gamma^2}{15}$	$\frac{\delta^2}{9}$	$\frac{\gamma^2}{15}$	$\frac{3\gamma^2+5\delta^2}{45}$
	${}^nI\left(\frac{\pi}{2}\right)$	$\overline{P_{yo}^2}+\overline{P_{zo}^2}$	$\frac{a^2}{2}$	$\frac{13\gamma^2}{90}$	$\frac{\delta^2}{6}$	$\frac{45a^2+13\gamma^2}{90}$	$\frac{45a^2+13\gamma^2+15\delta^2}{90}$
	$\rho_n(\pi/2)$	$\frac{\overline{P_{zo}^2}}{\overline{P_{yo}^2}}$	0	$\frac{6}{7}$	2	$\frac{6\gamma^2}{45a^2+7\gamma^2}$	$\frac{6\gamma^2+10\delta^2}{45a^2+7\gamma^2+5\delta^2}$
$Oz(\theta=0)$ and $-Oz(\theta=\pi)$	${}^nI_\perp(0)$	$\overline{P_{yo}^2}$	$\frac{a^2}{2}$	$\frac{7\gamma^2}{90}$	$\frac{\delta^2}{18}$	$\frac{45a^2+7\gamma^2}{90}$	$\frac{45a^2+7\gamma^2+5\delta^2}{90}$
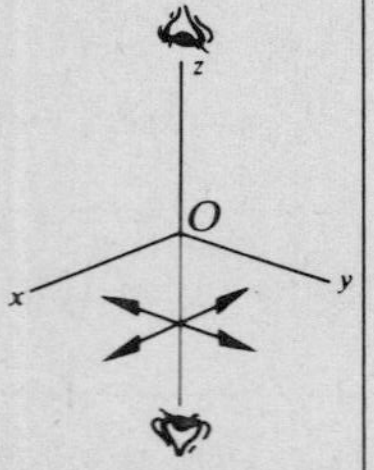	${}^nI_\parallel(0)$	$\overline{P_{xo}^2}$	$\frac{a^2}{2}$	$\frac{7\gamma^2}{90}$	$\frac{\delta^2}{18}$	$\frac{45a^2+7\gamma^2}{90}$	$\frac{45a^2+7\gamma^2+5\delta^2}{90}$
	${}^nI(0)$	$\overline{P_{xo}^2}+\overline{P_{yo}^2}$	a^2	$\frac{7\gamma^2}{45}$	$\frac{\delta^2}{9}$	$\frac{45a^2+7\gamma^2}{45}$	$\frac{45a^2+7\gamma^2+5\delta^2}{45}$
	$\rho_n(0)$	$\frac{\overline{P_{xo}^2}}{\overline{P_{yo}^2}}$	1	1	1	1	1

CRS 8

Table D $f(a^2, \gamma^2, \delta^2, \theta)$ for ${}^{\circledR}I_{\circledR}(\theta)$, $(= {}^{\textcircled{L}}I_{\textcircled{L}}(\theta))$, ${}^{\circledR}I_{\textcircled{L}}(\theta)(= {}^{\textcircled{L}}I_{\circledR}(\theta))$; $\mathscr{P}(\theta)$, and ${}^{\circledR}\mathscr{C}(\theta) = -{}^{\textcircled{L}}\mathscr{C}(\theta)$; $\theta = \theta$, $\theta = \pi/2$, $\theta = 0$, $\theta = \pi$.

Observation direction	Quantity observed	$f(a^2, \theta)$, $\mathscr{P}$ or $\mathscr{C}$ (scalar part)	$f(\gamma^2, \theta)$, $\mathscr{P}$ or $\mathscr{C}$ (anisotropic part)	$f(\delta^2, \theta)$, $\mathscr{P}$ or $\mathscr{C}$ (antisymmetric part)	$f(a^2, \gamma^2, \theta)$, $\mathscr{P}$ or $\mathscr{C}$ (scalar + anisotropic parts)	$f(a^2, \gamma^2, \delta^2, \theta)$, $\mathscr{P}$ or $\mathscr{C}$ (total)
$Ow(\theta = \theta)$ 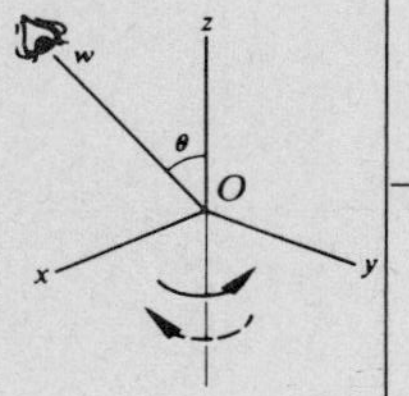	${}^{\circledR}I_{\circledR}(\theta) = {}^{\textcircled{L}}I_{\textcircled{L}}(\theta)$	$\frac{a^2(1+2c+c^2)}{4}$	$\frac{\gamma^2(13-10c+c^2)}{180}$	$\frac{\delta^2(3+2c-c^2)}{36}$	$\frac{45a^2(1+2c+c^2)+\gamma^2(13-10c+c^2)}{180}$	$\frac{45a^2(1+2c+c^2)+\gamma^2(13-10c+c^2)+5\delta^2(3+2c-c^2)}{180}$
	${}^{\circledR}I_{\textcircled{L}}(\theta) = {}^{\textcircled{L}}I_{\circledR}(\theta)$	$\frac{a^2(1-2c+c^2)}{4}$	$\frac{\gamma^2(13+10c+c^2)}{180}$	$\frac{\delta^2(3-2c-c^2)}{36}$	$\frac{45a^2(1-2c+c^2)+\gamma^2(13+10c+c^2)}{180}$	$\frac{45a^2(1-2c+c^2)+\gamma^2(13+10c+c^2)+5\delta^2(3-2c-c^2)}{180}$
	$\mathscr{P}(\theta)$	$\tan^4\frac{\theta}{2}$	$\frac{13+10c+c^2}{13-10c+c^2}$	$\frac{3-2c-c^2}{3+2c-c^2}$	$\frac{45a^2(1-2c+c^2)+\gamma^2(13+10c+c^2)}{45a^2(1+2c+c^2)+\gamma^2(13-10c+c^2)}$	$\frac{45a^2(1-2c+c^2)+\gamma^2(13+10c+c^2)+5\delta^2(3-2c-c^2)}{45a^2(1+2c+c^2)+\gamma^2(13-10c+c^2)+5\delta^2(3+2c-c^2)}$
	${}^{\circledR}\mathscr{C}(\theta) = -{}^{\textcircled{L}}\mathscr{C}(\theta)$	$\frac{2c}{1+c^2}$	$\frac{-10c}{13+c^2}$	$\frac{2c}{3-c^2}$	$\frac{90a^2c-10\gamma^2c}{45a^2(1+c^2)+\gamma^2(13+c^2)}$	$\frac{90a^2c-10\gamma^2c+10\delta^2c}{45a^2(1+c^2)+\gamma^2(13+c^2)+5\delta^2(3-c^2)}$
$Ox(\theta = \pi/2)$ 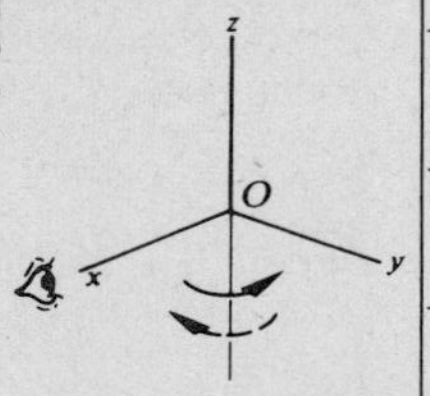	${}^{\circledR}I_{\circledR}\left(\frac{\pi}{2}\right) = {}^{\textcircled{L}}I_{\textcircled{L}}\left(\frac{\pi}{2}\right)$	$\frac{a^2}{4}$	$\frac{13\gamma^2}{180}$	$\frac{\delta^2}{12}$	$\frac{45a^2+13\gamma^2}{180}$	$\frac{45a^2+13\gamma^2+15\delta^2}{180}$
	${}^{\circledR}I_{\textcircled{L}}\left(\frac{\pi}{2}\right) = {}^{\textcircled{L}}I_{\circledR}\left(\frac{\pi}{2}\right)$	$\frac{a^2}{4}$	$\frac{13\gamma^2}{180}$	$\frac{\delta^2}{12}$	$\frac{45a^2+13\gamma^2}{180}$	$\frac{45a^2+13\gamma^2+15\delta^2}{180}$
	$\mathscr{P}\left(\frac{\pi}{2}\right)$	1	1	1	1	1
	${}^{\circledR}\mathscr{C}\left(\frac{\pi}{2}\right) = -{}^{\textcircled{L}}\mathscr{C}\left(\frac{\pi}{2}\right)$	0	0	0	0	0
$Oz(\theta = 0)$ and $-Oz(\theta = \pi)$ 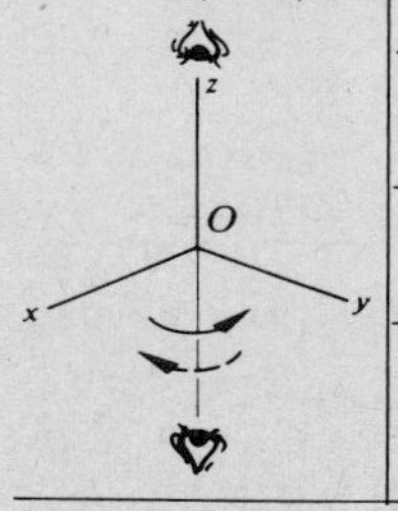	${}^{\circledR}I_{\circledR}(0) = {}^{\textcircled{L}}I_{\textcircled{L}}(0) = {}^{\circledR}I_{\textcircled{L}}(\pi) = {}^{\textcircled{L}}I_{\circledR}(\pi)$	a^2	$\frac{\gamma^2}{45}$	$\frac{\delta^2}{9}$	$\frac{45a^2+\gamma^2}{45}$	$\frac{45a^2+\gamma^2+5\delta^2}{45}$
	${}^{\circledR}I_{\textcircled{L}}(0) = {}^{\textcircled{L}}I_{\circledR}(0) = {}^{\circledR}I_{\circledR}(\pi) = {}^{\textcircled{L}}I_{\textcircled{L}}(\pi)$	0	$\frac{2\gamma^2}{15}$	0	$\frac{2\gamma^2}{15}$	$\frac{2\gamma^2}{15}$
	$\mathscr{P}(0) = \mathscr{P}^{-1}(\pi)$	0	6	0	$\frac{6\gamma^2}{45a^2+\gamma^2}$	$\frac{6\gamma^2}{45a^2+\gamma^2+5\delta^2}$
	${}^{\circledR}\mathscr{C}(0) = -{}^{\textcircled{L}}\mathscr{C}(0) = -{}^{\circledR}\mathscr{C}(\pi) = {}^{\textcircled{L}}\mathscr{C}(\pi)$	1	$-\frac{5}{7}$	1	$\frac{45a^2-5\gamma^2}{45a^2+7\gamma^2}$	$\frac{45a^2-5\gamma^2+5\delta^2}{45a^2+7\gamma^2+5\delta^2}$

$c = \cos\theta$, $s = \sin\theta$.

CRS 9

Table E $F(a^2, \gamma^2, \delta^2, \theta, P, \chi, \psi)$ for Stokes parameters for $\theta = \pi/2$, $\theta = 0$ and $\theta = \pi$; general and special polarization states of incident radiation

Observation direction	Stokes parameter	$P = 1 \chi = 0, \psi = 0$ (i.e., $E_x \neq 0$)	$P = 1, \chi = 0, \psi = \pi/2$ (i.e., $E_y \neq 0$)	$P = 0$ (i.e., $E_x = E_y \neq 0$)	$P = 1, \chi = \pi/4, \psi = 0$ (i.e., $E_y = -iE_x$)	$P = P, \chi = \chi, \psi = \psi$ (general polarization)
$Ox\left(\theta = \frac{\pi}{2}\right)$ 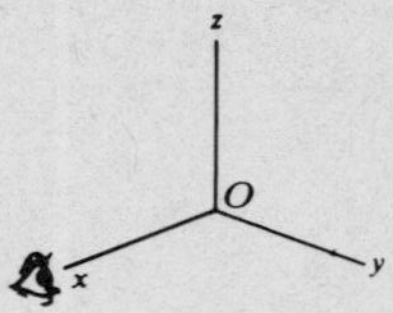	$S_0\left(\frac{\pi}{2}\right)$	$\frac{6\gamma^2 + 10\delta^2}{45}$	$\frac{45a^2 + 7\gamma^2 + 5\delta^2}{45}$	$\frac{45a^2 + 13\gamma^2 + 15\delta^2}{90}$	$\frac{45a^2 + 13\gamma^2 + 15\delta^2}{90}$	$\frac{(45a^2 + 13\gamma^2 + 15\delta^2) - (45a^2 + \gamma^2 - 5\delta^2)P \cos 2\chi \cos 2\psi}{90}$
	$S_1\left(\frac{\pi}{2}\right)$	0	$\frac{45a^2 + \gamma^2 - 5\delta^2}{45}$	$\frac{45a^2 + \gamma^2 - 5\delta^2}{90}$	$\frac{45a^2 + \gamma^2 - 5\delta^2}{90}$	$\frac{(45a^2 + \gamma^2 - 5\delta^2)(1 - P \cos 2\chi \cos 2\psi)}{90}$
	$S_2\left(\frac{\pi}{2}\right)$	0	0	0	0	0
	$S_3\left(\frac{\pi}{2}\right)$	0	0	0	0	0
$Oz(\theta = 0)$ and $-Oz(\theta = \pi)$ 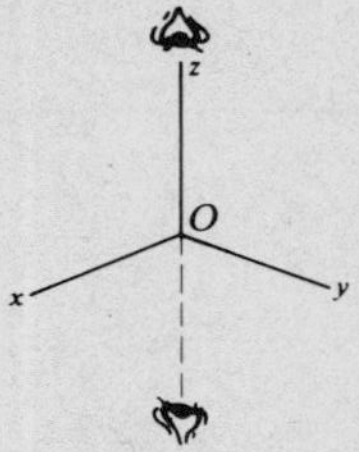	$S_0(0) = S_0(\pi)$	$\frac{45a^2 + 7\gamma^2 + 5\delta^2}{45}$	$\frac{45a^2 + 7\gamma^2 + 5\delta^2}{45}$	$\frac{45a^2 + 7\gamma^2 + 5\delta^2}{45}$	$\frac{45a^2 + 7\gamma^2 + 5\delta^2}{45}$	$\frac{45a^2 + 7\gamma^2 + 5\delta^2}{45}$
	$S_1(0) = S_1(\pi)$	$\frac{45a^2 + \gamma^2 - 5\delta^2}{45}$	$-\left\{\frac{45a^2 + \gamma^2 - 5\delta^2}{45}\right\}$	0	0	$\left\{\frac{45a^2 + \gamma^2 - 5\delta^2}{45}\right\} P \cos 2\chi \cos 2\psi$
	$S_2(0) = -S_2(\pi)$	0	0	0	0	$\left\{\frac{45a^2 + \gamma^2 - 5\delta^2}{45}\right\} P \cos 2\chi \sin 2\psi$
	$S_3(0) = -S_3(\pi)$	0	0	0	$\frac{45a^2 - 5\gamma^2 + 5\delta^2}{45}$	$\left\{\frac{45a^2 - 5\gamma^2 + 5\delta^2}{45}\right\} P \sin 2\chi$
		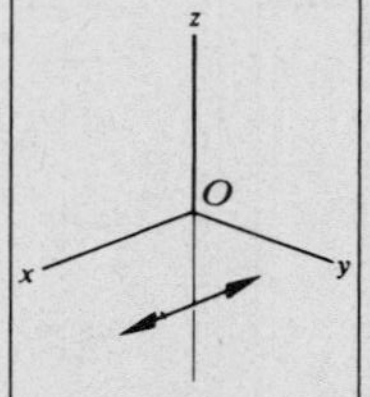	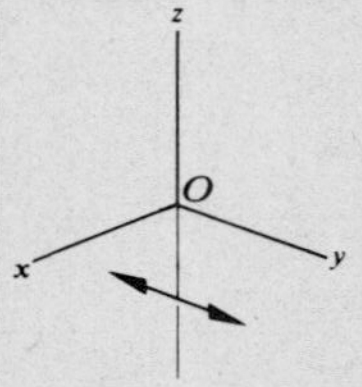	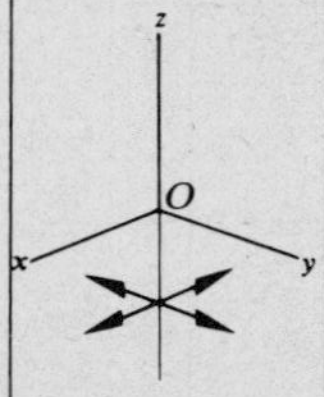	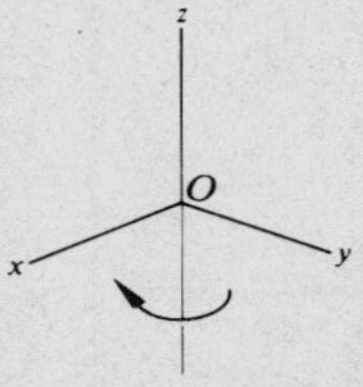	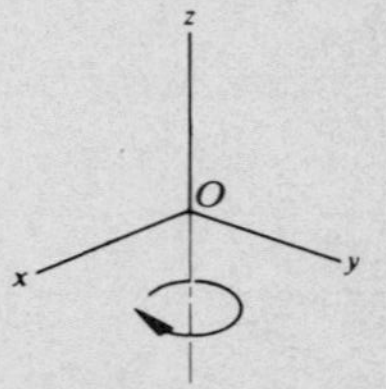

Table F $f(a^2, \gamma^2, \delta^2, \theta, P, \chi, \psi)$ for ${}^{p_i}I_{\|}(\theta)$, ${}^{p_i}I_{\perp}(\theta)$, ${}^{p_i}I_{\circledR}(\theta)$, ${}^{p_i}I_{\copyright}(\theta)$; $\rho(\theta)$, $\mathscr{P}(\theta)$, ${}^{p_i}\mathscr{C}(\theta)$; $\theta = 0$, $\theta = \pi/2$, $\theta = \pi$

Observation direction	Quantity observed	$f(a^2, \gamma^2, \delta^2, \theta, P, \chi, \psi), \rho, \mathscr{P}$ or $\mathscr{C}$
$Oz(\theta = 0)$ and $-Oz(\theta = \pi)$	${}^{p_i}I_{\|}(0) = {}^{p_i}I_{\|}(\pi)$	$\dfrac{(45a^2+7\gamma^2+5\delta^2)+(45a^2+\gamma^2-5\delta^2)P\cos 2\chi \cos 2\psi}{90}$
	${}^{p_i}I_{\perp}(0) = {}^{p_i}I_{\perp}(\pi)$	$\dfrac{(45a^2+7\gamma^2+5\delta^2)-(45a^2+\gamma^2-5\delta^2)P\cos 2\chi \cos 2\psi}{90}$
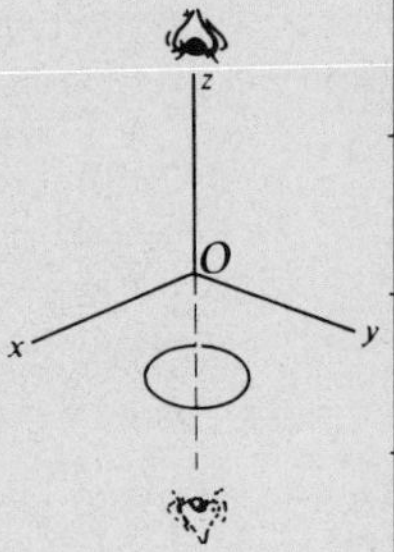	${}^{p_i}I_{\circledR}(0) = {}^{p_i}I_{\copyright}(\pi)$	$\dfrac{(45a^2+7\gamma^2+5\delta^2)+(45a^2-5\gamma^2+5\delta^2)P\sin 2\chi}{90}$
	${}^{p_i}I_{\copyright}(0) = {}^{p_i}I_{\circledR}(\pi)$	$\dfrac{(45a^2+7\gamma^2+5\delta^2)-(45a^2-5\gamma^2+5\delta^2)P\sin 2\chi}{90}$
	† $\dfrac{{}^{p_i}I_{\|}(0)}{{}^{p_i}I_{\perp}(0)}$	$\dfrac{(45a^2+7\gamma^2+5\delta^2)+(45a^2+\gamma^2-5\delta^2)P\cos 2\chi \cos 2\psi}{(45a^2+7\gamma^2+5\delta^2)-(45a^2+\gamma^2-5\delta^2)P\cos 2\chi \cos 2\psi}$
	* ${}^{p_i}\mathscr{P}(0) = {}^{p_i}\mathscr{P}^{-1}(\pi)$	$\dfrac{(45a^2+7\gamma^2+5\delta^2)-\lvert(45a^2-5\gamma^2+5\delta^2)P\sin 2\chi\rvert}{(45a^2+7\gamma^2+5\delta^2)+\lvert(45a^2-5\gamma^2+5\delta^2)P\sin 2\chi\rvert}$
	${}^{p_i}\mathscr{C}(0) = -{}^{p_i}\mathscr{C}(\pi)$	$\left(\dfrac{45a^2-5\gamma^2+5\delta^2}{45a^2+7\gamma^2+5\delta^2}\right)P\sin 2\chi$
$Ox(\theta = \pi/2)$	${}^{p_i}I_{\|}\left(\dfrac{\pi}{2}\right)$	$\dfrac{3\gamma^2+5\delta^2}{45}$
	${}^{p_i}I_{\perp}\left(\dfrac{\pi}{2}\right)$	$\dfrac{(45a^2+7\gamma^2+5\delta^2)-(45a^2+\gamma^2-5\delta^2)P\cos 2\chi \cos 2\psi}{90}$
	${}^{p_i}I_{\circledR}\left(\dfrac{\pi}{2}\right)$	$\dfrac{(45a^2+13\gamma^2+15\delta^2)-(45a^2+\gamma^2-5\delta^2)P\cos 2\chi \cos 2\psi}{180}$
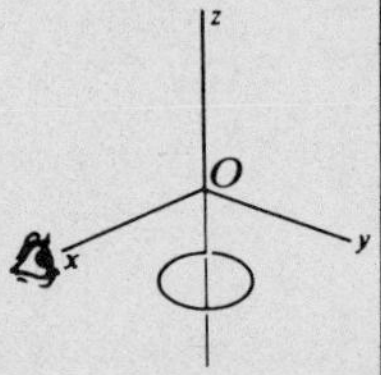	${}^{p_i}I_{\copyright}\left(\dfrac{\pi}{2}\right)$	$\dfrac{(45a^2+13\gamma^2+15\delta^2)-(45a^2+\gamma^2-5\delta^2)P\cos 2\chi \cos 2\psi}{180}$
	‡ $\dfrac{{}^{p_i}I_{\|}(\pi/2)}{{}^{p_i}I_{\perp}(\pi/2)}$	$\dfrac{6\gamma^2+10\delta^2}{(45a^2+7\gamma^2+5\delta^2)-(45a^2+\gamma^2-5\delta^2)P\cos 2\chi \cos 2\psi}$
	* ${}^{p_i}\mathscr{P}\left(\dfrac{\pi}{2}\right)$	1
	${}^{p_i}\mathscr{C}\left(\dfrac{\pi}{2}\right)$	0

* ${}^{\circledR}\mathscr{P}(o) = {}^{\copyright}\mathscr{P}(o)$

† This yields $\rho_{\perp}(0)$, $\rho_n(0)$ and $\{\rho_{\|}(0)\}^{-1}$; $\rho(0) = \rho(\pi)$.

‡ This yields $\rho_{\perp}(\pi/2)$, $\rho_n(\pi/2)$ and $\{\rho_{\|}(\pi/2)\}^{-1}$.

General formulae relating polarization properties to Stokes parameters will be found on pages 29–31; and general formulae relating scattered intensities to Stokes parameters on page 87.

Table G $F(a^2, \gamma^2, aG', \gamma_G^2, \gamma_A^2, \theta, P, \chi, \psi)$ for Stokes parameters $\theta = \pi/2$; and $f(a^2, \gamma^2, aG', \gamma_G^2, \gamma_A^2, \theta, P, \chi, \psi)$ for ${}^{pi}I_\parallel(\pi/2)$, ${}^{pi}I_\parallel(\pi/2)$, ${}^{\circledR}I_\parallel(\pi/2)$, ${}^{\textcircled{L}}I_\parallel(\pi/2)$, ${}^{\circledR}I_\perp(\pi/2)$, ${}^{\textcircled{L}}I_\perp(\pi/2)$; and $\Delta_\parallel(\pi/2)$, $\Delta_\perp(\pi/2)$, $\Delta(\pi/2)$, ${}^{\parallel}\mathscr{C}(\pi/2)$, and ${}^{\perp}\mathscr{C}(\pi/2)$

Stokes parameter	$F(a^2, \gamma^2, aG', \gamma_G^2, \gamma_A^2, \theta, P, \chi, \psi)$
$S_0\left(\frac{\pi}{2}\right)$	$\dfrac{(45a^2+13\gamma^2)-(45a^2+\gamma^2)P\cos 2\chi\cos 2\psi+2/c(45aG'+13\gamma_G^2-\gamma_A^2/3)P\sin 2\chi}{90}$
$S_1\left(\frac{\pi}{2}\right)$	$\dfrac{(45a^2+\gamma^2)(1-P\cos 2\chi\cos 2\psi)+(2/c)(45aG'+\gamma_G^2+\gamma_A^2)P\sin 2\chi}{90}$
$S_2\left(\frac{\pi}{2}\right)$	0
$S_3\left(\frac{\pi}{2}\right)$	$\dfrac{(2/c)(45aG'+13\gamma_G^2-\gamma_A^2/3)-(2/c)(45aG'+\gamma_G^2+\gamma_A^2)P\cos 2\chi\cos 2\psi}{90}$

Quantity observed	$f(a^2, \gamma^2, aG', \gamma_G^2, \gamma_A^2, \theta, P, \chi, \psi)$, Δ or $\mathscr{C}$
${}^{pi}I_\parallel\left(\frac{\pi}{2}\right)$	$\dfrac{6\gamma^2+(4/c)(3\gamma_G^2-\gamma_A^2/3)P\sin 2\chi}{90}$
${}^{pi}I_\perp\left(\frac{\pi}{2}\right)$	$\dfrac{(45a^2+7\gamma^2)-(45a^2+\gamma^2)P\cos 2\chi\cos 2\psi+(2/c)\left(45aG'+7\gamma_G^2+\frac{\gamma_A^2}{3}\right)P\sin 2\chi}{90}$
${}^{\circledR}I_\parallel\left(\frac{\pi}{2}\right)$	$\dfrac{6\gamma^2+(4/c)(3\gamma_G^2-\gamma_A^2/3)}{90}$
${}^{\textcircled{L}}I_\parallel\left(\frac{\pi}{2}\right)$	$\dfrac{6\gamma^2-(4/c)(3\gamma_G^2-\gamma_A^2/3)}{90}$
${}^{\circledR}I_\perp\left(\frac{\pi}{2}\right)$	$\dfrac{45a^2+7\gamma^2+(2/c)(45aG'+7\gamma_G^2+\gamma_A^2/3)}{90}$
${}^{\textcircled{L}}I_\perp\left(\frac{\pi}{2}\right)$	$\dfrac{45a^2+7\gamma^2-(2/c)(45aG'+7\gamma_G^2+\gamma_A^2/3)}{90}$
† $\Delta_\parallel\left(\frac{\pi}{2}\right)=\dfrac{{}^{\circledR}I_\parallel-{}^{\textcircled{L}}I_\parallel}{{}^{\circledR}I_\parallel+{}^{\textcircled{L}}I_\parallel}$	$\dfrac{2(3\gamma_G^2-\gamma_A^2/3)}{3c\gamma^2}$
† $\Delta_\perp\left(\frac{\pi}{2}\right)=\dfrac{{}^{\circledR}I_\perp-{}^{\textcircled{L}}I_\perp}{{}^{\circledR}I_\perp+{}^{\textcircled{L}}I_\perp}$	$\dfrac{2(45aG'+7\gamma_G^2+\gamma_A^2/3)}{c(45a^2+7\gamma^2)}$
† $\Delta\left(\frac{\pi}{2}\right)=\dfrac{{}^{\circledR}I-{}^{\textcircled{L}}I}{{}^{\circledR}I+{}^{\textcircled{L}}I}$	$\dfrac{2(45aG'+13\gamma_G^2-\gamma_A^2/3)}{c(45a^2+13\gamma^2)}$
${}^{\parallel}\mathscr{C}\left(\frac{\pi}{2}\right)=\dfrac{{}^{\parallel}S_3(\pi/2)}{{}^{\parallel}S_0(\pi/2)}$	$\dfrac{2(3\gamma_G^2-\gamma_A^2/3)}{3c\gamma^2}$
${}^{\perp}\mathscr{C}\left(\frac{\pi}{2}\right)=\dfrac{{}^{\perp}S_3(\pi/2)}{{}^{\perp}S_0(\pi/2)}$	$\dfrac{2(45aG'+7\gamma_G^2+\gamma_A^2/3)}{c(45a^2+7\gamma^2)}$

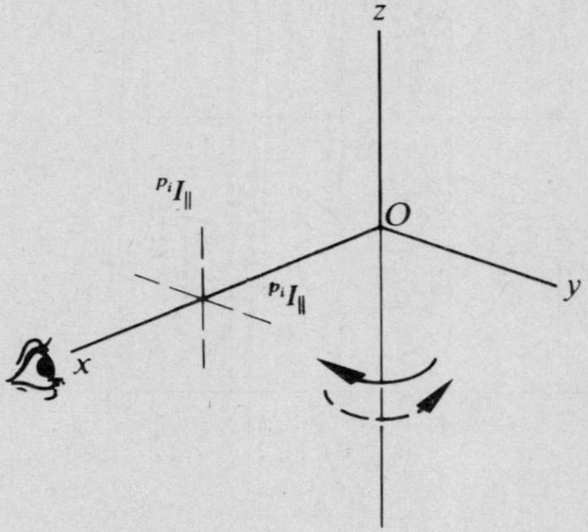

† ${}^{\circledR}I_\parallel = {}^{\circledR}I_\parallel(\pi/2)$ etc. for brevity.
c = speed of light.

Table H Vibration transition polarizability components $[\alpha_{xy}]_{v^f v^i}$ for fundamental, first overtone, binary combination, and difference transitions (assuming the Placzek polarizability theory is applicable)

Final vibration state	Initial vibration state	Selection rule	$[\alpha_{xy}]_{v^f v^i}$
$[v], v_k+1$	$[v], v_k$	$\Delta v_k = +1$	$(v_k+1)^{\frac{1}{2}}\left(\frac{\partial \alpha_{xy}}{\partial Q_k}\right)_0 b_{v_k}$
$[v], v_k-1$	$[v], v_k$	$\Delta v_k = -1$	$(v_k)^{\frac{1}{2}}\left(\frac{\partial \alpha_{xy}}{\partial Q_k}\right)_0 b_{v_k}$
$[v], v_k+2$	$[v], v_k$	$\Delta v_k = +2$	$\frac{1}{2}\{(v_k+1)(v_k+2)\}^{\frac{1}{2}}\left(\frac{\partial^2 \alpha_{xy}}{\partial Q_k^2}\right)_0 b_{v_k}^2$
$[v], v_k-2$	$[v], v_k$	$\Delta v_k = -2$	$\frac{1}{2}\{(v_k-1)(v_k)\}^{\frac{1}{2}}\left(\frac{\partial^2 \alpha_{xy}}{\partial Q_k^2}\right)_0 b_{v_k}^2$
$[v], v_k+1, v_l+1$	$[v], v_k, v_l$	$\Delta v_k = +1, \Delta v_l = +1$	$\frac{1}{2}\{(v_k+1)(v_l+1)\}^{\frac{1}{2}}\left(\frac{\partial^2 \alpha_{xy}}{\partial Q_k \partial Q_l}\right)_0 b_{v_k} b_{v_l}$
$[v], v_k-1, v_l-1$	$[v], v_k, v_l$	$\Delta v_k = -1, \Delta v_l = -1$	$\frac{1}{2}\{(v_k)(v_l)\}^{\frac{1}{2}}\left(\frac{\partial^2 \alpha_{xy}}{\partial Q_k \partial Q_l}\right)_0 b_{v_k} b_{v_l}$
$[v]$	$[v]$	$\Delta v = 0$	$(\alpha_{xy})_0$

$$b_{v_k} = \left(\frac{h}{8\pi^2 c\tilde{\nu}_k}\right)^{1/2} = \left(\frac{h}{4\pi\omega_k}\right)^{1/2}$$

$[v]$ denotes those vibrational quantum numbers which are unchanged in the transition. Derivatives higher than the second are not considered. c = speed of light.

Table I Space-averaged squares of transition polarizability tensor components for diatomic molecules for rotation and vibration–rotation transitions.

Upper state $v'J'$	Lower state $v''J''$	Selection rules[a]	$\overline{[\alpha_{ii}]^2_{v',J':v'',J''}}$	$\overline{[\alpha_{ij}]^2_{v',J':v'',J''}}$	$\overline{[\alpha_{ii}]_{v',J':v'',J''}[\alpha_{jj}]_{v',J':v'',J''}}$
v, J	v, J	$\Delta v = 0, \Delta J = 0$	$(a)_0^2 + \frac{4}{45} b_{J,J} (\gamma)_0^2$	$\frac{1}{15} b_{J,J} (\gamma)_0^2$	$(a)_0^2 - \frac{2}{45} b_{J,J} (\gamma)_0^2$
$v, J+2$	v, J	$\Delta v = 0, \Delta J = +2$	$\frac{4}{45} b_{J+2,J} (\gamma)_0^2$	$\frac{1}{15} b_{J+2,J} (\gamma)_0^2$	$\frac{2}{45} b_{J+2,J} (\gamma)_0^2$
[b] $v+1, J$	v, J	$\Delta v = +1, \Delta J = 0$	$(a')^2 + \frac{4}{45} b_{J,J} (\gamma')^2$	$\frac{1}{15} b_{J,J} (\gamma')^2$	$(a')^2 - \frac{2}{45} b_{J,J} (\gamma')^2$
[b] $v+1, J\pm2$	v, J	$\Delta v = +1, \Delta J = \pm 2$	$\frac{4}{45} b_{J,J} (\gamma')^2$	$\frac{1}{15} b_{J\pm2,J} (\gamma')^2$	$\frac{2}{45} b_{J\pm2,J} (\gamma')^2$

$$b_{J,J} = \frac{J(J+1)}{(2J-1)(2J+3)} \qquad b_{J+2,J} = \frac{3(J+1)(J+2)}{2(2J+1)(2J+3)} \qquad b_{J-2,J} = \frac{3J(J-1)}{2(2J+1)(2J-1)}$$

[a] Assumes zero angular momentum of electrons about bond axis ($\Lambda = 0$) and zero resultant electron spin ($S = 0$).

[b] Assumes applicability of Placzek polarizability theory; factor $(v+1)b_v^2$ has been omitted (see Table H). i, j represent x, y, z.

Table J $\Phi(a^2, \gamma^2, \theta)$ for ${}^{\perp}I_{\perp}(\pi/2)$, ${}^{\perp}I_{\parallel}(\pi/2)$, ${}^{\textcircled{R}}I_{\textcircled{R}}(0)(={}^{\textcircled{L}}I_{\textcircled{L}}(0))$, ${}^{\textcircled{R}}I_{\textcircled{L}}(0)(={}^{\textcircled{L}}I_{\textcircled{R}}(0))$; $\rho_{\perp}(\pi/2)$; and $\mathscr{P}(0)(=\mathscr{P}^{-1}(\pi))$ for diatomic molecules

Type of scattering	Branch	Selection rules		$\Phi(a^2, \gamma^2, \theta)$			$\Phi(a^2, \gamma^2, \theta)$		
		Δv	ΔJ	${}^{\perp}I_{\perp}\left(\frac{\pi}{2}\right)$	${}^{\perp}I_{\parallel}\left(\frac{\pi}{2}\right)$	$\rho_{\perp}\left(\frac{\pi}{2}\right)$	${}^{\textcircled{R}}I_{\textcircled{R}}(0)={}^{\textcircled{L}}I_{\textcircled{L}}(0)$	${}^{\textcircled{R}}I_{\textcircled{L}}(0)={}^{\textcircled{L}}I_{\textcircled{R}}(0)$	$\mathscr{P}(0)=\mathscr{P}^{-1}(\pi)$
Rayleigh	Q	0	0	$(a)_0^2+\frac{4}{45}b_{J,J}(\gamma)_0^2$	$\frac{1}{15}b_{J,J}(\gamma)_0^2$	$\frac{3b_{J,J}(\gamma)_0^2}{45(a)_0^2+4b_{J,J}(\gamma)_0^2}$	$(a)_0^2+\frac{b_{J,J}}{45}(\gamma)_0^2$	$\frac{2b_{J,J}(\gamma)_0^2}{15}$	$\frac{6b_{J,J}(\gamma)_0^2}{45(a)_0^2+b_{J,J}(\gamma)_0^2}$
Raman (rotation)	S	0	+2	$\frac{4}{45}b_{J+2,J}(\gamma)_0^2$	$\frac{1}{15}b_{J+2,J}(\gamma)_0^2$	$\frac{3}{4}$	$\frac{b_{J+2,J}}{45}(\gamma)_0^2$	$\frac{2b_{J+2,J}}{15}(\gamma)_0^2$	6
Raman[a] (vibration–rotation)	O	±1	−2	$\frac{4}{45}b_{J-2,J}(\gamma')^2$	$\frac{1}{15}b_{J-2,J}(\gamma')^2$	$\frac{3}{4}$	$\frac{b_{J-2,J}}{45}(\gamma)_0^2$	$\frac{2b_{J-2,J}}{15}(\gamma)_0^2$	6
	Q	±1	0	$(a')^2+\frac{4}{45}b_{J,J}(\gamma')^2$	$\frac{1}{15}b_{J,J}(\gamma')^2$	$\frac{3b_{J,J}(\gamma')^2}{45(a')^2+4b_{J,J}(\gamma')^2}$	$(a')^2+\frac{b_{J,J}(\gamma')^2}{45}$	$\frac{2b_{J,J}(\gamma')^2}{15}$	$\frac{6b_{J,J}(\gamma')^2}{45(a')^2+b_{J,J}(\gamma')^2}$
	S	±1	+2	$\frac{4}{45}b_{J+2,J}(\gamma')^2$	$\frac{1}{15}b_{J+2,J}(\gamma')^2$	$\frac{3}{4}$	$\frac{b_{J+2,J}}{45}(\gamma')^2$	$\frac{2b_{J+2,J}}{15}(\gamma')^2$	6

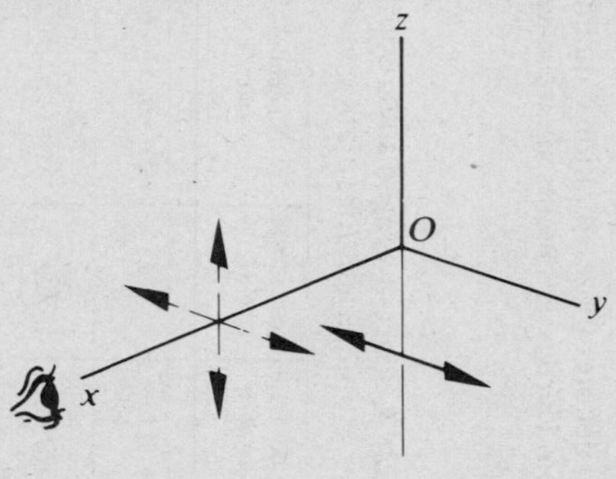

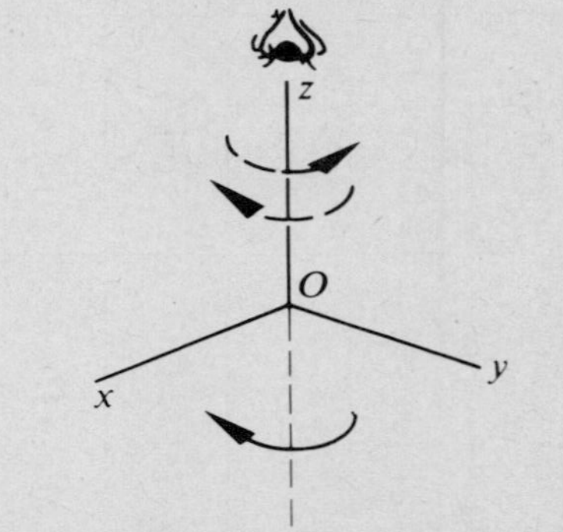

[a] The factors $(v+1)b_v^2$ (for $\Delta v=+1$) and vb_v^2 (for $\Delta v=-1$) have been omitted. For definitions of $b_{J,J}$ and $b_{J\pm2,J}$ see Table I.

CRS 14

Table K $\Phi(a^2, \gamma^2, \theta)$ for ${}^{\perp}I\,(\pi/2)$, ${}^{\perp}I_{\parallel}(\pi/2)$, ${}^{\circledR}I_{\circledR}(0)(={}^{\copyright}I_{\copyright}(0))$, ${}^{\circledR}I_{\copyright}(0)(={}^{\copyright}I_{\circledR}(0))$; $\rho_{\perp}(\pi/2)$ and $\mathscr{P}(0)(=\mathscr{P}^{-1}(\pi))$ for rotation transitions in symmetric top molecules

Type of scattering	Branch	Selection rules			$\Phi(a^2, \gamma^2, \theta)$		$\rho_{\perp}\left(\frac{\pi}{2}\right)$	$\Phi(a^2, \gamma^2, \theta)$		$\mathscr{P}(0)=\mathscr{P}^{-1}(\pi)$
		Δv	ΔJ	ΔK	${}^{\perp}I_{\perp}\left(\frac{\pi}{2}\right)$	${}^{\perp}I_{\parallel}\left(\frac{\pi}{2}\right)$		${}^{\circledR}I_{\circledR}(0)={}^{\copyright}I_{\copyright}(0)$	${}^{\circledR}I_{\copyright}(0)={}^{\copyright}I_{\circledR}(0)$	
Rayleigh	Q	0	0	0	$(a)_0^2+\frac{4}{45}b_{J,K:J,K}(\gamma)_0^2$	$\frac{1}{15}b_{J,K:J,K}(\gamma)_0^2$	$\frac{3b_{J,K:J,K}(\gamma)_0^2}{45(a)_0^2+4(\gamma)_0^2}$	$(a)_0^2+\frac{b_{J,K:J,K}(\gamma)_0^2}{45}$	$\frac{2b_{J,K:J,K}(\gamma)_0^2}{15}$	$\frac{6b_{J,K:J,K}(\gamma)_0^2}{45(a)_0^2+b_{J,k:J,k}(\gamma)_0^2}$
Raman (rotation)	O	0	+1	0	$\frac{4}{45}b_{J+1,K:J,K}(\gamma)_0^2$	$\frac{1}{15}b_{J+1,K:J,K}(\gamma)_0^2$	$\frac{3}{4}$	$\frac{b_{J+1,K:J,K}(\gamma)_0^2}{45}$	$\frac{2b_{J+1,K:J,K}(\gamma)_0^2}{15}$	6
	S	0	+2	0	$\frac{4}{45}b_{J+2,K:J,K}(\gamma)_0^2$	$\frac{1}{15}b_{J+2,K:J,K}(\gamma)_0^2$	$\frac{3}{4}$	$\frac{b_{J+2,K:J,K}(\gamma)_0^2}{45}$	$\frac{2b_{J+2,K:J,K}(\gamma)_0^2}{15}$	6

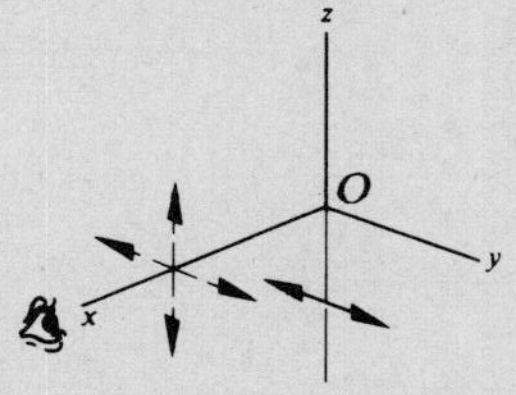

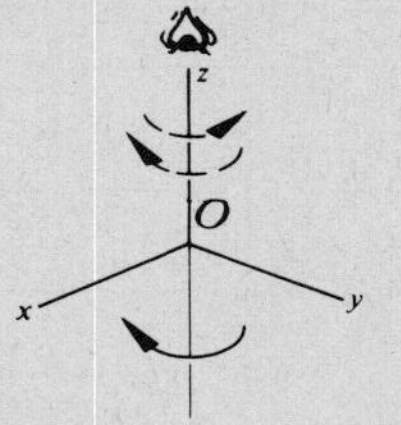

$$b_{J,K:J,K}=\frac{[J(J+1)-3K^2]^2}{J(J+1)(2J-1)(2J+3)} \qquad b_{J+1,K:J,K}=\frac{3K^2[(J+1)^2-K^2]}{J(J+1)(J+2)(2J+1)} \qquad b_{J+2,K:J,K}=\frac{3[(J+1)^2-K^2][(J+2)^2-K^2]}{2(J+1)(J+2)(2J+1)(2J+3)}$$

Table L General forms of $\Phi(a^2, \gamma^2, \theta)$ for ${}^{\perp}I_{\perp}(\pi/2)$, ${}^{\perp}I_{\parallel}(\pi/2)$, ${}^{\circledR}I_{\circledR}(0)(={}^{\textcircled{L}}I_{\textcircled{L}}(0))$, ${}^{\circledR}I_{\textcircled{L}}(0)(={}^{\textcircled{L}}I_{\circledR}(0))$; $\rho_{\perp}(\pi/2)$, $\mathscr{P}(0)=\mathscr{P}^{-1}(\pi)$ for rotation and vibration–rotation transitions

Selection rules		$\Phi(a^2, \gamma^2, \theta)$			$\Phi(a^2, \gamma^2, \theta)$		
		${}^{\perp}I_{\perp}\left(\frac{\pi}{2}\right)$	${}^{\perp}I_{\parallel}\left(\frac{\pi}{2}\right)$	$\rho_{\perp}\left(\frac{\pi}{2}\right)$	${}^{\circledR}I_{\circledR}(0)={}^{\textcircled{L}}I_{\textcircled{L}}(0)$	${}^{\circledR}I_{\textcircled{L}}(0)={}^{\textcircled{L}}I_{\circledR}(0)$	$\mathscr{P}(0)=\mathscr{P}^{-1}(\pi)$
$\Delta v=0$	$\Delta R=0$	$(a)_0^2+\frac{4}{45}b_{R^fR^i}(\gamma)_0^2$	$\frac{1}{15}b_{R^fR^i}(\gamma)_0^2$	$\frac{3b_{J,J}(\gamma)_0^2}{45(a)_0^2+4b_{R^fR^i}(\gamma)_0^2}$	$(a_0)^2+\frac{b_{R^fR^i}}{45}(\gamma)_0^2$	$\frac{2b_{R^fR^i}(\gamma)_0^2}{15}$	$\frac{6b_{R^fR^i}(\gamma')^2}{45(a)_0^2+b_{R^fR^i}(\gamma)_0^2}$
	$\Delta R\neq 0$	$\frac{4}{45}b_{R^fR^i}(\gamma)_0^2$	$\frac{1}{15}b_{R^fR^i}(\gamma)_0^2$	$\frac{3}{4}$	$\frac{b_{R^fR^i}}{45}(\gamma)_0^2$	$\frac{2b_{R^fR^i}}{15}(\gamma)_0^2$	6
$\Delta v=\pm 1$[a]	$\Delta R=0$	$(a')^2+\frac{4}{45}b_{R^fR^i}(\gamma')^2$	$\frac{1}{15}b_{R^fR^i}(\gamma')^2$	$\frac{3b_{J,J}(\gamma')^2}{45(a')^2+4b_{R^fR^i}(\gamma')^2}$	$(a')^2+\frac{b_{R^fR^i}(\gamma')^2}{45}$	$\frac{2b_{R^fR^i}(\gamma')^2}{15}$	$\frac{6b_{R^fR^i}(\gamma')^2}{45(a')^2+b_{R^fR^i}(\gamma')^2}$
	$\Delta R\neq 0$	$\frac{4}{45}b_{R^fR^i}(\gamma')^2$	$\frac{1}{15}b_{R^fR^i}(\gamma')^2$	$\frac{3}{4}$	$\frac{b_{R^fR^i}(\gamma')^2}{45}$	$\frac{2b_{R^fR^i}}{15}(\gamma')^2$	6

The factors $(v+1)b_v^2$ (for $\Delta v=+1$) and vb_v^2 (for $\Delta v=-1$) have been omitted. For definitions of $b_{R^f}b_{R^i}$ see Table M.

Table M Factors $b_{R^fR^i}$ giving relative anisotropic intensities of individual rotational lines for the transition $J^f, K^f \leftarrow J, K$

K^f \ J^f	J	$J+1$[d]	$J+2$[e]
K[a]	$\frac{[J(J+1)-3K^2]^2}{J(J+1)(2J-1)(2J+3)}$	$\frac{3K^2[(J+1)^2-K^2]}{J(J+1)(J+2)(2J+1)}$	$\frac{3[(J+1)^2-K^2][(J+2)^2-K^2]}{2(J+1)(J+2)(2J+1)(2J+3)}$
$K\pm 1$[b,c]	$\frac{3(2K\pm 1)^2(J\mp K)(J\pm K+1)}{2J(J+1)(2J-1)(2J+3)}$	$\frac{(J\mp 2K)^2(J\pm K+1)(J\pm K+2)}{2J(J+1)(J+2)(2J+1)}$	$\frac{[(J+1)^2-K^2](J\pm K+2)(J\pm K+3)}{(J+1)(J+2)(2J+1)(2J+3)}$
$K\pm 2$[c]	$\frac{3[J^2-(K\pm 1)^2][(J+1)^2-(K\pm 1)^2]}{2J(J+1)(2J-1)(2J+3)}$	$\frac{[(J+1)^2-(K\pm 1)^2](J\pm K+1)(J\pm K+3)}{2J(J+1)(J+2)(2J+1)}$	$\frac{(J\pm K+1)(J\pm K+2)(J\pm K+3)(J\pm K+4)}{4(J+1)(J+2)(2J+1)(2J+3)}$

[a] Values in this row are applicable to the pure rotational spectra and to the totally symmetric bands ($l=0$) of linear molecules by putting $K=0$.
[b] Values in this row are applicable to degenerate bands ($l=1$) of linear molecules by putting $K=0$.
[c] For $K+1$ (and $K+2$) the upper of the two signs $\pm$ or $\mp$ must be used; for $K-1$ (and $K-2$) the lower sign must be used.
[d] Values in this column are applicable to R branch lines ($\Delta J=+1$). For P branch lines ($\Delta J=-1$) replace J by $J-1$.
[e] Values in this column are applicable to S branch lines ($\Delta J=+2$). For O branch lines ($\Delta J=-2$) replace J by $J-2$.

Similarly, for Rayleigh scattering we have

$$[\boldsymbol{P}^{(1)}]_{ii} = [\tilde{\boldsymbol{P}}_0^{(1)}]_{ii} \exp\{-i\omega_0 t\} + [\tilde{\boldsymbol{P}}_0^{(1)}]_{ii}^* \exp\{i\omega_0 t\} \tag{5.29}$$

where the x components of the transition moment amplitudes are given by

$$[\tilde{P}_{xo}^{(1)}]_{ii} = \sum_y [\tilde{\alpha}_{xy}]_{ii} \tilde{E}_{yo} \quad \text{and} \quad [\tilde{P}_{xo}^{(1)*}]_{ii} = \sum_y [\tilde{\alpha}_{xy}^*]_{ii} \tilde{E}_{yo}^* \tag{5.30}$$

and the complex transition polarizabilities are given by

$$[\tilde{\alpha}_{xy}]_{ii} = \frac{1}{\hbar} \sum_r \left\{ \frac{[\tilde{P}_y]_{ir}[\tilde{P}_x]_{ri}}{\omega_{ri} + \omega_0} + \frac{[\tilde{P}_x]_{ir}[\tilde{P}_y]_{ri}}{\omega_{ri} - \omega_0} \right\} \tag{5.31}$$

and

$$[\tilde{\alpha}_{xy}^*]_{ii} = \frac{1}{\hbar} \sum_r \left\{ \frac{[\tilde{P}_x]_{ir}[\tilde{P}_y]_{ri}}{\omega_{ri} + \omega_0} + \frac{[\tilde{P}_y]_{ir}[\tilde{P}_x]_{ri}}{\omega_{ri} - \omega_0} \right\} \tag{5.32}$$

The tensor relationship between the electric field and the transition moment has now clearly emerged. However, while this relationship is of the same general form as that of the equations of (4.4), the transition polarizability components (or components of the polarizability matrix element) can now be complex and thus no longer necessarily have the symmetry properties assumed in the treatment in chapter 4.

Equations (5.27), (5.28), (5.31), and (5.32) are of fundamental importance, for they relate the transition polarizability to the energy levels and wave functions of the scattering system. In principle, if all the energy levels and wave functions of the system were known, the transition polarizability could be evaluated, but in practice such complete knowledge is usually denied us. Despite this, these expressions for the transition polarizabilities give us a very considerable insight into the nature and mechanism of light scattering at the molecular level, as the following sections will show.

5.4 The transition polarizability

We shall now discuss a number of aspects of the transition polarizability, its symmetry properties and the conditions under which it can assume a simpler form.

5.4.1 *The role of the states* r

We see that the numerators of the transition polarizability components always contain products of components of transition moments of the types $[P_x]_{ri}$, involving the initial state and the state r, and $[P_y]_{fr}$, involving the final state and the state r. Since these transition moments always occur as products, if the scattering tensor component is to be non-zero, there must exist in the system at least one state r which has a non-zero dipole transition moment to both the initial state and the final state. It must be stressed that this requirement does not mean that the transition from state i to state f, which is associated with Raman scattering, occurs in two distinguishable stages via a transition to an 'intermediate' state r. We note also that incident radiation of energy $\hbar\omega_0$ is not absorbed, since there is no requirement that the energy $\hbar\omega_0$ corresponds to an energy difference between two discrete states of the system, i.e., $\hbar\omega_0$ does not have to be equal to $\hbar\omega_r$. Nor does the condition for existence of a real

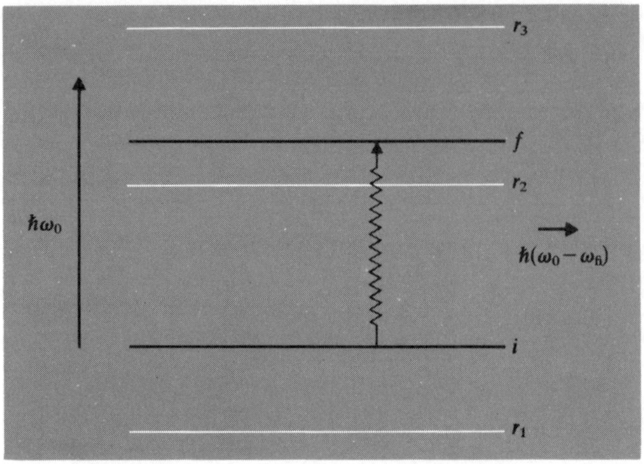

Fig. 5.1 The states r for the Stokes transition $f \leftarrow i$

transition moment (eq. 5.20) impose any restriction upon the states r relative to ω_0. Indeed, the state r can lie above the final state f, below the initial state i, or between i and f (see Fig. 5.1). The dependence of the transition polarizabilities on the states r arises because quantum mechanical language describes the reaction of a system to a perturbation in terms of the eigenfunctions and energies of *all* the possible states of the system. It is perhaps a little misleading that the states r are sometimes referred to as intermediate states, and we shall avoid that description here.

5.4.2 *Coherence properties of the scattered radiation*

In the mathematical development of section 5.2 and section 5.3, the phase factors in the time-independent wave functions ψ_i, ψ_r, and ψ_f were disregarded. When these are taken into account, the transition moment $[\tilde{\boldsymbol{P}}]_{ri}$ will have the phase factor $\exp\{-i(\delta_r - \delta_i)\}$ and the transition moment $[\tilde{\boldsymbol{P}}]_{fr}$ will have the phase factor $\exp\{-i(\delta_f - \delta_r)\}$. Thus, the Raman transition moment $[\tilde{\boldsymbol{P}}]_{fi}$ will have the phase factor $\exp\{-i(\delta_f - \delta_i)\}$ which will vary arbitrarily from one scattering molecule to the next, and so Raman scattering will be incoherent. In the case of the Rayleigh transition moment $[\tilde{\boldsymbol{P}}]_{ii}$, the phase factor is unity since $\exp\{-i(\delta_i - \delta_i)\} = 1$, and so Rayleigh scattering is coherent, provided the state i is non-degenerate. If the state i is degenerate, then in addition to coherent Rayleigh scattering involving the same state i there is an incoherent contribution involving transitions between the various degenerate states i.

The general consequences of the difference in coherence properties of Rayleigh and Raman scattering have already been discussed on page 57 and will not be repeated here.

5.4.3 *Selection rules*

We have already seen that the existence of at least one level r which has a non-zero dipole transition moment with both the initial *and* final states of the system is a necessary condition for the transition polarizability to be non-zero. The selection rules can, however, be cast in a form which involves only the properties of the initial and final states and not the properties of the states *r*.

It can be shown that $[\tilde{\alpha}_{xy}]_{fi}$ has the same transformation properties as

$$\langle f|xy|i\rangle \tag{5.33}$$

Thus, the general condition for $[\tilde{\alpha}_{xy}]_{fi}$ to be non-zero is that the product $\psi_f xy \psi_i$ belongs to a representation which contains the totally symmetric species. The selection rules we have discussed in chapter 4 are special cases of this rule. It should be noted that the general condition we have just stated, while a necessary one, is not a sufficient one. Although a transition polarizability satisfies the symmetry requirements, it may still have a near-zero value.

5.4.4 *Frequency dependence*

The denominators of the transition polarizability components contain frequency terms of the type $\omega_{rf}+\omega_0$ and $\omega_{ri}-\omega_0$. Thus, the magnitude of the transition dipole moment is, in general, dependent on the excitation frequency ω_0; and the intensity of scattering will not depend simply on the fourth power of the frequency of the scattered radiation since eq. (2.110) is based on a dipole whose amplitude is frequency-independent. Only if the excitation frequency ω_0 is much smaller than the transition frequencies ω_{rf} and ω_{ri} can the transition polarizability be regarded as independent of ω_0, so that the intensity of scattering follows a fourth power law.

However, as ω_0 begins to approach a transition frequency, the intensity of scattering will depart from the fourth power law. In particular, if ω_0 is reasonably close to ω_{ri}, the frequency of an absorption band of the system, the transition polarizability components will have values which are greatly enhanced compared with the values when ω_0 is remote from ω_{ri}. The anomalously intense Raman scattering that results is termed resonance Raman scattering. Enhancement of this kind is found in many optical phenomena; a familiar example is the dependence of optical rotation on the wavelength of the radiation used to measure it.

It is easy to see that, in principle, by appropriate choice of the exciting frequency ω_0, resonance Raman scattering could be exploited to increase dramatically the intensity of Raman scattering in almost any system. The recent availability of lasers continuously tuneable over a wide frequency range now makes this possible, and resonance Raman scattering is now an important spectroscopic technique. Intensity enhancements up to $\times 10^5$ are possible and, as a result, scattering species in very low concentrations can be quite readily studied.

In addition, resonance Raman scattering can exhibit selection rules and polarization phenomena which are different from ordinary Raman scattering. This is because, as we shall see subsequently, when ω_0 is much smaller than any transition frequency the transition polarizability tensor acquires symmetry properties that are not possessed by the general tensor. The lower symmetry of the general tensor can mean, for example, that transitions inactive in normal Raman scattering become active in resonance Raman scattering.

When $\omega_0 \approx \omega_{ri}$, it would appear that the transition polarizability will tend to infinity. However, the formulae we have derived here do not apply in this limiting case. In a more general treatment, the lifetimes of the states *r* have to be taken into account. The states *r* are assumed to decay exponentially with time so that the stationary state r, defined by a form of eq. (5.1), becomes a quasi-stationary state given by

$$\Psi_r = \psi_r \exp\{-i(\omega_r - \tfrac{1}{2}i\Gamma_r)t\} \qquad (5.34)$$

where Γ_r is inversely proportional to the lifetime of the state r. The finite lifetimes of the states r can therefore be incorporated into our formulae simply by changing to complex energies, i.e., replacing $\hbar\omega_r$ by $\hbar\omega_r - \frac{1}{2}i\hbar\Gamma_r$. Thus, the frequency denominators in the transition polarizability will now be of the type $(\omega_{ri} - \omega_0 - i\Gamma_{ri})$ where $\Gamma_{ri} = \frac{1}{2}(\Gamma_r - \Gamma_i)$ and is related to the width of the $r \leftarrow i$ transition. Clearly the new frequency denominators do not become zero when $\omega_0 \approx \omega_{ri}$.

5.4.5 *Symmetry properties of the transition polarizability tensor*

We shall consider, first, the symmetry properties of the complex Rayleigh scattering tensor with components $[\tilde{\alpha}_{xy}]_{ii}$. Using the Hermitian property of the transition moment, namely, that

$$[\tilde{P}_x]_{ri} = [\tilde{P}_x^*]_{ir} \qquad (5.35)$$

it is readily shown that

$$[\tilde{\alpha}_{xy}]_{ii} = [\tilde{\alpha}_{yx}^*]_{ii} \qquad (5.36)$$

and thus the tensor itself is Hermitian. If the components of the tensor are real then the tensor is symmetric, since

$$[\alpha_{xy}]_{ii} = [\alpha_{yx}]_{ii} \qquad (5.37)$$

If the components of the tensor are complex we may write

$$[\tilde{\alpha}_{xy}]_{ii} = [\alpha_{xy}]_{ii} - i[\alpha'_{xy}]_{ii} \qquad (5.38)$$

where $[\alpha_{xy}]_{ii}$ is real and $i[\alpha'_{xy}]_{ii}$ imaginary. Now, using eq. (5.37), we may write

$$[\tilde{\alpha}_{yx}^*]_{ii} = [\alpha_{yx}]_{ii} + i[\alpha'_{yx}]_{ii} \qquad (5.39)$$

It follows from eq. (5.36) that $[\alpha'_{xy}]_{ii} = -[\alpha'_{yx}]_{ii}$. Thus, the imaginary part of the complex tensor is always antisymmetric, whereas the real part is always symmetric. For Rayleigh scattering, there can be no antisymmetric real part and no symmetric imaginary part of the scattering tensor. Now a complex tensor requires complex wave functions and these can only exist for systems in the presence of external magnetic fields or where there are internal magnetic perturbations, as in the case of spin–orbit interactions. Thus, in the absence of magnetic perturbation, the Rayleigh scattering tensor is real and symmetric. It should be noted that there is no restriction on the frequency denominators; the symmetry properties of the Rayleigh scattering tensor are the same for transparent and absorbing samples.

For the complex Raman scattering tensor with components $[\tilde{\alpha}_{xy}]_{fi}$, it is easily shown that

$$[\tilde{\alpha}_{xy}]_{fi} \neq [\tilde{\alpha}_{yx}^*]_{fi} \qquad (5.40)$$

and thus, in general, the scattering tensor is not Hermitian and so is not symmetric, even if the tensor components are real.

Our subsequent analysis of the Raman scattering tensor will be in two stages. Initially, we shall develop the Placzek polarizability theory which will reveal under what conditions the tensor can be first Hermitian and then symmetric. We shall then examine the properties of the general Raman scattering tensor in particular situations. This will involve us in the properties of antisymmetric tensors.

5.5 The Placzek[1] polarizability theory

The wave functions ψ_i, ψ_r, and ψ_f involved in the transition polarizabilities are the time-independent total wave functions of the system. We now separate the wave functions into their electronic, vibrational, and rotational parts. We write, for ψ_i, for example,

$$\psi_i = \psi_{e^i}\psi_{v^iR^i} \tag{5.41}$$

In eq. (5.41), ψ_{e^i} is the electronic part of the wave function with the set of electronic quantum numbers e^i, and is a function of the electronic and nuclear coordinates; and $\psi_{v^iR^i}$ is the vibrational and rotational part of the wave function with the set of vibrational quantum numbers v^i and the set of rotational quantum numbers R^i. The wave function $\psi_{v^iR^i}$ can be expanded as

$$\psi_{v^iR^i} = \phi_{v^i}\Theta_{R^i} \tag{5.42}$$

where ϕ_{v^i} is the vibrational wave function which is a function of the normal coordinates of vibration and Θ_{R^i} is the rotational wave function which is a function of the Euler angles which define the orientation of the molecule with respect to space-fixed axes. In what follows we shall be concerned mainly with the separation of the electronic part from the vibrational and rotational parts, and it will not be necessary to introduce eq. (5.42) into eq. (5.41) at this stage.

If the expansions of ψ_i, ψ_r, and ψ_f are introduced into eq. (5.27), the resulting expression for the transition polarizability is rather complicated, and the electronic and nuclear contributions appear to be inextricably mixed. However, if certain special conditions are satisfied, the scattering tensor associated with a transition in which only the vibrational and/or rotational quantum numbers change and the system starts and finishes in the ground electronic state is given by

$$[\tilde{\alpha}_{xy}]_{fi} = \langle v^fR^f|[\tilde{\alpha}_{xy}]_{e^0e^0}|v^iR^i\rangle \tag{5.43}$$

where

$$[\tilde{\alpha}_{xy}]_{e^0e^0} = \frac{1}{\hbar}\sum_r\left\{\frac{[\tilde{P}_y]_{e^0e^r}[\tilde{P}_x]_{e^re^0}}{\bar{\omega}_{e^re^0}+\omega_0}+\frac{[\tilde{P}_x]_{e^0e^r}[\tilde{P}_y]_{e^re^0}}{\bar{\omega}_{e^re^0}-\omega_0}\right\} \tag{5.44}$$

In eq. (5.44), e^0 and e^r refer, respectively, to the *electronic* quantum numbers of the ground state and an intermediate state *r*, and the summation is over all electronic states; $\bar{\omega}_{e^re^0}$ is the average frequency separation of the ground electronic state and the electronic state *r*; and the transition moment components in eq. (5.44) relate to transitions between *electronic* states, in equilibrium nuclear geometry so that

$$[\tilde{P}_y]_{e^re^0} = \langle e^r|P_{y_e}|e^0\rangle, \text{etc.} \tag{5.45}$$

where P_{y_e} is the *y*-component of the electronic part of the electric dipole operator (i.e., in the definition of the dipole moment operator given by eq. (5.4) the summation is now confined to the electrons).

The rather extensive manipulations that are necessary to arrive at eqs. (5.43) and (5.44) will not be given here. We shall content ourselves with noting the special conditions that have to be introduced in the course of the manipulations. These are:

(a) The excitation frequency ω_0 must be much larger than the frequency associated with any vibration or rotation transition of the system.
(b) The excitation frequency ω_0 must be much less than any electronic transition frequency $\bar{\omega}_{e^r e^o}$ of the system.
(c) The ground electronic state must not normally be degenerate (although there may be special cases where this is not a necessary condition).

The special form of the transition polarizability given by eq. (5.43) is Hermitian, since $[\tilde{\boldsymbol{\alpha}}]_{e^o e^o}$ is Hermitian. Thus, when the conditions in (a), (b), and (c) are satisfied, the Raman transition polarizability has the same symmetry properties as the Rayleigh scattering tensor, i.e., it has only a real symmetric part and an antisymmetric imaginary part.

In the absence of magnetic perturbations, the electronic wave functions of the ground state are real, $[\boldsymbol{\alpha}]_{e^o e^o}$ is real and hence symmetric and so $[\alpha_{xy}]_{fi}$ is real and symmetric. We then have in place of eqs. (5.43) and (5.44)

$$[\alpha_{xy}]_{fi} = \langle v^f R^f | [\alpha_{xy}]_{e^o e^o} | v^i R^i \rangle \tag{5.46}$$

and

$$[\alpha_{xy}]_{e^o e^o} = \frac{1}{\hbar} \sum_r \left\{ \frac{[P_y]_{e^o e^r}[P_x]_{e^r e^o}}{\bar{\omega}_{e^r e^o} + \omega_0} + \frac{[P_x]_{e^o e^r}[P_y]_{e^r e^o}}{\bar{\omega}_{e^r e^o} - \omega_0} \right\} \tag{5.47}$$

$[\alpha_{xy}]_{e^o e^o}$ is the electronic polarizability in the ground electronic state.

In addition to the symmetry properties with which $[\alpha_{xy}]_{e^o e^o}$ is endowed, it is a function of the nuclear coordinates only, and not of the electron coordinates.

Thus, it is only when these several conditions are met that the polarizability can be regarded as a frequency-independent, symmetric real tensor which is a function simply of the nuclear coordinates, and the formulae developed in chapters 3 and 4 are applicable. Fortunately, these conditions are sufficiently well satisfied in many cases, but the availability of tuneable laser sources makes it necessary to examine more closely the resonance and near-resonance cases.

5.6 Resonance Raman scattering

5.6.1 *General considerations*

We now examine the case where $\omega_0 \approx \omega_{ri}$ and resonance Raman scattering occurs. The simplifications of the Placzek theory no longer apply, and we have to deal in general with a Raman scattering tensor which is both unsymmetric and complex. However, in the absence of magnetic perturbation, the tensor becomes real although remaining unsymmetric, and we shall therefore concentrate our attention on this case. Explicit expressions for the tensor components in terms of the electronic, vibration, and rotation wave functions and energy levels of the system may be obtained, for example, by application of vibronic coupling theory.[2] We shall not consider such developments here, but confine ourselves to a closer examination of the general properties of the tensor.

5.6.2 *The unsymmetric real scattering tensor*

To simplify the notation we shall consider a polarizability tensor $\boldsymbol{a}$ with components α_{xy} rather than a transition polarizability tensor $[\boldsymbol{a}]_{fi}$ with components $[\alpha_{xy}]_{fi}$. The results are, of course, equally applicable to both cases.

In section 3.3.5, we showed that a symmetric tensor can always be written as the sum of two other symmetric tensors: an isotropic tensor $\boldsymbol{a}_{\text{iso}}$, and an anisotropic tensor $\boldsymbol{a}_{\text{aniso}}$. If the tensor is unsymmetric, i.e., $\alpha_{xy} \neq \alpha_{yx}$, it may be written as the sum of three tensors, $\boldsymbol{a}_{\text{iso}}$, $\boldsymbol{a}_{\text{aniso}}$, and $\boldsymbol{a}_{\text{anti}}$, where $\boldsymbol{a}_{\text{iso}}$ is still defined by eq. (3.26) but $\boldsymbol{a}_{\text{aniso}}$, although still symmetric, is given by

$$\boldsymbol{a}_{\text{aniso}}: \quad \begin{matrix} \alpha_{xx} - a & \dfrac{\alpha_{xy} + \alpha_{yx}}{2} & \dfrac{\alpha_{xz} + \alpha_{zx}}{2} \\ \dfrac{\alpha_{yx} + \alpha_{xy}}{2} & \alpha_{yy} - a & \dfrac{\alpha_{yz} + \alpha_{zy}}{2} \\ \dfrac{\alpha_{zx} + \alpha_{xz}}{2} & \dfrac{\alpha_{zy} + \alpha_{yz}}{2} & \alpha_{zz} - a \end{matrix} \tag{5.48}$$

and $\boldsymbol{a}_{\text{anti}}$, an antisymmetric tensor with $\alpha_{ij} = -\alpha_{ji}$, is defined by

$$\boldsymbol{a}_{\text{anti}}: \quad \begin{matrix} 0 & \left(\dfrac{\alpha_{xy} - \alpha_{yx}}{2}\right) & \left(\dfrac{\alpha_{xz} - \alpha_{zx}}{2}\right) \\ -\left(\dfrac{\alpha_{xy} - \alpha_{yx}}{2}\right) & 0 & \left(\dfrac{\alpha_{yz} - \alpha_{zy}}{2}\right) \\ -\left(\dfrac{\alpha_{xz} - \alpha_{zx}}{2}\right) & -\left(\dfrac{\alpha_{yz} - \alpha_{zy}}{2}\right) & 0 \end{matrix} \tag{5.49}$$

It is clear that

$$\boldsymbol{a} = \boldsymbol{a}_{\text{iso}} + \boldsymbol{a}_{\text{aniso}} + \boldsymbol{a}_{\text{anti}} \tag{5.50}$$

where $\boldsymbol{a}$ is now an unsymmetric matrix given by

$$\boldsymbol{a}: \quad \begin{matrix} \alpha_{xx} & \alpha_{xy} & \alpha_{xz} \\ \alpha_{yx} & \alpha_{yy} & \alpha_{yz} \\ \alpha_{zx} & \alpha_{zy} & \alpha_{zz} \end{matrix} \tag{5.51}$$

with $\alpha_{xy} \neq \alpha_{yx}$. Of course, if $\alpha_{xy} = \alpha_{yx}$, $\boldsymbol{a}_{\text{anti}}$ is zero and the definition of $\boldsymbol{a}_{\text{aniso}}$ given by eq. (5.48) becomes identical with that of eq. (3.27).

The trace of the product of the tensor $\boldsymbol{a}_{\text{aniso}}$ with itself is given by

$$\text{Trace}\{\boldsymbol{a}_{\text{aniso}} : \boldsymbol{a}_{\text{aniso}}\} = \tfrac{2}{3}\gamma^2 \tag{5.52}$$

This is of the same form as eq. (3.30), but γ^2 is now defined by

$$\gamma^2 = \tfrac{1}{2}[(\alpha_{xx} - \alpha_{yy})^2 + (\alpha_{yy} - \alpha_{zz})^2 + (\alpha_{zz} - \alpha_{xx})^2 + \tfrac{3}{2}\{(\alpha_{xy} + \alpha_{yx})^2 + (\alpha_{yz} + \alpha_{zy})^2 + (\alpha_{zx} + \alpha_{xz})^2\}] \tag{5.53}$$

and not by eq. (3.19). Of course, eq. (5.53) reduces to eq. (3.19) if $\alpha_{xy} = \alpha_{yx}$. The trace of the product of the tensor $\boldsymbol{\alpha}_{\text{anti}}$ with itself is given by

$$\text{Trace}\,\{\boldsymbol{\alpha}_{\text{anti}} : \boldsymbol{\alpha}_{\text{anti}}\} = \tfrac{2}{3}\delta^2 \tag{5.54}$$

where

$$\delta^2 = \tfrac{3}{4}\{(\alpha_{xy} - \alpha_{yx})^2 + (\alpha_{yz} - \alpha_{zy})^2 + (\alpha_{zx} - \alpha_{xz})^2\} \tag{5.55}$$

which reduces to zero if $\alpha_{xy} = \alpha_{yx}$. The trace of the product of $\boldsymbol{\alpha}_{\text{iso}}$ with itself is still given by eq. (3.29), and the definition of a is still given by eq. (3.18).

Since both $\boldsymbol{\alpha}_{\text{iso}}$ and $\boldsymbol{\alpha}_{\text{aniso}}$ are symmetric tensors, we may rewrite eq. (5.50) in the form

$$\boldsymbol{\alpha} = \boldsymbol{\alpha}_{\text{sym}} + \boldsymbol{\alpha}_{\text{anti}} \tag{5.56}$$

where

$$\boldsymbol{\alpha}_{\text{sym}} = \boldsymbol{\alpha}_{\text{iso}} + \boldsymbol{\alpha}_{\text{aniso}} \tag{5.57}$$

The intensity of scattering depends on α^2, i.e., upon

$$\{\alpha_{\text{sym}} + \alpha_{\text{anti}}\}^2 \tag{5.58}$$

In a freely orienting system, we have to form space averages, and it can be shown that the space averages of cross terms involving $\alpha_{\text{sym}}\alpha_{\text{anti}}$ are always zero. Thus, very conveniently, the contribution to the scattered intensity from the antisymmetric tensor is simply additive. The necessary space averages are given below:

$$\overline{\left(\frac{\alpha_{xy} - \alpha_{yx}}{2}\right)^2} = \overline{\left(\frac{\alpha_{xz} - \alpha_{zx}}{2}\right)^2} = \overline{\left(\frac{\alpha_{yz} - \alpha_{zy}}{2}\right)^2} = \frac{\delta^2}{9} \tag{5.59}$$

It happens that the symmetry properties of the antisymmetric tensor contribution to the scattered radiation are the same as those of magnetic dipole radiation, and hence Placzek[1] designated such scattering as magnetic dipole scattering. This label is somewhat misleading and we shall not use it here.

5.6.3 *Antisymmetric tensor contributions to intensities and polarization properties of scattered radiation*

Because of the properties discussed in section 5.6.2 we may readily graft on to the formulae for intensities and polarization properties for randomly oriented, non-absorbing systems, obtained in chapters 3 and 4, the additional contributions from the antisymmetric part of the scattering tensor.

We need only consider a few cases by way of examples. We adhere to the nomenclature of chapters 3 and 4 where the radiation is incident along the z-axis and the scatter plane is the xz-plane. Then, for incident radiation plane polarized with the electric vector perpendicular to the scatter plane ($E_y \neq 0$), the only contribution from the anisotropic tensor to the intensity observed along x is by way of an *additional* contribution to $\overline{[P_{xo}^{(1)}]^2}$, and this is simply

$$\overline{[P_{xo}^{(1)}]^2_{\text{anti}}} = \frac{\delta^2}{9} E_{yo}^2 \tag{5.60}$$

Hence,

$$\rho_\perp(\pi/2) = \frac{3\gamma^2 + 5\delta^2}{45a^2 + 4\gamma^2} \tag{5.61}$$

For incident radiation plane polarized with the electric vector parallel to the scatter plane ($E_x \neq 0$), $\overline{[P_{zo}^{(1)}]^2}$ and $\overline{[P_{yo}^{(1)}]^2}$ have equal contributions from the antisymmetric tensor, and hence $\rho_\parallel(\pi/2)$ is unchanged, i.e.,

$$\rho_\parallel(\pi/2) = 1{\cdot}0 \tag{5.62}$$

For incident natural illumination ($E_x = E_y \neq 0$), there will be additional contributions to both $\overline{[P_{xo}^{(1)}]^2}$ (from E_{yo} only) and to $\overline{[P_{zo}^{(1)}]^2}$ (from E_{xo} and E_{yo}), and these additional contributions are

$$\overline{[P_{xo}^{(1)}]^2_{\text{anti}}} = \frac{\delta^2}{9} E_{yo}^2 \tag{5.63}$$

and

$$\overline{[P_{zo}^{(1)}]^2_{\text{anti}}} = \frac{\delta^2}{9} (E_{xo}^2 + E_{yo}^2) \tag{5.64}$$

Hence, remembering that for natural radiation, $E_{x_o}^2 = E_{yo}^2 = \mathscr{I}/c\varepsilon_0$

$$\rho_n(\pi/2) = \frac{6\gamma^2 + 10\delta^2}{45a^2 + 7\gamma^2 + 5\delta^2} \tag{5.65}$$

For circularly polarized incident radiation, the additional contributions are not quite so easily identified. It is necessary to repeat the type of calculations presented in section 3.4.3 without the assumption that $\overline{\alpha_{xy} = \alpha_{yx}}$. It is then found that for incident right circularly polarized radiation, there is a contribution from the anisotropic tensor to the intensity observed along z (0° scattering) by way of an additional contribution to $\overline{(P_0)_{®}(P_0)^*_{®}}$ only; and this is given by

$$\overline{[(P_0)_{®}(P_0)^*_{®}]_{\text{anti}}} = \frac{\delta^2}{9} \{E_{xo}^2 + E_{yo}^2\} \tag{5.66}$$

Referring to page 61 we see that the reversal coefficients are now given by

$$\mathscr{P}(0) = \frac{6\gamma^2}{45a^2 + \gamma^2 + 5\delta^2} = \mathscr{P}(\pi)^{-1} \tag{5.67}$$

and the degrees of circularity by

$$^{®}\mathscr{C}(0) = \frac{45a^2 - 5\gamma^2 + 5\delta^2}{45a^2 + 7\gamma^2 + 5\delta^2} = -{}^{©}\mathscr{C}(0) \tag{5.68}$$

The antisymmetric tensor contributions to the scattered intensities and polarization characteristics for different types of incident radiation are included in Tables A–D in the central reference section.

The consequences of contributions to the scattered radiation from the antisymmetric tensor are interesting, especially in relation to polarization properties. For

Raman scattering arising from non-totally symmetric modes where $a=0$, eq. (5.61) becomes

$$\rho_{\perp}(\pi/2)=\frac{3}{4}+\frac{5}{4}\frac{\delta^2}{\gamma^2} \tag{5.69}$$

If $\delta^2=0$, we have 'normal' polarization with $\rho_{\perp}(\pi/2)=\frac{3}{4}$, but if $\delta^2\neq 0$, we have 'anomalous' polarization with $\rho_{\perp}(\pi/2)>\frac{3}{4}$ if $\gamma^2\neq 0$, and inverse polarization with $\rho_{\perp}(\pi/2)=\infty$ if $\gamma^2=0$.

For totally symmetric modes where $a\neq 0$, if $\delta^2=0$, eq. (5.61) may be written as

$$\rho_{\perp}(\pi/2)=\frac{3}{(45a^2/\gamma^2+4)} \tag{5.70}$$

and $0\leqslant\rho_{\perp}(\pi/2)<\frac{3}{4}$ according as $\infty\geqslant a^2/\gamma^2>0$. If, however, $\delta^2\neq 0$, then we must revert to the general formula (5.61). We note, however, that 'anomalous' polarization will only arise if $\delta^2>\frac{27}{4}a^2$.

We see from eqs. (5.67) and (5.68) that for inverse polarization ($\rho_{\perp}(\pi/2)=\infty$, $a=\gamma=0$, $\delta\neq 0$) $\mathscr{P}(0)=0$, ${}^{\circledR}\mathscr{C}(0)=1$ and ${}^{\copyright}\mathscr{C}(0)=-1$.

We may appeal to group theory to determine which invariants are non-zero for particular vibrational modes in molecules of a given symmetry. In principle, this requires a knowledge of the symmetry properties of the general polarizability tensor as listed in Appendix II. For example, consider a vibration of the a_2 symmetry class of the point group C_{4v}. From the symmetry properties of the general polarizability tensor, we find that the a_2 class contains two non-vanishing components, α_{xy} and α_{yx}, which are equal in magnitude and opposite in sign, i.e., $\alpha_{xy}=-\alpha_{yx}$. It follows from eqs. (3.18), (5.53), and (5.55) that $a=0$, $\gamma^2=0$ and $\delta^2=\frac{9}{4}(\alpha_{xy}-\alpha_{yx})^2=9\alpha_{xy}{}^2$. Hence, this mode will show inverse polarization. It is useful to note that an irreducible representation containing a rotation component also contains the corresponding components of the antisymmetric tensor $\boldsymbol{\alpha}_{\text{anti}}$. Thus, in the point group C_{4v}, R_z belongs to the a_2 representation and thus α_{xy} and $-\alpha_{yx}$ will also belong to this representation.

We have noted previously (page 84) that for a symmetric polarizability tensor an intensity measurement and one polarization measurement suffice to determine the magnitudes of the invariants a and γ. For a non-symmetric tensor where, in general, three invariants a, γ, and δ are involved, it is necessary to measure not only the intensity but also *two* polarization properties, e.g., $\rho_{\perp}(\pi/2)$ and $\mathscr{P}(0)$, to characterize completely the scattered radiation.

Since the shapes of Raman bands generated by isotropic, anisotropic, and antisymmetric scattering will be different, it has been suggested that in principle the relative contributions to a given Raman band could be determined from 90° scattering by decomposing the band into its isotropic, anisotropic, and antisymmetric components.[3]

5.6.4 *Antisymmetric tensor contributions to the Stokes parameters of scattered radiation*

The *additional* contributions to the Stokes parameters for 0°, 90° and 180° scattering from the *antisymmetric* tensor may be shown to be as follows:

$$S_0(0)=S_0(\pi)=KE_0^2\left(\frac{\delta^2}{9}\right) \tag{5.71}$$

$$S_1(0) = S_1(\pi) = -KE_0^2\left(\frac{\delta^2}{9}\right) P \cos 2\chi \cos 2\psi \tag{5.72}$$

$$S_2(0) = -S_2(\pi) = -KE_0^2\left(\frac{\delta^2}{9}\right) P \cos 2\chi \sin 2\psi \tag{5.73}$$

$$S_3(0) = -S_3(\pi) = KE_0^2\left(\frac{\delta^2}{9}\right) P \sin 2\chi \tag{5.74}$$

$$S_0(\pi/2) = \tfrac{1}{2}KE_0^2\left(\frac{\delta^2}{9}\right)(3 + P \cos 2\chi \cos 2\psi) \tag{5.75}$$

$$S_1(\pi/2) = -\tfrac{1}{2}KE_0^2\left(\frac{\delta^2}{9}\right)(1 - P \cos 2\chi \cos 2\psi) \tag{5.76}$$

$$S_2(\pi/2) = 0 \tag{5.77}$$

$$S_3(\pi/2) = 0 \tag{5.78}$$

where K is given by eq. (4.43).

Hence for *pure antisymmetric* scattering at 90°

$$\frac{{}^{p_i}I_{\|}(\pi/2)}{{}^{p_i}I_{\perp}(\pi/2)} = \frac{2}{1 + P \cos 2\chi \cos 2\psi} \tag{5.79}$$

whence

$$\rho_n(\pi/2)(\text{anti}) = 2 \tag{5.80}$$

$$\rho_{\perp}(\pi/2)(\text{anti}) = \infty \tag{5.81}$$

$$\rho_{\|}(\pi/2)(\text{anti}) = 1 \tag{5.82}$$

The reversal coefficients and the degrees of circularity for pure antisymmetric scattering are

$$\mathscr{P}(0)(\text{anti}) = \frac{1 - |P \sin 2\chi|}{1 + |P \sin 2\chi|} = \mathscr{P}^{-1}(\pi) \tag{5.83}$$

and

$${}^{p_i}\mathscr{C}(0)(\text{anti}) = P \sin 2\chi = -{}^{p_i}\mathscr{C}(\pi) \tag{5.84}$$

Hence, for circularly polarized incident radiation, $\mathscr{P}(0) = 0$, ${}^{®}\mathscr{C}(0) = 1$, ${}^{©}\mathscr{C}(0) = -1$; $\mathscr{P}(\pi) = \infty$, ${}^{®}\mathscr{C}(\pi) = -1$, ${}^{©}\mathscr{C}(\pi) = +1$; and so in the forward direction the pure antisymmetric scattered radiation has completely the same sense and in the backward direction the scattered radiation has completely the reverse sense. Tables E and F include the antisymmetric tensor contributions to the Stokes parameters, ρ, $\mathscr{P}$ and $\mathscr{C}$.

5.6.5 *Frequency dependence of resonance scattering*

We now consider briefly the frequency dependence of scattering when resonance effects are important. A general treatment of this problem is hardly possible since so many of the levels of the system can, in principle, be involved. However, if one or two levels are considered to play an overwhelmingly predominant role, then some progress can be made. For example Albrecht and Hutley[4] have shown that, if only one electronic level is important, then the total frequency (wavenumber) dependence of the *scattered intensity*, including both the normal fourth power law dependence and the frequency (wavenumber) dependence of the scattering tensor, is given by

$$F_A^2 = \left\{ \frac{\tilde{\nu}'^2(\tilde{\nu}^2_{e^r e^o} + \tilde{\nu}_0^2)}{(\tilde{\nu}^2_{e^r e^o} - \tilde{\nu}_0^2)^2} \right\}^2 \tag{5.85}$$

where $\tilde{\nu}_{e^r e^o}$ is the wavenumber associated with the transition from the ground electronic state to the electronic state in question, and

$$\tilde{\nu}' = \tilde{\nu}_0 \pm \tilde{\nu}_M \tag{5.86}$$

If, however, two electronic levels are important, then the total frequency (wavenumber) dependence is given by

$$F_B^2 = \left\{ \frac{2\tilde{\nu}'^2(\tilde{\nu}_{e^r e^o}\tilde{\nu}_{e^{r'} e^o} + \tilde{\nu}_0^2)}{(\tilde{\nu}^2_{e^r e^o} - \tilde{\nu}_0^2)(\tilde{\nu}^2_{e^{r'} e^o} - \tilde{\nu}_0^2)} \right\}^2 \tag{5.87}$$

If scattered intensities are measured using a series of excitation frequencies (wavenumbers), these expressions can be used for extrapolation of scattered intensities (or scattering tensors) to $\tilde{\nu}_0 = 0$ (or $\tilde{\nu}_{e^r e^o} = \tilde{\nu}_{e^{r'} e^o} = \infty$) to obtain 'non-resonant' values.

5.6.6 *The imaginary polarizability tensor*

In absorbing systems, when magnetic perturbations are present the symmetric and antisymmetric parts of the imaginary polarizability can contribute to scattering. These are rather esoteric effects and can be more conveniently discussed in section 5.7, where the complex tensor is considered in more detail in relation to Rayleigh and Raman optical activity.

5.7 Time-dependent perturbation treatment including magnetic dipole and electric quadrupole terms[5]

5.7.1 *General considerations*

In this section, we shall present, without derivation, the results obtained when the magnetic dipole and electric quadrupole terms in the interaction Hamiltonian are taken into account. To simplify the nomenclature, we shall continue the practice introduced in section 5.6.2 and write, for the transition polarizability tensor $[\boldsymbol{\alpha}]_{fi}$ with components $[\alpha_{xy}]_{fi}$, simply $\boldsymbol{\alpha}$ and α_{xy}, respectively, and similarly for all other transition tensors and their invariants.

For irradiation with a plane wave of frequency ω_0 propagating along the z-axis with complex electric field amplitudes $\tilde{E}_{x_0}$ and $\tilde{E}_{y_0}$, it is found that $[\tilde{P}^{(1)}_{x_0}]_{fi}$, the x component of the amplitude of the complex electric dipole transition moment, is

$$\begin{aligned}[\tilde{P}^{(1)}_{x_0}]_{fi} &= \left\{ \tilde{\alpha}_{xx} + \frac{1}{c}\tilde{G}_{xy} + \frac{i\omega_0}{3c}\tilde{A}_{xzx} \right\} \tilde{E}_{x_0} \\ &\quad + \left\{ \tilde{\alpha}_{xy} + \frac{1}{c}\tilde{G}_{xz} + \frac{i\omega_0}{3c}\tilde{A}_{xzy} \right\} \tilde{E}_{y_0}\end{aligned} \tag{5.88}$$

and so on, instead of equations like (5.26).

$[\tilde{M}^{(1)}_{xo}]_{fi}$, the x component of the amplitude of the complex magnetic dipole transition moment, is given by

$$[\tilde{M}^{(1)}_{xo}]_{fi} = \tilde{G}^*_{xx}\tilde{E}_{xo} + \tilde{G}^*_{yx}\tilde{E}_{yo} \tag{5.89}$$

and so on.

$[\hat{\Theta}^{(1)}_{xyo}]_{fi}$, the xy component of the amplitude of the complex electric quadrupole transition moment, is given by

$$[\tilde{\Theta}^{(1)}_{xyo}]_{fi} = \tilde{A}^*_{xxy}\tilde{E}_{xo} + \tilde{A}^*_{yxy}\tilde{E}_{yo} \tag{5.90}$$

We recall that $\tilde{\alpha}_{xx}$, $\tilde{\alpha}_{xy}$ are components of a complex second-order tensor $\tilde{\boldsymbol{\alpha}}$, and that time-dependent perturbation theory shows that $\tilde{\alpha}_{xy}$ involves a summation over the products of two electric dipole terms of the type

$$\langle f|P_x|r\rangle\langle r|P_y|i\rangle \tag{5.91}$$

with frequency denominators of the type $\omega_{rf}+\omega_0$ and $\omega_{ri}-\omega_0$. P_x and P_y are components of the dipole moment operator defined by eq. (5.4).

$\tilde{G}_{xy}$, $\tilde{G}_{xz}$, etc., are the components of a second-order complex tensor $\tilde{\boldsymbol{G}}$. Time-dependent perturbation theory shows that $\tilde{G}_{xy}$ involves a summation over the products of an electric dipole term and a magnetic dipole term of the type

$$\langle f|P_x|r\rangle\langle r|M_y|i\rangle \tag{5.92}$$

and frequency denominators of the type $\omega_{rf}+\omega_0$ and $\omega_{ri}-\omega_0$. M_y is a component of the magnetic dipole moment operator defined by

$$M_y = \sum_i \frac{e_i}{2m_i}(z_i p_{x_i} - x_i p_{z_i}) \tag{5.93}$$

where the i-th particle has charge e_i, mass m_i, coordinates x_i, y_i, z_i and momentum components p_{x_i}, p_{y_i}, p_{z_i} (see page 36).

$\tilde{A}_{xzx}$, $\tilde{A}_{xxy}$, etc., are components of a complex third-order tensor $\tilde{\boldsymbol{A}}$. Time-dependent perturbation theory shows that $\tilde{A}_{xxy}$ involves a summation over the products of an electric dipole term and an electric quadrupole term of the type

$$\langle f|P_x|r\rangle\langle r|\Theta_{xy}|i\rangle \tag{5.94}$$

and frequency denominators of the type $\omega_{rf}+\omega_0$ and $\omega_{ri}-\omega_0$, where Θ_{xy} is a component of the electric quadrupole operator. From eq. (2.40)

$$\Theta_{lm} = \tfrac{1}{2}\sum_i e_i(3l_i m_i - r_i^2\delta_{lm}) \tag{5.95}$$

5.7.2 *Decomposition of the tensors* $\tilde{\boldsymbol{\alpha}}$, $\tilde{\boldsymbol{G}}$, *and* $\tilde{\boldsymbol{A}}$: *Rayleigh and Raman optical activity*

In general, the tensors $\tilde{\boldsymbol{\alpha}}$, $\tilde{\boldsymbol{G}}$, and $\tilde{\boldsymbol{A}}$ can all be complex and so may be written as

$$\tilde{\boldsymbol{\alpha}} = \boldsymbol{\alpha} - i\boldsymbol{\alpha}' \tag{5.96}$$

$$\tilde{\boldsymbol{G}} = \boldsymbol{G} - i\boldsymbol{G}' \tag{5.97}$$

and

$$\tilde{\boldsymbol{A}} = \boldsymbol{A} - i\boldsymbol{A}' \tag{5.98}$$

In the absence of magnetic perturbations $\boldsymbol{a}'$, $\boldsymbol{A}'$ and $\boldsymbol{G}$ are all zero. In this case, when the square of the dipole transition moment is formed to calculate the scattered intensity, there will be non-zero terms of the following five types:

$$\alpha\alpha\,;\,\alpha A\,;\,\alpha G'\,;\,AA\,;\,G'G'$$

The most important contribution to Rayleigh and Raman scattering arises from $\alpha\alpha$; the contributions from αA and $\alpha G'$ are of the order of 10^{-3} that from $\alpha\alpha$; and the contributions from AA and $G'G'$ are of the order of 10^{-6} that from $\alpha\alpha$. Furthermore, the products αA and $\alpha G'$ are only non-zero in chiral systems.

The additional contributions from chiral molecules to the intensity of Rayleigh and Raman scattering arising from the terms αA and $\alpha G'$ although very small, can be distinguished from the strong scattering originating in the term $\alpha\alpha$ because they exhibit different polarization effects. For example, the scattered intensity is different according to whether the incident radiation is right or left circularly polarized; and for incident radiation which is either linearly polarized or natural, a circularly polarized component is produced.

The difference in scattered intensity resulting from incident left or right circularly polarized radiation is termed the circular intensity differential (CID), which can be appropriately defined by the dimensionless quantity $\Delta_{p_s}(\pi/2)$ given by

$$\Delta_{p_s}(\pi/2) = \frac{{}^{\circledR}I_{p_s}(\pi/2) - {}^{\copyright\!\!\!\!\text{L}}I_{p_s}(\pi/2)}{{}^{\circledR}I_{p_s}(\pi/2) + {}^{\text{Ⓛ}}I_{p_s}(\pi/2)} \tag{5.99}$$

where ${}^{\circledR}I_{p_s}(\pi/2)$ and ${}^{\text{Ⓛ}}I_{p_s}(\pi/2)$ are the intensities of the scattered radiation at 90° with polarization p_s produced with right and left circularly polarized incident light.

Conventional optical rotatory dispersion and circular dichroism measure electronic optical activity and so provide structural and stereochemical information by probing chromophores which are perturbed by the chiral arrangements of bonds in the molecule. Since vibrational motion extends over the whole molecule, vibrational optical activity is expected to yield new stereochemical information. In principle, vibrational optical activity can be measured by extending rotatory dispersion and circular dichroism into the infrared region. However, optical activity depends on the frequency of the exciting light, and thus vibrational optical activity is barely accessible through infrared rotatory dispersion. However, it has recently been detected through infrared circular dichroism. Vibrational optical activity is rather more readily accessible through Raman CID measurements, since the mechanism of the Raman effect results in scattering at $\omega_0 \pm \omega_M$, and if ω_0 is in the visible region so also is $\omega_0 \pm \omega_M$.

Although Raman CID measurements make considerable demands on experimental techniques (see chapter 7) the results are of considerable structural interest, since each vibrational band of a chiral molecule can be studied.

5.7.3 *Intensities and polarizations in Rayleigh and Raman optical activity*

We now give, without derivation, formulae for the intensity, depolarization ratios and circular intensity differential (CID) associated with Rayleigh and Raman optical activity. The formulae have been derived by considering only terms in α^2, $\alpha G'$, and

αA so that magnetic perturbation effects are excluded. Furthermore, the scattering system is considered to be isotropic and non-absorbing.

For incident radiation of irradiance $\mathcal{I}$ propagating along the z-axis partially polarized with a degree of polarization P and characterized by an azimuth ψ and an ellipticity χ, the *additional* contributions to the Stokes parameters of the radiation scattered at 90° to z, along x, are given by

$$S_0\left(\frac{\pi}{2}\right)=\frac{1}{2}KE_0^2\left\{\frac{2}{c}\left(\frac{45aG'+13\gamma_G^2-\frac{1}{3}\gamma_A^2}{45}\right)P\sin 2\chi\right\} \tag{5.100}$$

$$S_1\left(\frac{\pi}{2}\right)=\frac{1}{2}KE_0^2\left\{\frac{2}{c}\left(\frac{45aG'+\gamma_G^2+\gamma_A^2}{45}\right)P\sin 2\chi\right\} \tag{5.101}$$

$$S_2\left(\frac{\pi}{2}\right)=0 \tag{5.102}$$

$$S_3\left(\frac{\pi}{2}\right)=\frac{1}{2}KE_0^2\frac{2}{c}\left\{\frac{(45aG'+13\gamma_G^2-\frac{1}{3}\gamma_A^2)}{45}-\frac{(45aG'+\gamma_G^2+\gamma_A^2)P\cos 2\chi\cos 2\psi)}{45}\right\} \tag{5.103}$$

where K is defined by eq. (4.43), a is the mean polarizability defined by eq. (3.18), G' is the mean optical activity given by

$$G'=\tfrac{1}{3}(G'_{xx}+G'_{yy}+G'_{zz}) \tag{5.104}$$

and γ_G^2 and γ_A^2, the polarizability optical activity anisotropies, are given by

$$\begin{aligned}\gamma_G^2=\tfrac{1}{2}\{&(\alpha_{xx}-\alpha_{yy})(G'_{xx}-G'_{yy})+(\alpha_{yy}-\alpha_{zz})(G'_{yy}-G'_{zz})+(\alpha_{zz}-\alpha_{xx})(G'_{zz}-G'_{xx})\\&+3[\alpha_{xy}(G'_{xy}+G'_{yx})+\alpha_{yz}(G'_{yz}+G'_{zy})+\alpha_{zx}(G'_{zx}+G'_{xz})]\}\end{aligned} \tag{5.105}$$

$$\begin{aligned}\gamma_A^2=\tfrac{3}{2}\omega_0\{&A_{zxy}(\alpha_{yy}-\alpha_{xx})+A_{yzx}(\alpha_{xx}-\alpha_{zz})+A_{xyz}(\alpha_{zz}-\alpha_{yy})\\&+\alpha_{xy}(A_{yzy}-A_{zyy}+A_{zxx}-A_{xzx})+\alpha_{xz}(A_{yzz}-A_{zyz}+A_{xyx}-A_{yxx})\\&+\alpha_{yz}(A_{zxz}-A_{xzz}+A_{xyy}-A_{yxy})\}\end{aligned} \tag{5.106}$$

The *complete* Stokes parameters $S_0(\pi/2)$, $S_1(\pi/2)$, $S_2(\pi/2)$ and $S_3(\pi/2)$ for an isotropic, non-absorbing and chiral system are included in Table G in the central reference section.

It is the Stokes parameters S_0 and S_1 which determine the intensities of the components polarized parallel and perpendicular to the scattering plane, and it can be seen from eqs. (5.100) and (5.101) that the contributions of optically active scattering (i.e., the contributions from the terms $\alpha G'$ and αA) to the Stokes parameters S_0 and S_1 always involve $P\sin 2\chi$. Thus, such contributions are zero if the incident light is unpolarized ($P=0$) or linearly polarized ($\chi=0$). Also, it can be seen from eqs. (5.101) and (5.102) that the azimuth of the light scattered at 90° to the scattering plane is perpendicular to the scattering plane, and eq. (5.103) shows that optically active scattering produces a circularly polarized component in the scattered radiation.

The intensity components ${}^{p_i}I_{\parallel}(\pi/2)$ and ${}^{p_i}I_{\perp}(\pi/2)$ parallel and perpendicular to the scattering plane are given by

$$ {}^{p_i}I_{\parallel}(\pi/2) = \frac{1}{2}K'\{S_0(\pi/2) - S_1(\pi/2)\} $$

$$ = k_{\tilde{\nu}}(\nu')^4\left\{\frac{6\gamma^2 + 4/c(3\gamma_G^2 - \frac{1}{3}\gamma_A^2)P\sin 2\chi}{90}\right\}\mathscr{I} \tag{5.107} $$

with K' defined by eq. (4.61) and

$$ {}^{p_i}I_{\perp}(\pi/2) = \frac{1}{2}K'\{S_0(\pi/2) + S_1(\pi/2)\} $$

$$ = k_{\tilde{\nu}}(\nu')^4\left\{\frac{(45a^2 + 7\gamma^2) - (45a^2 + \gamma^2)P\cos 2\chi\cos 2\psi + 2/c(45aG' + 7\gamma_G^2 + \frac{1}{3}\gamma_A^2)P\sin 2\chi}{90}\right\}\mathscr{I} \tag{5.108} $$

The components of the circular intensity differential defined by eq. (5.99), which are polarized perpendicular and parallel to the scatter plane, are given by

$$ \Delta_{\perp}\left(\frac{\pi}{2}\right) = \frac{2(45aG' + 7\gamma_G^2 + \frac{1}{3}\gamma_A^2)}{c(45a^2 + 7\gamma^2)} \tag{5.109} $$

and

$$ \Delta_{\parallel}\left(\frac{\pi}{2}\right) = \frac{2(\gamma_G^2 - \frac{1}{9}\gamma_A^2)}{c(\gamma^2)} \tag{5.110} $$

$\Delta_{\perp}(\pi/2)$ and $\Delta_{\parallel}(\pi/2)$ are referred to as the polarized and depolarized CIDs. The numerator and denominator for the CID with no analyser in the scattered beam are obtained by adding the numerators and the denominators, respectively, in eqs. (5.109) and (5.110). We note that the degrees of circularity ${}^{\perp}\mathscr{C}(0)$ and ${}^{\parallel}\mathscr{C}(0)$ produced when linearly polarized incident light is used give the same information as the CID. The degree of circularity is measured by S_3/S_0, and this ratio is equal to $\Delta_{\perp}(\pi/2)$ if the incident radiation is linearly polarized perpendicular to the scattering plane and is equal to $\Delta_{\parallel}(\pi/2)$ if the incident radiation is polarized parallel to the scattering plane. Formulae for intensity components, CIDs and degrees of circularity for an isotropic, non-absorbing chiral system are included in Table G in the central reference section.

5.7.4 *Extension of the Placzek theory to Raman optical activity: vibrational selection rules*

Since we are dealing with non-absorbing systems, we may extend the Placzek polarizability theory to the tensors $\boldsymbol{G}'$ and $\boldsymbol{A}$ subject to the same conditions relating to the excitation frequency ω_0 and the degeneracy of the ground electronic state. We recall that when the transition polarizability is expanded as a function of the normal coordinates Q, then in the harmonic oscillator approximation the intensity of vibrational Raman scattering associated with the k-th normal coordinate is proportional to terms of the type

$$ \left(\frac{\partial\alpha_{xy}}{\partial Q_k}\right)_0^2 \langle v^f|Q_k|v^i\rangle^2 \tag{5.111} $$

Such terms are only non-zero if $v^f - v^i = \pm 1$ and $(\partial\alpha_{xy}/\partial Q)_0$ is non-zero. Similarly, the intensity of Raman scattering arising from optical activity through the terms $\alpha G'$ and αA is proportional to terms of the type

$$\left(\frac{\partial\alpha_{xy}}{\partial Q_k}\right)_0^*\left(\frac{\partial G'_{xy}}{\partial Q_k}\right)_0 |\langle v^f|Q_k|v^i\rangle|^2 \tag{5.112}$$

and

$$\left(\frac{\partial\alpha_{xy}}{\partial Q_k}\right)_0^*\left\{\left(\frac{\partial A_{yzy}}{\partial Q_k}\right)_0 - \left(\frac{\partial A_{zyy}}{\partial Q_k}\right)_0 + \left(\frac{\partial A_{zxx}}{\partial Q_k}\right)_0 - \left(\frac{\partial A_{xzx}}{\partial Q_k}\right)_0\right\}\langle v^f|Q_k|v^i\rangle^2 \tag{5.113}$$

We see that the same selection rule operates for the vibrational quantum numbers, namely, $v^f - v^i = \pm 1$. Also only those fundamental modes which are spanned by either α_{xy} and G'_{xy} or α_{xy} and the appropriate linear combinations of the components of $\boldsymbol{A}$ will show Raman CIDs.

Natural optical activity is exhibited only by asymmetric molecules or molecules that are disymmetric to the extent that they belong to pure rotation groups like C_n, D_n, O, T, and I. For these groups it happens that α_{xy}, G'_{xy} and the appropriate linear combinations of $\boldsymbol{A}$ have identical transformation properties. Thus, all Raman active vibrational modes in optically active molecules should show CIDs, and the Stokes and anti-Stokes CIDs of a particular vibrational mode should have the same sign and magnitude.

5.7.5 *Magnetic Rayleigh and Raman optical activity*

Magnetic Rayleigh and Raman optical activity can arise from terms involving unperturbed $\boldsymbol{a}$ and $\boldsymbol{a}'$ perturbed to first order in the magnetic field, and also unperturbed $\boldsymbol{a}'$ and $\boldsymbol{a}$ perturbed to first order. If the system is transparent, $\boldsymbol{a}$ is symmetric and $\boldsymbol{a}'$ antisymmetric; but if it is absorbing, both $\boldsymbol{a}$ and $\boldsymbol{a}'$ can contain symmetric and antisymmetric parts. Magnetic Raman activity has recently been observed in a resonance Raman spectrum (see chapter 7).

5.7.6 *Electric Rayleigh and Raman optical activity*

Electric Rayleigh and Raman optical activity can arise from terms involving unperturbed $\boldsymbol{a}$ and $\boldsymbol{G}'$ *plus* $\boldsymbol{A}$ perturbed to first order in the electric field, and unperturbed $\boldsymbol{G}'$ *plus* $\boldsymbol{A}$ and $\boldsymbol{a}$ perturbed to first order. Since electric Rayleigh and Raman optical activity is a perturbation of Rayleigh and Raman optical activity, it will be very weak and has not so far been detected.

References

1. G. Placzek, Rayleigh-Streuung und Raman-Effekt, in E. Marx (Ed.), *Handbuch der Radiologie*, vol. **VI**, 2, pp. 205–374, Akademische Verlag, Leipzig, 1934.
2. L. D. Barron, *Molec. Phys.*, **31**, 129, 1976.
3. L. D. Barron, in D. A. Long, R. F. Barrow and J. Sheridan (Eds.), *Chemical Society Specialist Reports*, No. 29, *Molecular Spectroscopy*, vol. 4, Chemical Society, London, 1976.
4. A. C. Albrecht and M. C. Hutley, *J. Chem. Phys.*, **55**, 4438, 1971.
5. L. D. Barron and A. D. Buckingham, *Ann. Rev. Phys. Chem.*, **26**, 381, 1975.

6 Experimental procedures

"Macht doch den zweiten Fensterladen auch auf, damit mehr Licht herein komme."

J. W. von Goethe

6.1 Introduction

The block diagram of Fig. 6.1 shows the components of the equipment necessary for the observation of Raman spectra. They are a source of monochromatic radiation, an appropriate sample device, a system for dispersion of the scattered radiation, and a detection device which may be either photographic or photoelectric in nature. We shall consider each of these briefly, in turn, concentrating on describing for each component the important requirements and how far these can be realized.

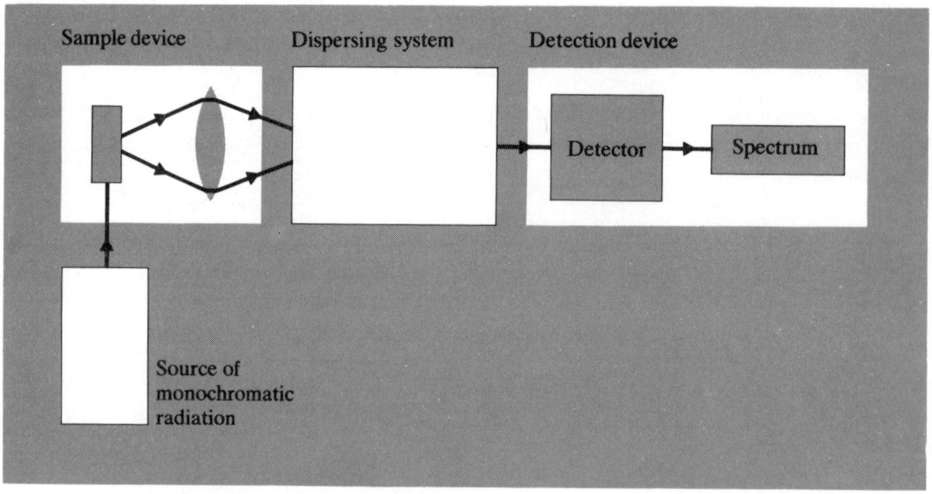

Fig. 6.1 Block diagram of equipment for observation of Raman spectra

6.2 Sources of monochromatic radiation

The essential requirements of a source for the excitation of Raman spectra are that it should be highly monochromatic (i.e., have a narrow line width) and be capable of giving a high irradiance at the sample. The gas laser meets these requirements perfectly and, in addition, provides radiation which is self-collimated and plane polarized. Most gas lasers can provide a number of discrete wavenumbers of varying power, and dye iasers can provide an excitation wavenumber which is continuously

Table 6.1(a) Wavelengths (and wavenumbers) available from typical gas lasers

λ_{air} (Å)	$\tilde{\nu}_{vac}$ (cm^{-1})	Relative output powers (mW)			
		He/Ne	Kr	Ar/Kr	Ar
7993·2	12 507		30		
7931·4	12 605		10		
7525·5	13 285		100		
6764·4	14 779		120	20	
6470·9	15 450		500	200	
6328·2	15 798	70			
5681·9	17 595		150	80	
5308·7	18 832		200	80	
5208·3	19 195		70	20	
5145·3	19 430			200	800
5017·2	19 926			20	140
4965·1	20 135			50	300
4879·9	20 487			200	700
4825·2	20 719		30	10	
4764·9	20 981			60	300
4762·4	20 992		50		
4726·9	21 150				60
4657·9	21 463				50
4579·4	21 831			20	150
4545·1	21 996				20
3637·9 3511·1	27 481 28 473				20
3564·2 3507·4	28 049 28 503		40		

Notes: (i) See page 40 for a discussion of λ_{air} and $\tilde{\nu}_{vac}$.
(ii) The powers for the krypton, argon/krypton and argon lasers relate to typical commercial lasers described as having a nominal output of about 2 W.

Table 6.1(b) Wavelength (and wavenumber) ranges covered by typical dye lasers

Dye	Wavelength range (Å)	Wavenumber range (cm^{-1})
Nile Blue perchlorate	7100–8000	14 100–12 500
Cresyl Violet perchlorate	6700–7100	14 900–14 100
Rhodamine B	5900–6900	16 900–14 500
Rhodamine 6G	5600–6600	17 900–15 200
Rhodamine 110	5300–6200	18 900–16 100
Sodium Fluorescein	5300–5800	18 900–17 200
Coumarin 6	5200–5600	19 200–17 900
Coumarin 102	4600–5200	21 700–19 200
Coumarin 2	4300–4800	23 300–20 800

Notes: (i) Output powers depend on pump laser power but typically range over 100–900 mW.
(ii) The wavelength ranges quoted above are approximate only; they vary somewhat with the solvent and pumping conditions.

variable over a limited range. Tables 6.1(a) and (b) summarize the characteristic properties of some typical currently available gas lasers and dye lasers. Plate 1, opposite, illustrates a dye laser in action.

Using a lens, a laser beam may be focused to produce a beam of much smaller diameter. The small-diameter beam extends over a short length before beginning to diverge again, and we may refer to the region in which the beam is most concentrated as the focal cylinder (see Fig. 6.2). Diffraction considerations prevent the volume of

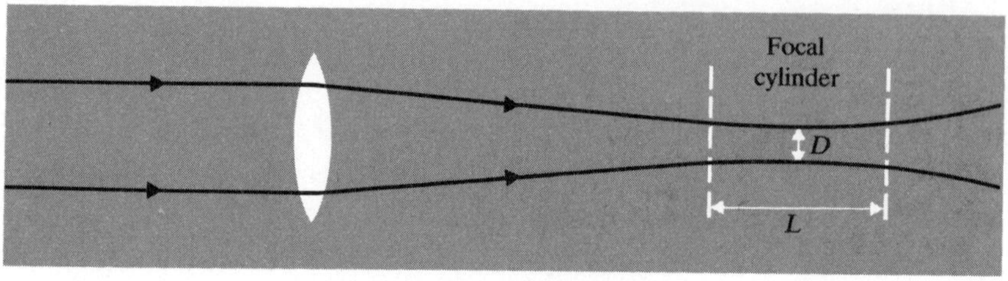

Fig. 6.2 The focal cylinder

this cylinder tending to zero. The following formulae give the diameter D and length L of the focal cylinder:

$$D=\frac{4\lambda f}{\pi d} \tag{6.1}$$

$$L=\frac{16\lambda f^2}{\pi d^2} \tag{6.2}$$

where λ is the wavelength of the laser radiation, d the diameter of the unfocused laser beam, and f the focal length of the focusing lens. For $\lambda = 5000$ Å (500·0 nm) and $d = 0{\cdot}15$ cm, the formulae can be written in the useful forms:

$$D/\text{cm} = 4.24 \times 10^{-4}(f/\text{cm}) \tag{6.3}$$

$$L/\text{cm} = 1.13 \times 10^{-2}(f/\text{cm})^2 \tag{6.4}$$

Using a lens with $f = 10$ cm, $D = 4{\cdot}24 \times 10^{-3}$ cm and $L = 1{\cdot}13$ cm, and the volume of the focal cylinder is thus about 10^{-5} cm^3. The ratio $d/D \approx 36$, and thus the area of the focused beam is about 10^{-3} that of the unfocused beam and so the irradiance or power per unit area at the sample is increased by about 10^3. Focused laser beams are therefore normally used for the production of Raman spectra except in special circumstances where, for example, the high irradiance may be harmful to the sample.

Under normal conditions of operation, the output of a laser, nominally described as having a given wavenumber, actually consists of a number of modes of slightly different wavenumber which together form a band envelope whose width determines the observed width of that laser wavenumber. By the selection of one of these modes with appropriate optical devices, very much narrower laser line widths can be achieved. For example, the argon laser line at 5145·3 Å (514·53 nm, 19 430 cm^{-1}) normally has a width of 0·15 cm^{-1}, but using a mode-selecting etalon inside the cavity, a line width of 0·001 cm^{-1} can be achieved. The power in the single-mode output is then reduced to about 50 per cent of the multimode output. Single-mode operation of a laser is important for high resolution studies.

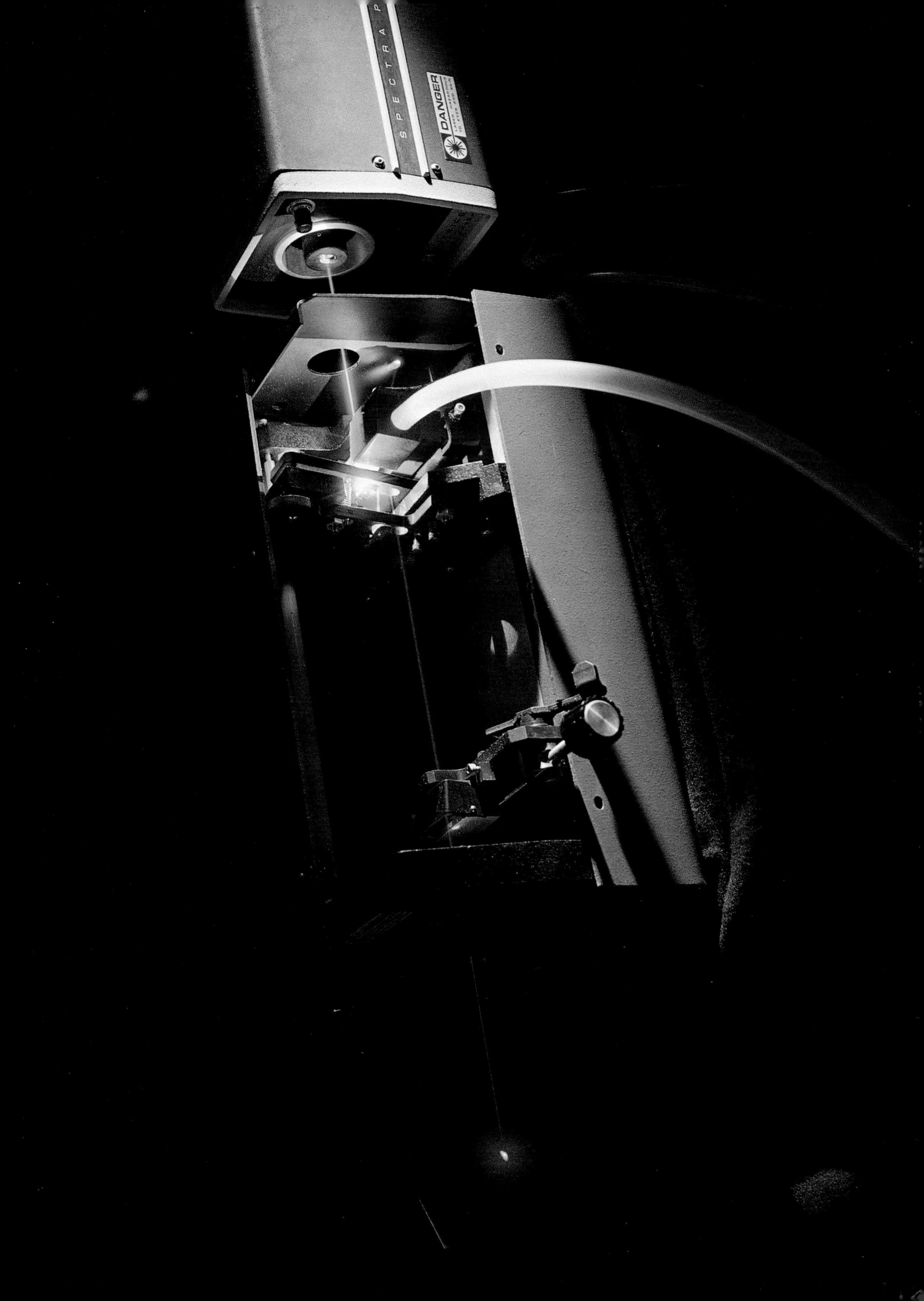
SPECTRA P
DANGER

6.3 Sample devices

Here the vital considerations are the most effective ways of illuminating the sample and collecting the scattered radiation for subsequent dispersion. In most cases the sample is placed outside the laser cavity, and then a typical arrangement is that shown in Fig. 6.3(a). The lens L_1 focuses the laser beam into the sample, and L_2 collects the optimum solid angle of scattered radiation and matches this to the collection angle of the dispersing system. Two additional simple optical devices shown in Fig. 6.3(a) can increase the observed intensity of scattering by about ten times. The concave mirror M_2 virtually doubles the solid angle of scattered radiation fed to the dispersing system. The concave mirror M_1 multipasses the laser beam through the focus in the sample, increasing the effective intensity of illumination by up to five times, provided the sample does not absorb the radiation.

The volume of the focal cylinder in which exists the high irradiance of the focused laser beam is only of the order of 10^{-5} cm^3, as we have seen. In principle, only sufficient sample to fill this focal cylinder is necessary, and for strong scatterers even smaller volumes could suffice. Good Raman spectra have indeed been obtained from liquid samples of the order of 10^{-6} cm^3; and spectra have been recorded from a cube of solid sample of side 10^{-4} cm (volume 10^{-12} cm^3). Although the capability of working with such samples is very important in special circumstances, problems of cell design usually result in somewhat bigger samples being used wherever possible.

The sample system of Fig. 6.3(a) is relatively simple and inexpensive. A number of other arrangements for sample illumination are shown in diagrammatic form in Figs. 6.3(b) to 6.3(j).

Only a fraction of the power inside the cavity of the laser itself passes out through the end mirror and, in principle, therefore, it would be advantageous to put the sample inside the laser cavity. In practice this is possible only for non-absorbing samples of excellent optical quality; otherwise the losses in the sample will stop the system lasing. An intra-cavity sample cell is sometimes used for the study of rotation and vibration–rotation Raman spectra of non-absorbent, stable gases. A typical arrangement is shown in Fig. 6.4. The sample is situated at the focus of the laser beam within the cavity; the power per unit area at the sample is of the order of 20 times greater than with the sample at a focus external to the cavity.

Filters and other optical devices may be inserted into the incident laser beam or the scattered radiation beam. For example, before focusing, the laser beam may pass through an interference filter to suppress unwanted laser or plasma frequencies (especially if these are likely to cause photochemical decomposition), or through a half-wave plate to rotate the plane of polarization through 90° if required; also, an iris diaphragm may be inserted in the beam to limit the beam diameter and reduce interference from the laser plasma. The scattered beam may pass through a polarization analyser and then through a crystal quartz wedge which acts as a polarization scrambler and ensures that natural light is incident on the slit of the spectrometer; in this way, possible errors in polarization measurements due to differential transmission by the spectrometer of light of different polarizations are avoided.

6.4 Dispersing systems

The characteristics of the dispersing system depend on whether it is to be used for the resolution of individual lines in rotation and vibration–rotation Raman bands, or for

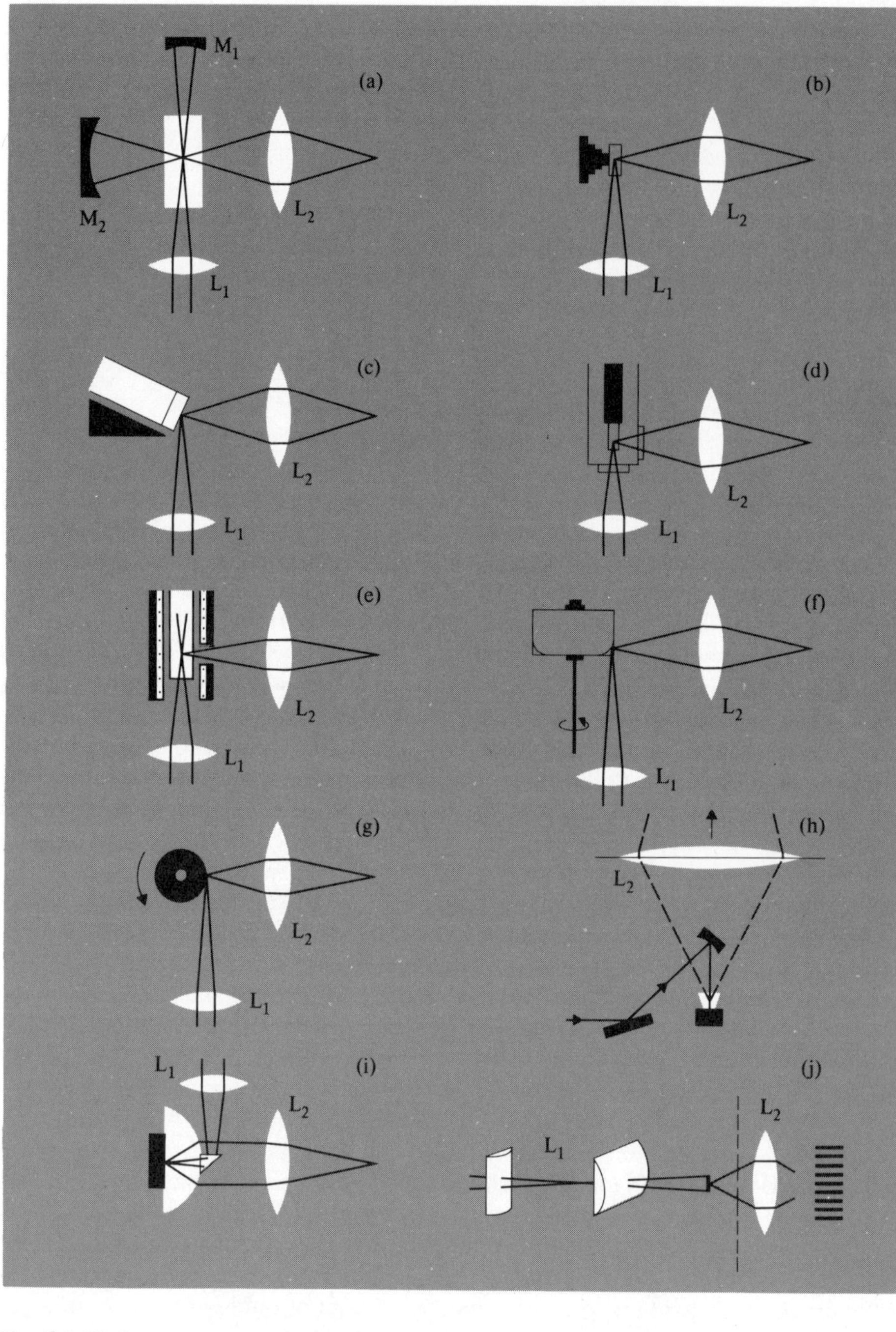

Fig. 6.3 Various arrangements for sample illumination: (a) sample extra-cavity, (b) single crystal mounted on a goniometer, (c) powdered solid, (d) low-temperature cryostat, (e) high-temperature cell, (f) spinning sample cell (liquids), (g) spinning reel for fibres, (h) diamond anvil for high pressure studies of crystals, (i) 180° scattering geometry, and (j) 0° scattering geometry

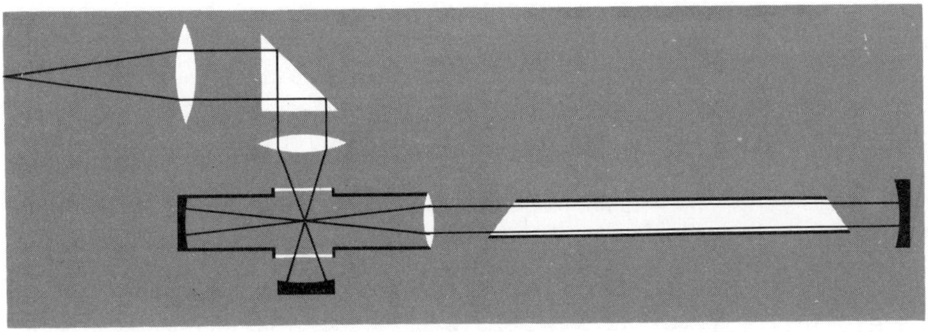

Fig. 6.4 Intra-cavity sample cell for gas samples

the study of vibrational bands under conditions of moderate resolution. In both types of investigation, the dispersing system is almost invariably based on a diffraction grating rather than a prism, but in most other respects the two kinds of study call for quite different design features.

The optical layout of a typical grating dispersing system for rotation and vibration–rotation Raman spectroscopy is shown in Fig. 6.5. Front-surfaced concave mirrors are used to collimate the radiation before dispersion and to refocus it, after dispersion, on the photographic plate which is the usual detector in such systems. For a fixed grating position, a wide range of wavenumbers reach the photographic plate. The dispersing system is then described as a *spectrograph*. The primary requirements are high resolving power and high reciprocal linear dispersion. A typical instrument would use a plane reflectance grating with 316 grooves mm^{-1} and

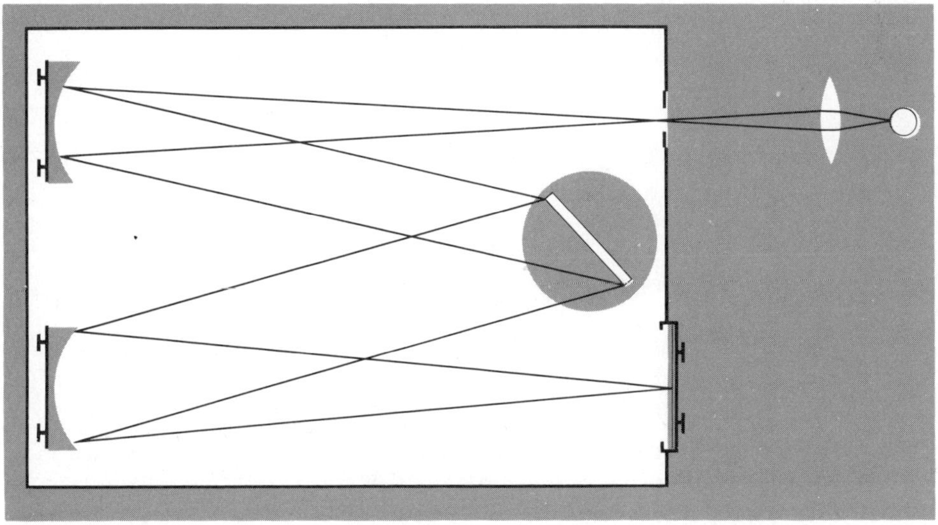

Fig. 6.5 Typical dispersing system for the study of rotation and vibration–rotation Raman spectra under high resolution

a ruled area of $25 \times 12\ cm^2$ in the thirteenth order of diffraction. The theoretical resolving power of such systems approaches 10^6 and the practical resolving power is usually 90 per cent of this. Thus, at 20 000 cm^{-1}, lines with a separation of about 0·02 cm^{-1} can be resolved. Since photographic plates of adequate speed will resolve only about 50 lines per mm, a reciprocal linear dispersion of 1 cm^{-1} mm^{-1} is required if lines with a separation of 0·02 cm^{-1} are to be distinguishable on the plate. To achieve this dispersion, long focal length mirrors must be used. For example, if the grating under discussion is used with a 300 cm focal length mirror, the reciprocal linear dispersion is about 1·4 cm^{-1} mm^{-1} at 20 000 cm^{-1}. Such long focal lengths mean that the system is 'slow' in the photographic sense; the *f*/number of the system discussed here is about *f*/25. Exposure times ranging from minutes to days are required, but the photographic plate has the compensating advantage that it simultaneously records a whole range of wavenumbers. To determine the wavenumbers of the Raman lines, a polynomial relating position on the plate to wavenumber is first established using a calibration spectrum (e.g., a hollow-cathode thorium source) whose wavenumbers are well known. The wavenumbers of the Raman lines may then be determined using this polynomial and measurements of the position of the Raman lines on the plate.

Interferometric techniques have also been used to achieve a resolution of the order of 0.02 cm^{-1} over a restricted wavenumber range. An interferometer crossed with a spectrograph has been used to measure the wavenumber of a Raman line to 0·003 cm^{-1}.

The optical layout of a double monochromator dispersing system for the study of

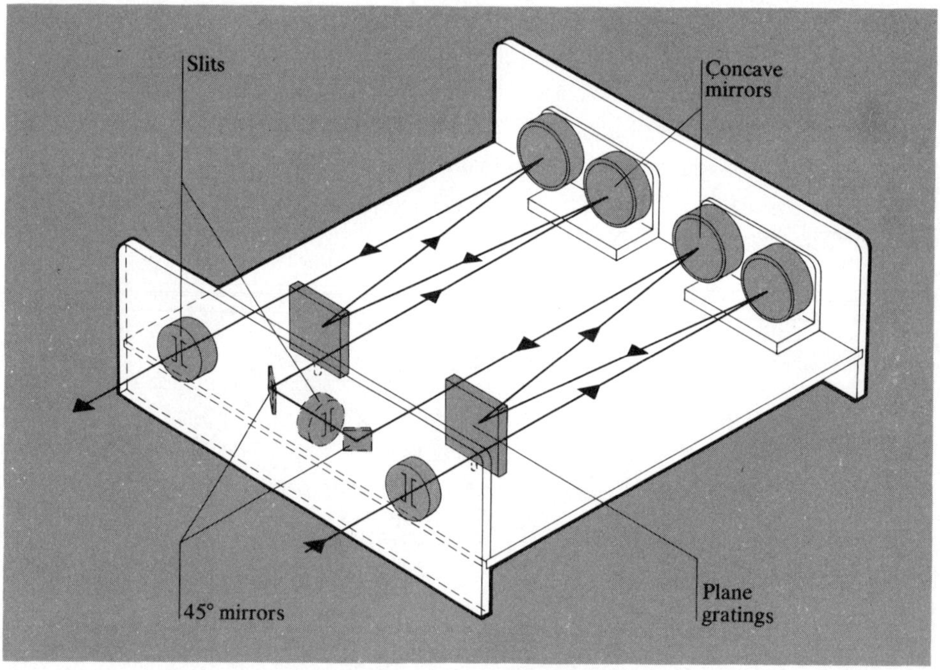

Fig. 6.6 Typical double monochromator grating dispersing system for the study of vibrational Raman spectra under medium resolution

vibrational Raman spectra under medium resolution is shown in Fig. 6.6. The dispersed radiation is detected photoelectrically. The exit slit allows only a narrow wavenumber band to reach the photomultiplier detector, and rotation of the diffraction gratings allows successive bands to reach the detector.

A most important property of any monochromator system used for Raman spectroscopy is its spectral purity. Essentially, this is its ability to distinguish radiation of the narrow wavenumber band $\tilde{\nu} \pm \delta\tilde{\nu}$, to which it is set, from radiation of other wavenumbers. When studying spectra in which the intensities of the lines are roughly comparable, the ability of a monochromator to distinguish between different wavenumbers depends essentially upon factors like the resolving power, dispersion, and slit width. In light scattering spectroscopy, other factors arise. Rayleigh scattering, which always occurs with Raman scattering, is of the order of 10^2 to 10^4 times more intense than Raman scattering; and optical shortcomings of the sample, as, for example, dust particles in a liquid or imperfections in a crystal, can still further increase the intensity of scattered light at the incident wavenumber relative to Raman scattering. Through defects in the dispersing system, mainly scattering at optical surfaces, particularly the grating, a small fraction of light of wavenumber $\tilde{\nu}_0$ will appear in the output focal plane at positions corresponding to other wavenumbers in the neighbourhood of $\tilde{\nu}_0$. If the intensity of $\tilde{\nu}_0$ is large enough, the intensity of the light masquerading as other wavenumbers may be comparable with, or even greater than, any Raman scattering whose true wavenumber is in this region. For example, in a typical single monochromator, about 10^{-5} of the radiation at $\tilde{\nu}_0$ will masquerade as $\tilde{\nu}_0 - 100$ cm^{-1}. For some samples (e.g., optically clear liquids or single crystals), this may be an order of magnitude less than any Raman scattering at $\tilde{\nu}_0 - 100$ cm^{-1}, but in other samples this can entirely swamp any Raman scattering at small wavenumber shifts. However, if two monochromators are used in series (as in Fig. 6.6), the resulting system will have only about $(10^{-5})^2 = 10^{-10}$ of the radiation at $\tilde{\nu}_0$ masquerading as $\tilde{\nu}_0 - 100$ cm^{-1}, and for three monochromators in series (a triple monochromator) this falls to $(10^{-5})^3 = 10^{-15}$. For most systems a double monochromator suffices, although the additional cost of a triple monochromator can be justified in some cases. The ratio of the amount of spurious or stray radiation at various wavenumbers to the radiation at $\tilde{\nu}_0$ is plotted as a function of wavenumber shift from $\tilde{\nu}_0$ for single, double, and triple monochromators, in Fig. 6.7. Selective filters placed before the entrance slit can also be used to reduce the intensity of the Rayleigh line, but filters are by no means as effective, since they do not have adequately narrow absorption characteristics.

Each monochromator system of a typical double monochromator for Raman spectroscopy would have a 10 cm × 10 cm plane reflectance grating, with 1200 grooves mm^{-1} used in the first order of diffraction, and collimating and focusing mirrors of 12 cm diameter and 75 cm focal length. With the two component monochromators coupled so that their dispersions are additive, then, overall, the double monochromator system would have a reciprocal linear dispersion of 20 cm^{-1} mm^{-1} and a resolution of 1 cm^{-1} in the region of 5000 Å for 50 μ slit width.

In scanning spectrometers, the wavenumber is usually read directly from a scale which translates the grating orientation into wavenumbers, but direct calibration with an emission spectrum with known wavenumbers is a desirable check. Standard scanning speeds usually range from 500 cm^{-1} to 0·5 cm^{-1} per minute, but much more rapid scan times are possible and are useful for the study of kinetic phenomena.

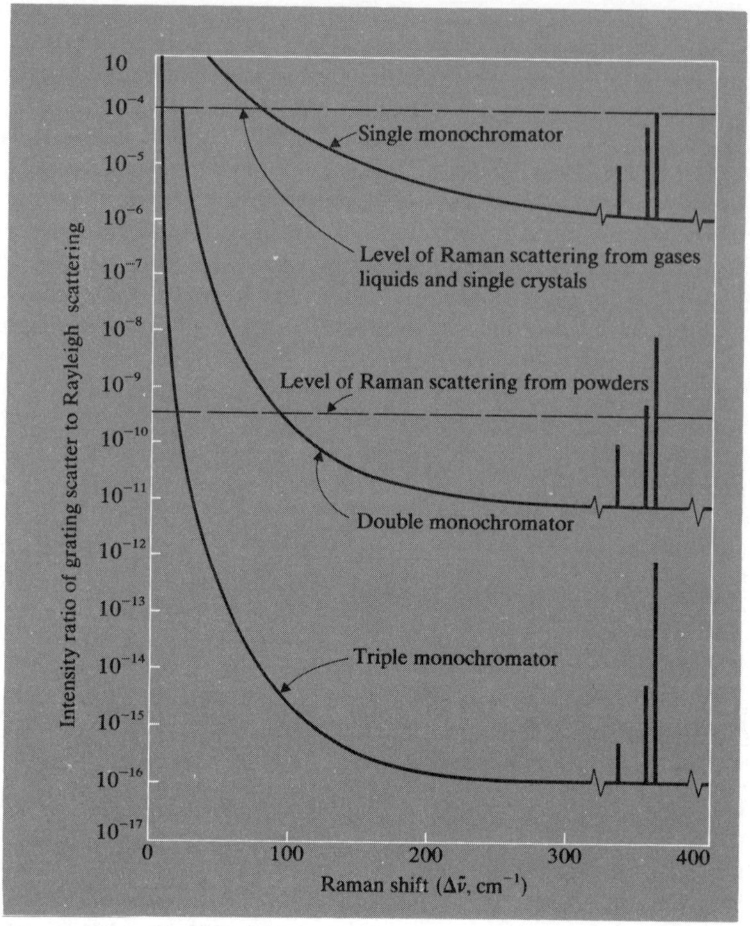

Fig. 6.7 Stray light rejection for single, double, and triple monochromators

6.5 Detection devices

In a spectrograph, the detector is a photographic plate. The choice of photographic plate is usually a compromise between speed and resolution. Fast emulsions tend to be coarse-grained and have relatively poor resolution, whereas fine-grained plates with better resolution are necessarily slower. Spectral response must also be considered; photographic plates are more sensitive in the blue and green regions of the spectrum (below 5300 Å) than in the yellow and red regions (beyond 5300 Å). Special attention must be paid to development of plates to prevent distortion of the image, through differential stretching of the emulsion during processing. This is particularly important in high resolution studies. The photographic plate acts as an integrator of weak light signals and covers a range of wavenumbers. Provided background radiation is very low, the photographic plate is a very effective detector for high resolution spectroscopy.

In spectrometers, the detector is almost invariably a special kind of photocell called a photomultiplier. When a photon falls on the photosensitive cathode of a photomultiplier, the photoelectron that is released is accelerated to a first dynode which is maintained at about 100 volts above the cathode. The accelerated photo-electron releases several electrons from the first dynode, and these electrons are then accelerated to the next dynode where further electrons are released. Since ten or more dynode stages are usually involved, there can emerge from the last stage of the photomultiplier a pulse of about 10^6 electrons associated with the arrival of one photon at the photocathode.

This electron multiplication process makes the photomultiplier very suitable for the detection of low-level light signals. It must be appreciated, however, that there will also be random emission of electrons from the photocathode and the dynodes, and these will also be multiplied and contribute a 'noise' element to the signal. If the signal is to be recognizable, the ratio of signal to noise must be substantially greater than unity. Thermal emission of electrons is a major cause of noise, but noise from this source can be substantially reduced by cooling the photomultiplier, and this is standard practice when dealing with very low light levels. Discrimination against noise can also be made in the course of processing the output pulses of electrons from the photomultiplier. A variety of methods may be used for processing the output electron pulses. The most commonly used are based on pulse counting, direct current measurement, or a.c. amplification. In each case, the ultimate objective is to produce a signal adequate for the operation of a pen recorder (analogue output) or for computer processing (digital output). We shall consider, briefly, each of these processing techniques in turn.

To appreciate the principles involved in photon counting and the circumstances in which it is most effective, it is necessary first to consider the response time of the basic photomultiplier circuit shown in Fig. 6.8. In this circuit, R is a load resistor across which a voltage is developed as a result of electrons from the photomultiplier anode flowing through it to ground. The magnitude of this voltage produced by a pulse of electrons is usually called the pulse height. The secondary emission characteristics of the dynode chain are such that each pulse of electrons produced by multiplication of a single primary photoelectron has a duration of about 10^{-8} s. The capacitance to ground of a photomultiplier is about 10 to 100 pF, and this represents the irreducible value of C in Fig. 6.8. Since there is no point in making the product CR (which determines the circuit response time) less than the minimum pulse width of 10^{-8} s, the minimum value of R is of the order of 100 Ω to 1 kΩ. With such R values, of the order of 10^7 pulses per second can be counted. This is more than adequate for Raman spectroscopy, where the rate of arrival of photons at the detector usually does not

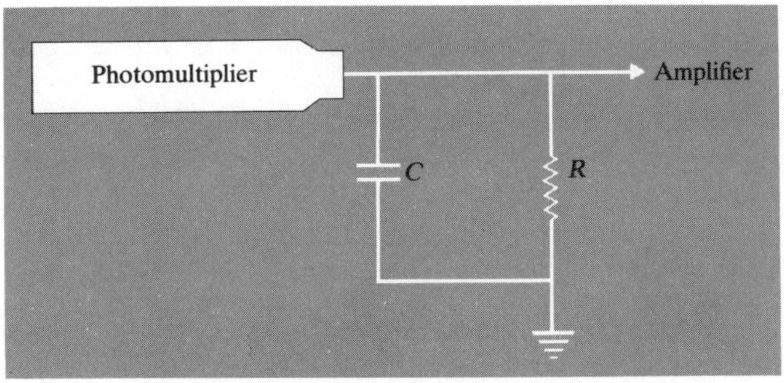

Fig. 6.8 Basic photomultiplier circuit diagram

exceed 10^4 photons per second. For such counting rates, R can be increased to 100 kΩ when the time response of the basic circuit is about 10^{-5} s. The pulse height varies somewhat from one photon to another, but in acceptable photomultipliers it lies in a range significantly above that for pulses originating from thermally emitted electrons. Thus, a pulse height discriminator circuit can be used to reject most of the noise pulses and pass the true signal pulses. Some noise pulses arise from cosmic rays and the radioactivity of the glass. Such pulses usually have much larger heights than the signal pulses and so can also be discriminated against. The pulses which pass the discriminator are then passed through pulse-shaping circuits which produce pulses of equal amplitude and duration to facilitate further processing. This pulse standardization procedure represents no loss of information. It is only the number of pulses which is significant, since each pulse represents the emissions of one photoelectron from the photocathode. The variation in pulse height is a statistical property of the secondary emission process, and conveys no information about the radiation incident on the photocathode. The standardized pulses may then be counted by an electronic counter for computer processing either directly (on-line) or indirectly via paper tape (off-line). More usually, the pulses are passed to a suitable RC network with time constants in the range 10^{-1} to 10 s. The slowly varying voltage developed in such a circuit is then amplified and displayed on a potentiometer pen recorder. Photon counting is particularly suitable for the detection of low light levels, because of the discrimination it affords against detector noise. Satisfactory Raman spectra can be recorded directly when the signal pulse arrival rate is as low as 10 per second. Even weaker signals can be dealt with by using signal averaging techniques. If the spectral region is scanned n times, then, since the signal is constant with time but the noise varies randomly, the average of n spectra will show an improvement in the signal–noise ratio relative to a single spectrum of $n^{1/2}$. When signal averaging is used, the output is most conveniently handled using a computer. Examples of how effectively signal averaging can extract weak signals from noise are shown in chapter 8. Photon counting is not appropriate for strong light levels since there is an upper limit to the number of pulses per second that such systems can cope with.

The response time of the photomultiplier circuit may be lengthened by increasing the load resistance R and increasing the capacitance C by adding a capacitor. If the reciprocal of the response time is made greater than the photon arrival rate, then the pulses corresponding to individual photons will no longer be distinguishable. Under these conditions, there is flowing through R effectively a continuous current, whose magnitude varies directly as the intensity of light falling on the photocathode. This current (the variations in which are relatively slow and, for a given spectrum, are determined by the scan speed of the spectrometer) may be measured with a suitable current meter, either directly, if it is large enough, or after d.c. amplification, if it is too small for direct measurement. Commercially available systems enable currents ranging from 10^{-10} to 10^{-4} A to be handled and displayed on a pen recorder. These currents correspond to pulse rates of 10 to 10^6 per second. Such systems usually incorporate a variable damping circuit so that the overall response time of the detection system is in the range 10^{-1} to 10^2 s.

A third method of dealing with the signal from a photomultiplier involves phase-sensitive or lock-in amplification. The slowly varying direct current which flows in the resistor R when the circuit response time is high enough may be converted into an alternating current with slowly varying amplitude if the radiation

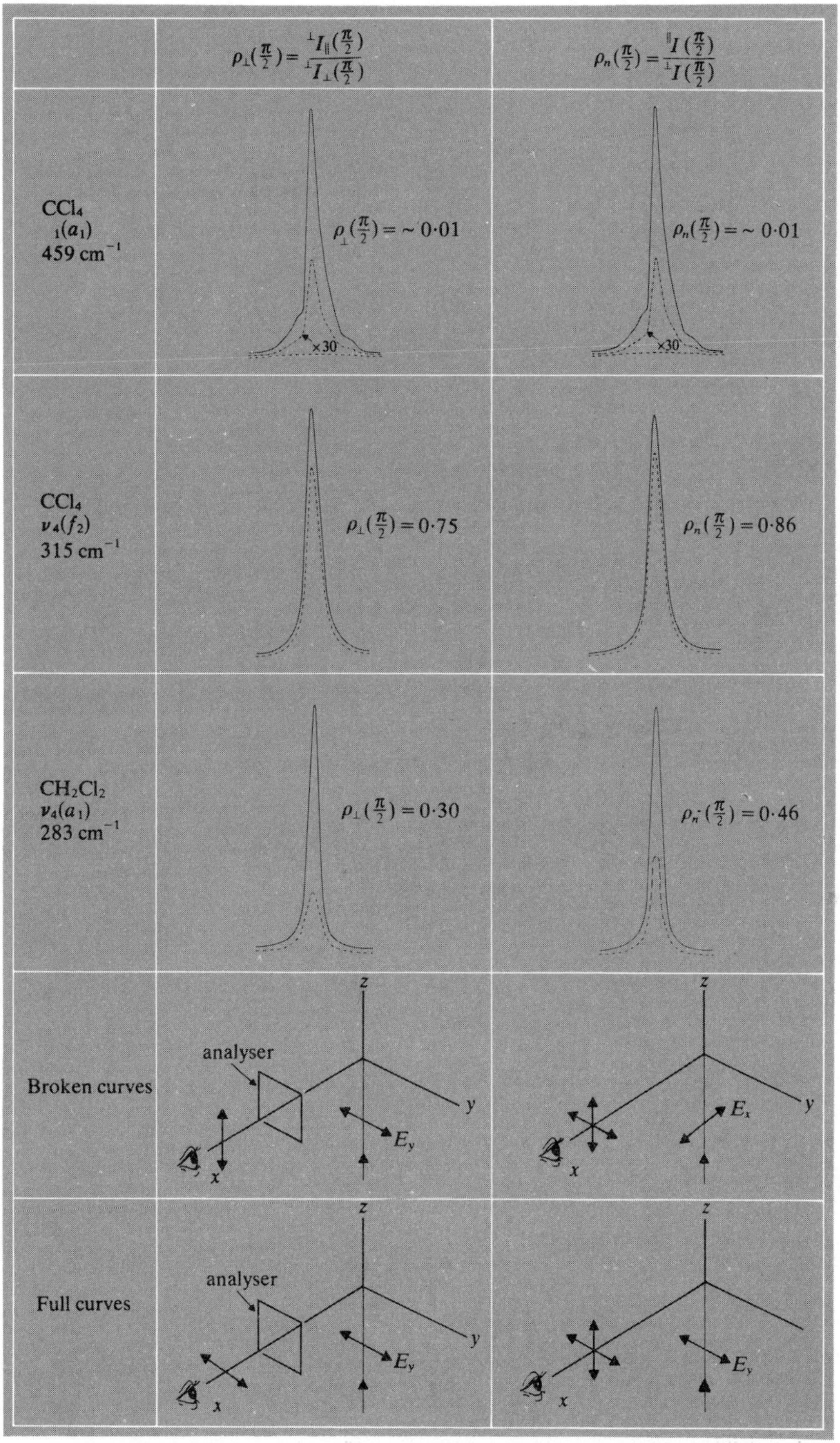

Fig. 6.9 Measurement of $\rho_{\perp}(\pi/2)$ and $\rho_{n}(\pi/2)$ for selected fundamental vibrations of CCl_4 and CH_2Cl_2

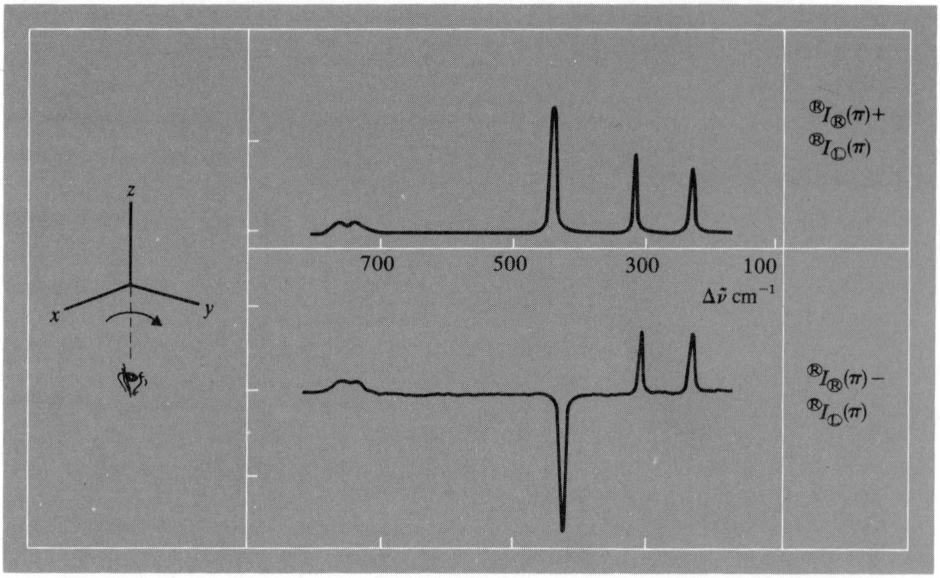

Fig. 6.10 Stokes Raman spectra of CCl_4: $^{®}I_{®}(\pi)+{}^{®}I_{Ⓛ}(\pi)$ and $^{®}I_{®}(\pi)-{}^{®}I_{Ⓛ}(\pi)$

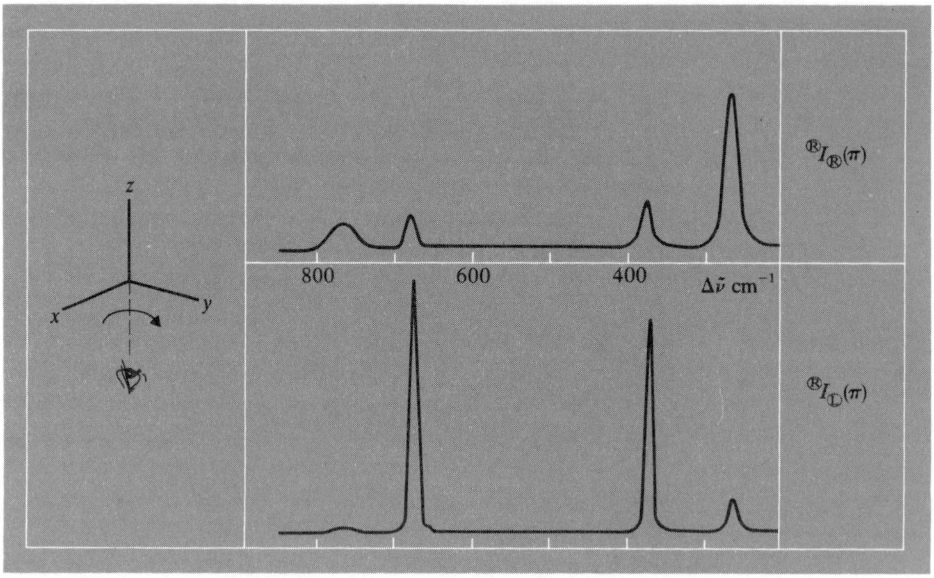

Fig. 6.11 Stokes Raman spectra of $CHCl_3$ ($\Delta\tilde{\nu}$, 200–800 cm^{-1}) $^{®}I_{®}(\pi)$ and $^{®}I_{Ⓛ}(\pi)$

incident on the photomultiplier is modulated. This is readily achieved by chopping the radiation with a mechanical chopper. A typical system would utilise, say, a ten-blade chopper rotating at 50 revolutions per second to impose a 500 cycle modulation on the radiation. The current in the load resistor R then has a frequency of 500 Hz and may be amplified with an a.c. amplifier tuned to respond only to $500 \pm \Delta$ Hz. Rectification is achieved by mixing (locking-in) the amplified a.c. signal with a fixed amplitude reference signal of the same frequency and correct phase. This reference signal is derived from the mechanical chopper system. Such a lock-in system also gives a significant improvement in signal relative to noise. The noise signal covers a wide frequency range, but only that small portion around 500 Hz will be processed by the system, whereas all the true signal will be passed on. It is particularly useful when the radiation falling on the entrance slit of the spectrometer includes some radiation which does not arise from scattering of light by the sample. For example, in the measurement of Raman scattering from a sample at a high temperature, thermal radiation from the hot sample as well as scattered radiation can reach the slit. If, however, the exciting laser radiation is modulated at, say, 500 Hz before it is incident on the sample, the Rayleigh and Raman scattering will also be modulated at the same frequency, but the unwanted thermal radiation will not. A lock-in or phase-sensitive system will then discriminate effectively against the unwanted signal produced by the thermal radiation.

Multichannel detection devices are now beginning to be used in Raman spectroscopy particularly in situations where signal averaging is required. Multichannel detection is considered briefly in chapter 8 page 225.

6.6 The measurement of polarization properties of Raman scattering

Many examples of Raman spectra from a variety of samples, under a wide range of conditions recorded in different ways, will be found in chapter 7. We conclude this chapter by presenting examples of the measurement of polarization properties of Raman scattering.

Figure 6.9 illustrates the measurement of $\rho_\perp(\pi/2)$ and $\rho_n(\pi/2)$ for selected fundamental vibrations of CCl_4 and CH_2Cl_2. For CCl_4 we see that for the spherically symmetric $\nu_1(a_1)$ vibration for which $(\gamma')_1 = 0$, the depolarization ratios tend to zero, whereas for the degenerate vibration $\nu_4(f_2)$ for which $(a')_4 = 0$, $\rho_\perp(\pi/2) = \frac{3}{4}$ and $\rho_n(\pi/2) = \frac{6}{7}$. For the $\nu_4(a_1)$ vibration of CH_2Cl_2 for which $(a')_4 \neq 0$ and $(\gamma')_4 \neq 0$, $0 < \rho_\perp(\pi/2) < \frac{3}{4}$ and $0 < \rho_n(\pi/2) < \frac{6}{7}$.

Figures 6.10 and 6.11 illustrate Raman spectra obtained with incident circularly polarized radiation. From Fig. 6.10 we see that for CCl_4, ${}^{\circledR}I_{\circledR}(\pi) - {}^{\circledR}I_{\copyright}(\pi)$ is negative for $\nu_1(a_1)$ for which $(\gamma')_1 = 0$ and positive for the other vibrations for which $(a') = 0$. From Fig. 6.11 we see that for $CHCl_3$, ${}^{\circledR}I_{\copyright}(\pi) > {}^{\circledR}I_{\circledR}(\pi)$ for the a_1 vibrations ($\tilde{\nu}_3 = 364$ cm^{-1} and $\tilde{\nu}_2 = 667$ cm^{-1}) whereas ${}^{\circledR}I_{\circledR}(\pi) > {}^{\circledR}I_{\copyright}(\pi)$ for the e vibrations ($\tilde{\nu}_6 = 260$ cm^{-1} and $\tilde{\nu}_5 = 760$ cm^{-1}).

7 Some examples of the application of Raman spectroscopy

"You ain't heard nothin' yet folks."

Al Jolson

7.1 Introduction

The applications of Raman spectroscopy are widespread and range over the chemical, physical, biological, and medical sciences. Such applications vary from the purely qualitative to the highly quantitative. Quite often, the Raman spectrum is used simply to identify a chemical species. In such cases, the fact that 'each different scattering molecular species gives its own characteristic Raman spectrum—writes (as it were) its autograph . . .'[1] is utilized. In other cases, Raman spectroscopy, usually in conjunction with infrared spectroscopy, is used to deduce the symmetry of the scattering species and to assign wavenumbers to modes of vibration. Here, use is made of the relation between molecular symmetry, selection rules, and band contours (in gases), or polarization characteristics (in gases and liquids), or the orientation dependence of scattered intensity (in single crystals). In favourable instances, vibrational assignments enable quantitative information about intramolecular and intermolecular forces to be obtained and thermodynamic functions to be calculated. A number of other applications depend on the use of vibrational Raman intensities to measure the concentration of scattering species. In still other applications, changes in the wavenumber, intensity, and band contour of Raman bands may be used to study relaxation phenomena and the effect of environment, temperature, and pressure on chemical species, and, in selected cases, resonance effects may be invoked to enhance the sensitivity of the Raman scattering to such factors. Continuing the imagery introduced above, we may say that it is possible to 'deduce character from the handwriting'! Also bond lengths and other structural parameters in the ground and excited vibrational states of many molecules may be obtained with good precision from the quantitative analysis of pure rotation and vibration–rotation Raman spectra of gases. Such measurements are particularly useful in the case of non-polar molecules which do not normally exhibit rotation spectra in absorption.

The early classical work in Raman spectroscopy, utilizing mercury arc excitation and photographic detection, contains examples of many of the applications broadly outlined above. However, the systems that could be studied had to be chosen carefully to be amenable to investigation with the experimental techniques then available. While the availability of laser sources and photoelectric detection has opened up some quite new applications of Raman spectroscopy, the main consequence of these new experimental techniques is that now the Raman spectrum of almost any type of material under almost any physical conditions may be obtained. The potential of this form of spectroscopy can therefore be very much more widely exploited. The material to be studied does not now need to be virtually colourless or of good optical quality, and only very small quantities are necessary; the sample may be in any physical state and may be investigated under a wide range of temperature and pressure; small wavenumber shifts may be measured with the same facility as large wavenumber shifts, and even weak scattering is unlikely to escape detection. Of particular importance is that the wavenumber shift is no longer the only characteristic parameter of a Raman spectrum which may be measured with relative ease; the intensity and contour of a Raman band, and its state of polarization, are now also susceptible to precise measurement, and, in the case of crystals, fruitful orientation studies may be made relatively easily. Also, because of the availability of a wide range of exciting frequencies from laser sources, resonance effects may be more fully utilized. For example, applications to a wide range of problems in biological systems have now become possible. In the investigation of the fine structure of rotation and vibration–rotation Raman bands, the narrow line widths of laser sources have removed the severe restrictions which were imposed by the substantial width of the mercury arc lines.

The many examples which follow have been chosen to illustrate this wide applicability of Raman spectroscopy, both as regards the nature of the samples that can be studied and the information that can be obtained. They also show some of the differences between Raman spectroscopy and infrared spectroscopy. For example, Raman spectroscopy may be used for the study of aqueous solutions over the whole range of vibrational wavenumbers, whereas infrared spectroscopy is restricted to certain wavenumber ranges; Raman spectroscopy yields bands of low wavenumber as readily as bands of high wavenumber, in contrast to infrared spectroscopy; Raman bands are often sharper than infrared bands; Raman spectra contain fewer overtone and combination bands than infrared spectra; and the intensity of Raman scattering is normally linearly related to the concentration of the scattering species, whereas in infrared spectra the absorption is a logarithmic function of the concentration.

The examples chosen for discussion do not represent every facet of Raman spectroscopy; the body of published work is too considerable. The reader should look upon what follows as an aperitif and turn to the guided reading list in Appendix I for more solid fare.

7.2 Rotation and vibration–rotation Raman spectra

7.2.1 *Introduction*

The study of rotation and vibration–rotation Raman spectra leads to precise information concerning molecular parameters. Laser sources have greatly extended the range of systems that can be studied. Only very small amounts of a material are now required, and so isotopically substituted molecules can be much more easily investigated. Also, the spectra from relatively heavy molecules with small interline spacings can now be resolved.

In the main, such spectra are recorded photographically using specially designed grating spectrographs of high resolving power. Fabry–Perot interferometers are also used, sometimes in conjunction with a spectrograph.

7.2.2 *Rotational Raman spectra of diatomic molecules*

The pure rotation Raman spectra of the isotopically pure homonuclear diatomic molecules $^{18}O_2$, $^{14}N_2$, $^{19}F_2$, and $^{35}Cl_2$ are shown in Figs. 7.1,[2] 7.2,[3] 7.3,[4] and 7.4,[5] respectively. The experimental conditions, which vary from spectrum to spectrum, are summarized in the figures. We shall consider how these spectra may be analysed, first for nuclear spin information and then for bond lengths.

Table 4.1 shows that if the nuclear spin is zero, then every other rotation line will be missing, whereas for other nuclear spin values, an intensity alternation will be observed. The spectrum of $^{19}F_2$ shows clearly that the intensities of lines of even and

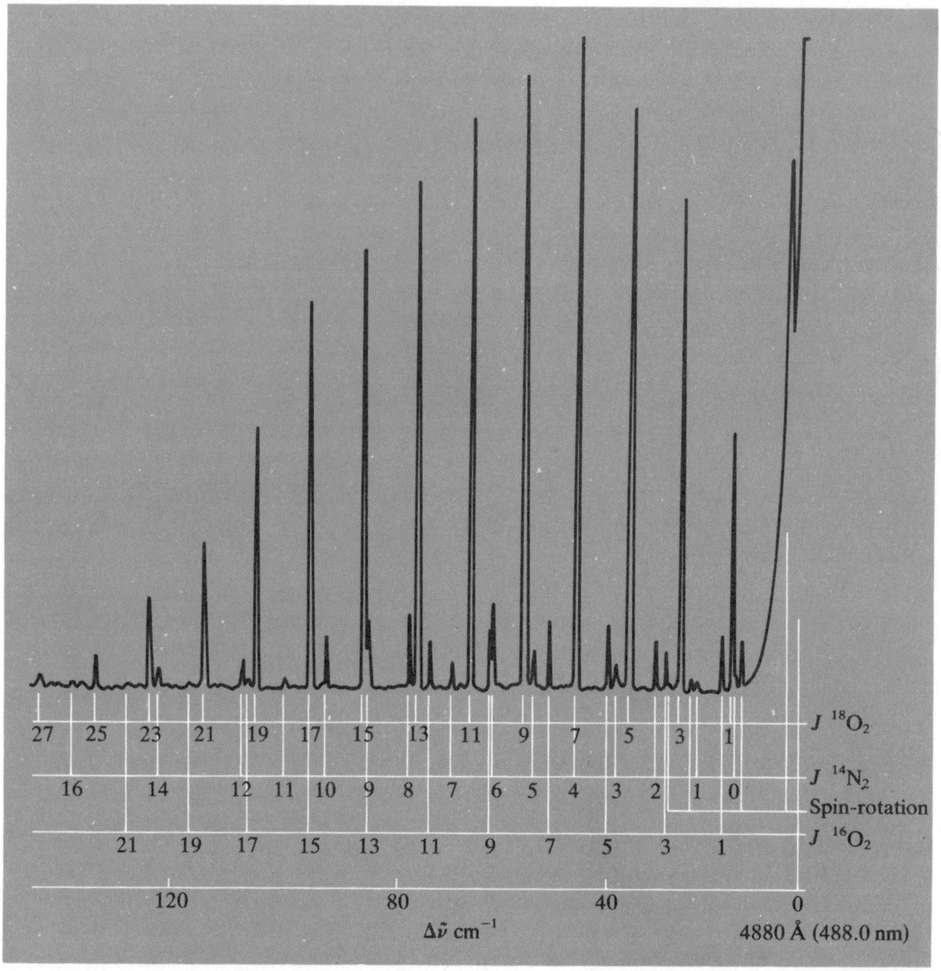

Fig. 7.1 A microdensitometer trace of the pure rotation Raman spectrum (Stokes region) of $^{18}O_2$, pressure 760 torr, temperature 288 K, 4880 Å (488·0 nm) excitation

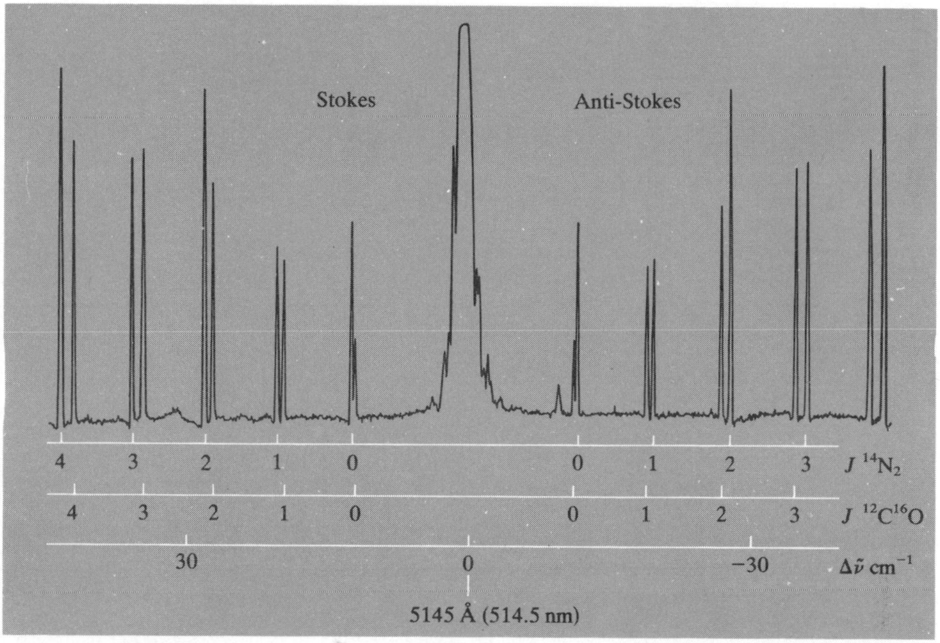

Fig. 7.2 A microdensitometer trace of part of the pure rotation Raman spectrum of $^{14}N_2$, pressure 30 torr, temperature 288 K, 5145 Å (514·5 nm) excitation. The spectrum contains part of the pure rotation Raman spectrum of $^{12}C^{16}O$ (pressure 150 torr) which was used as an internal wavenumber calibrant

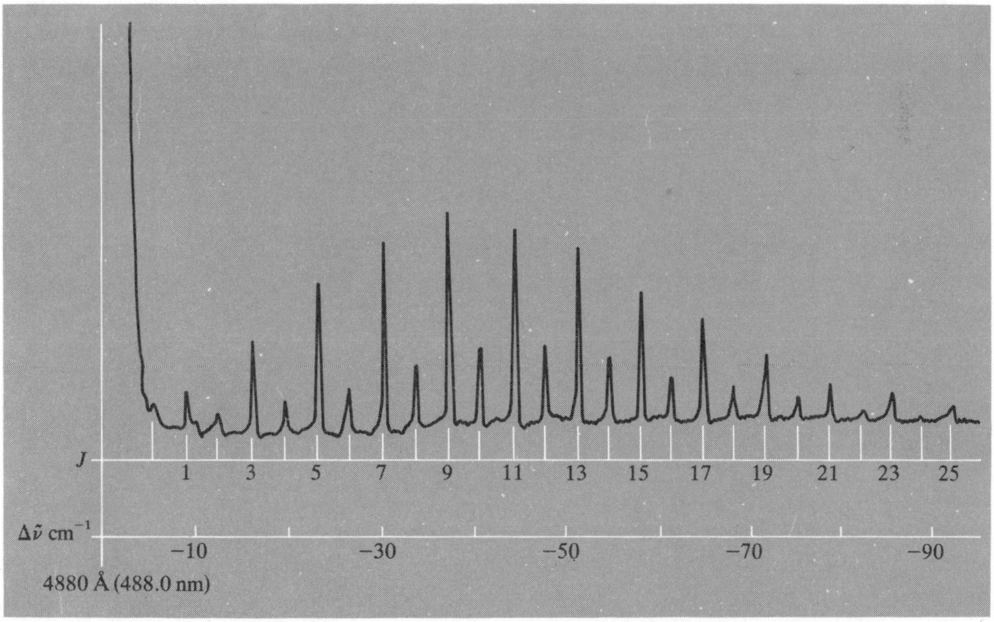

Fig. 7.3 Microdensitometer trace of the pure rotation Raman spectrum (anti-Stokes region) of $^{19}F_2$, pressure 550 torr, temperature 288 K, 4880 Å (488·0 nm) excitation

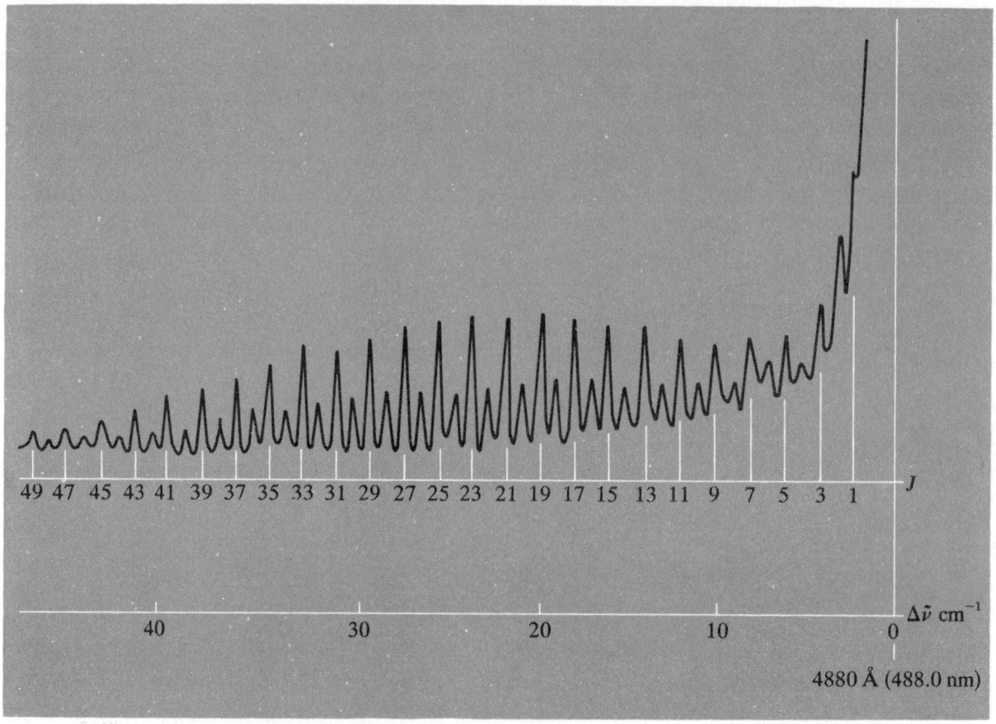

Fig. 7.4 Microdensitometer trace of the pure rotation Raman spectrum (Stokes region) of $^{35}Cl_2$, pressure 600 torr, temperature 288 K, 4880 Å (488·0 nm) excitation

odd J values are in the ratio 1:3. Thus, the ^{19}F nucleus has a nuclear spin of $\frac{1}{2}$ and obeys Fermi statistics. In the spectrum of $^{14}N_2$ the intensities of lines of even and odd J values are in the ratio 2:1, and thus the ^{14}N nucleus has a spin of 1 and obeys Bose statistics. The spectrum of $^{35}Cl_2$ shows that the intensities of lines of odd and even J values are in the ratio 5:3, and hence the ^{35}Cl nucleus has a spin of $\frac{3}{2}$ and obeys Fermi statistics. The spectrum of $^{18}O_2$ does not appear to show an intensity alternation, but, if the spectrum is analysed on the basis of every possible rotation line being present, the B value is about twice that expected, suggesting that alternate lines are missing. The ratio of the wavenumber separation of the first rotation lines in the Stokes and anti-Stokes regions to the wavenumber separation of successive rotation lines in either the Stokes or anti-Stokes region may be used to determine if alternate lines are missing. If no lines are absent, the two separations are, in the rigid rotor approximation, $12B$ and $4B$, and stand in the ratio 6:2. If lines of odd J values are missing, the two separations are $12B$ and $8B$, and stand in the ratio 3:2. If lines of even J values are missing, the two separations are $20B$ and $8B$, and stand in the ratio 5:2. For $^{18}O_2$, the ratio is 5:2, and thus even J values are absent, the nuclear spin of ^{18}O is zero, and the nuclei obey Bose statistics.

As an example of an analysis of such spectra to give bond lengths, we shall consider the case of $^{18}O_2$. Figure 7.5 shows a plot of $|\Delta\tilde{\nu}|/(J+\frac{3}{2})$ against $(J+\frac{3}{2})^2$ for this molecule. It can be seen from eq. (4.106) that such a plot should be linear, with the

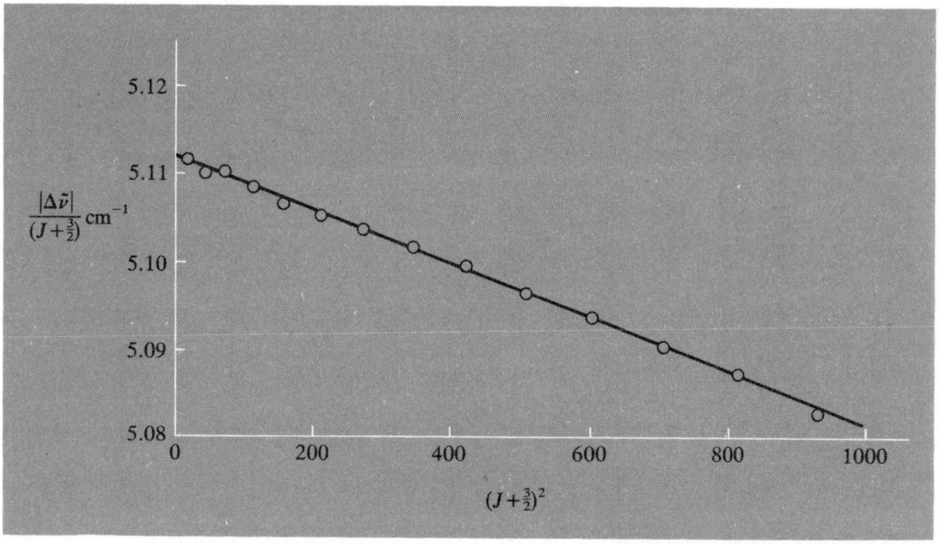

Fig. 7.5 Plot of $|\Delta\tilde{\nu}|/(J+\frac{3}{2})$ against $(J+\frac{3}{2})^2$ for $^{18}O_2$

intercept on the ordinate axis equal to $(4B_0 - 6D_0)$ and the slope equal to $-8D_0$. A least squares analysis of the data[2] gave $B_0 = 1{\cdot}278\,03 \pm 0{\cdot}000\,05$ cm^{-1} and $10^6 D_0 = 4{\cdot}7 \pm 0{\cdot}2$ cm^{-1}. The value of r_0 calculated from B_0, using eq. (4.98), is $1{\cdot}210\,64 \pm 0{\cdot}000\,04$ Å.

Combination of this r_0 value for $^{18}O_2$ with the r_0 values of $^{16}O_2$ and $^{16}O^{18}O$ obtained from similar analyses of their pure rotation Raman spectra gives $r_e = 1{\cdot}207\,4 \pm 0{\cdot}000\,1$ Å. This is within one standard deviation of the microwave result, but oxygen is a favourable case since it is a relatively light molecule, and so has large interline spacings. (The $^3\Sigma$ electronic ground state of $^{16}O_2$ allows magnetic dipole rotational transitions which fall in the microwave region.)

Similar analyses yield the following parameters for $^{14}N_2$, $^{19}F_2$ and $^{35}Cl_2$:

	B_0/cm^{-1}	$10^6 D_0$/cm^{-1}	r_0/Å
$^{14}N_2$	$1{\cdot}989\,57 \pm 0{\cdot}000\,01$	$5{\cdot}76 \pm 0{\cdot}03$	$1{\cdot}100\,873 \pm 0{\cdot}000\,007$
$^{19}F_2$	$0{\cdot}883\,31 \pm 0{\cdot}000\,04$	$3{\cdot}48 \pm 0{\cdot}06$	$1{\cdot}417\,44 \pm 0{\cdot}000\,06$
$^{35}Cl_2$	$0{\cdot}243\,10 \pm 0{\cdot}000\,02$	$1{\cdot}43 \pm 0{\cdot}08$	$1{\cdot}991\,5 \pm 0{\cdot}000\,1$

Similar precision in the determination of the rotational constants from Raman spectra can be achieved using a Fabry–Perot étalon with a medium-resolution spectrograph. The results obtained[6] in this way for $^{16}O_2$ are: $B_0 = 1{\cdot}437\,682 \pm 0{\cdot}000\,009$ cm^{-1}, $10^6\, D_0 = 4{\cdot}852 \pm 0{\cdot}012$ cm^{-1}, and $r_0 = 1{\cdot}210\,85 \pm 0{\cdot}000\,03$ Å.

The pure rotational Raman spectrum of natural sulphur dimer, S_2, at 900 K is shown in Fig. 7.6. Analysis of this spectrum[7] gave $B_0 = 0{\cdot}294\,43 \pm 0{\cdot}000\,05$ cm^{-1} and $10^8 D_0 = 19{\cdot}0 \pm 0{\cdot}5$ cm^{-1}. This is an interesting illustration of the possibility of studying species at high temperatures using Raman spectroscopy.

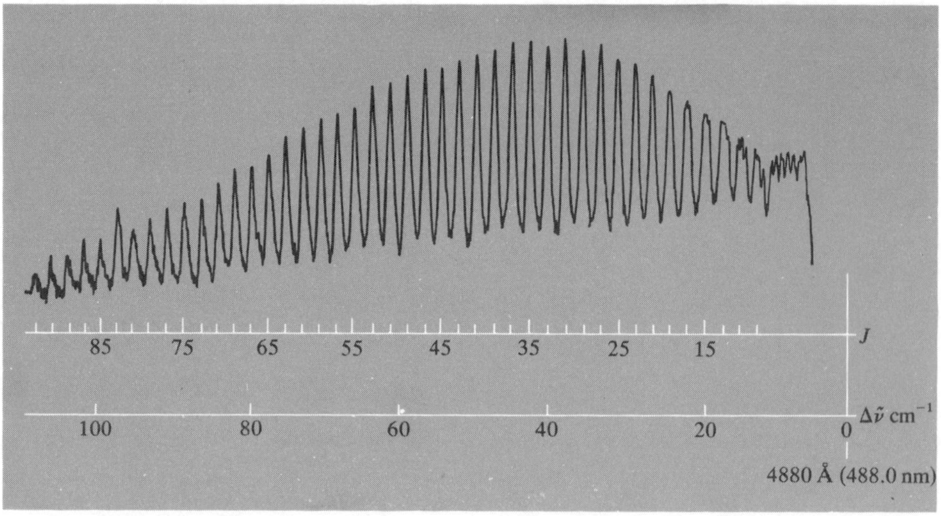

Fig. 7.6 Microdensitometer trace of the pure rotation Raman spectrum (Stokes region) of S_2 pressure 38 torr, temperature 900 K, 4880 Å (488·0 nm) excitation

7.2.3 *Rotational Raman spectra of linear molecules*

The pure rotation Stokes Raman spectrum[8] of the linear molecule $^{16}O^{12}C^{16}O$ is shown in Fig. 7.7. In the ground vibrational state, the nuclear statistics require lines of odd J to be missing, and the interline separation is thus $8B_0$. The strong lines in the spectrum in Fig. 7.7 are fully consistent with this. However, the spectrum also shows

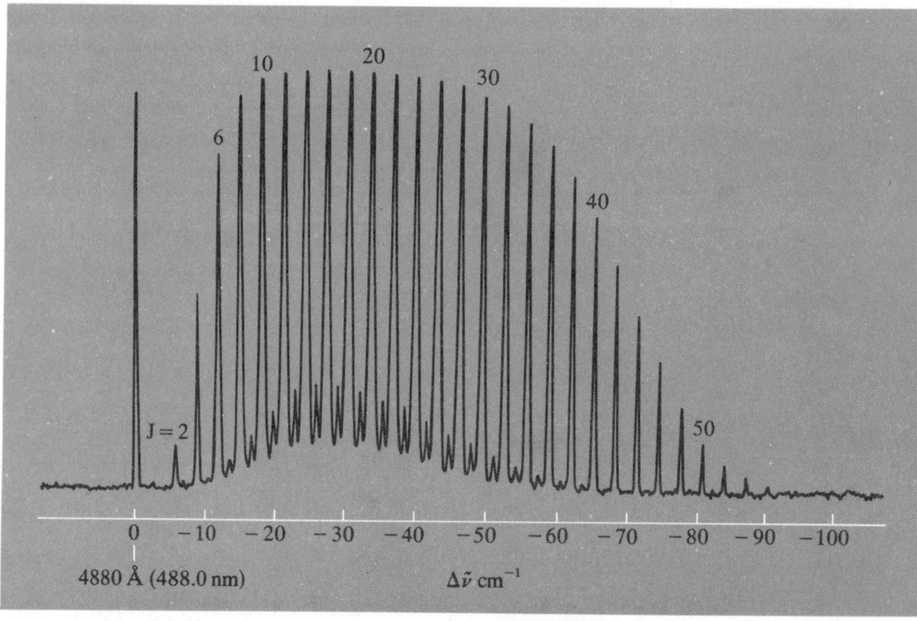

Fig. 7.7 Microdensitometer trace of the pure rotation Raman spectrum (anti-Stokes region) of $^{12}C^{16}O_2$, pressure 760 torr, temperature 298 K, 4880 Å (488·0 nm) excitation

much weaker lines at odd J. These lines are associated with the pure rotation spectrum of CO_2 in a thermally populated excited vibration state. The vibrational mode in question is the bending vibration ($\tilde{\nu}_M = 667\ cm^{-1}$) and in this vibrational state all rotational states have the same nuclear spin weighting; thus, rotational Raman lines associated with both even and odd J values will be observed with an interline spacing of approximately $4B$. The lines associated with even J values coincide with the strong lines arising from the CO_2 molecule in its ground vibrational state.

Another example of the pure rotational Raman spectrum of a linear molecule is shown in Fig. 7.8, where the spectrum of $^{12}C_2{}^{14}N_2$ is presented.[9] The intensity

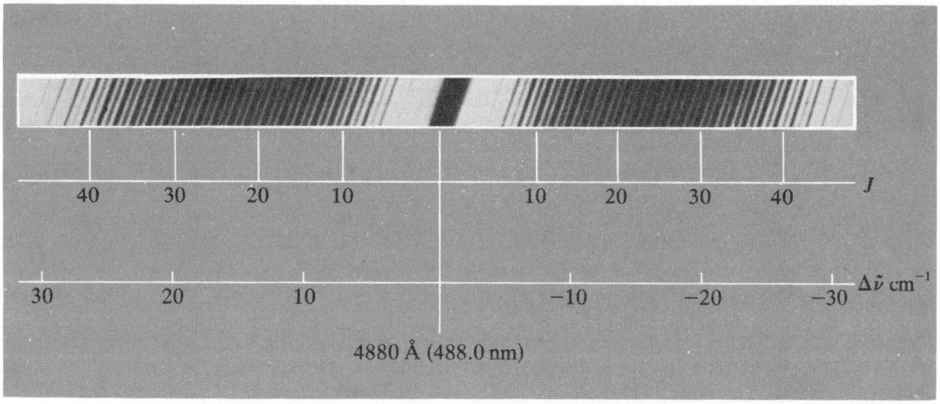

Fig. 7.8 The pure rotation Raman spectrum of $^{12}C_2{}^{14}N_2$, pressure 380 torr, temperature 288 K, 4880 Å (488·0 nm) excitation

alternation is consistent with a nuclear spin of 1 for ^{14}N. Analysis of the rotational Raman spectra of $^{12}C_2{}^{14}N_2$ and $^{12}C_2{}^{15}N_2$ leads to the following values of the bond lengths: r_0(C—C) = 1·40 ± 0·03 Å and r_0(C—N) = 1·15 ± 0·02 Å. The precision of these results is somewhat lower since vibrational levels associated with bending modes are substantially populated at normal temperatures.

7.2.4 *Rotational Raman spectra of symmetric top molecules*

The pure rotation Raman spectrum of benzene, C_6H_6, provides a good example of the study of a symmetric top molecule. The spectrum in Fig. 7.9(a) was obtained using laser excitation at $\lambda = 4880$ Å (488·0 nm), and the R branch lines ($\Delta J = 1, \Delta K = 0$) and the S branch lines ($\Delta J = 2, \Delta K = 0$) are clearly resolved. To illustrate the improvement in the quality of the spectra obtained with laser excitation, Fig. 7.9(b) shows the pure rotation Raman spectrum of C_6H_6 obtained using mercury arc excitation.[10] All the R branch lines and S branch lines out to $J = 30$ are obscured. For a symmetric top molecule, a plot of $|\Delta\tilde{\nu}|/(J+\frac{3}{2})$ against $(J+\frac{3}{2})^2$ should be linear with an intercept on the ordinate axis of $(4B_0 - 6D_J - 4D_{JK}K^2)$ and a slope of $-8D_J$. If D_{JK} is neglected, which is justifiable since the spectra show no evidence of K dependence, then B_0 and D_J may be evaluated. In this way, the following values of B_0 and I_{B_0} were obtained for C_6H_6 and C_6D_6, and *sym*-$C_6H_3D_3$:

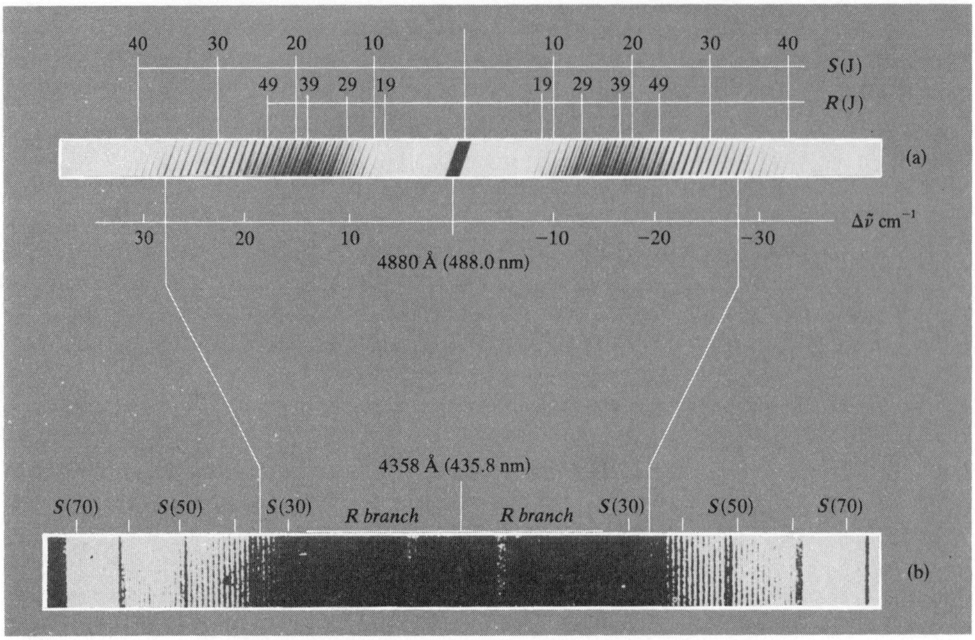

Fig. 7.9 Photographically recorded pure rotation Raman spectrum of C_6H_6: (a) pressure 70 torr, temperature 288 K, 4880 Å (488·0 nm) excitation; and (b) pressure 380 torr, temperature 333 K, 4358 Å (435·8 nm) excitation

	B_0/cm^{-1}	$10^{40}I_{B_0}$/g cm^2
C_6H_6	0·189 50±0·000 01	147·609±0·008
C_6D_6	0·156 85±0·000 02	178·36±0·02
$C_6H_3D_3$	0·171 77±0·000 01	162·97±0·01

On the assumption that the benzene molecule is planar and the carbon ring is a regular hexagon (D_{6h} symmetry), I_{B_0}, the moment of inertia about any axis in the plane of the molecule and passing through its centre, may be related to r_0(C—H) (and/or r_0(C—D)) and r_0(C—C). For example, for C_6H_6,

$$I_{B_0} = 3\{m_C r^2(\text{C—C}) + m_H[r(\text{C—C}) + r(\text{C—H})]^2\} \tag{7.1}$$

Assuming that r_0(C—H) = r_0(C—D), the experimentally determined values of I_{B_0} for C_6H_6, *sym*-$C_6H_3D_3$, and C_6D_6 enable values of r_0(C—C) and r_0(C—H) to be calculated. The best values were[11] r_0(C—C) = 1·397 9±0·000 2 Å and r_0(C—H) = 1·079±0·001 Å.

7.2.5 *Vibration–rotation Raman spectra*

Figure 7.10 shows the $Q(\Delta J = 0)$ branch of the $1 \leftarrow 0$ vibrational transition in the Raman spectrum of tritium, $^3H_2$. Seven lines corresponding to $J = 0$ to $J = 6$ were observed. The intensities of lines of even and odd J values are in the ratio 1:3 and show that the ^{3}H nucleus has a spin of $\frac{1}{2}$ and obeys Fermi statistics. When eq. (4.114)

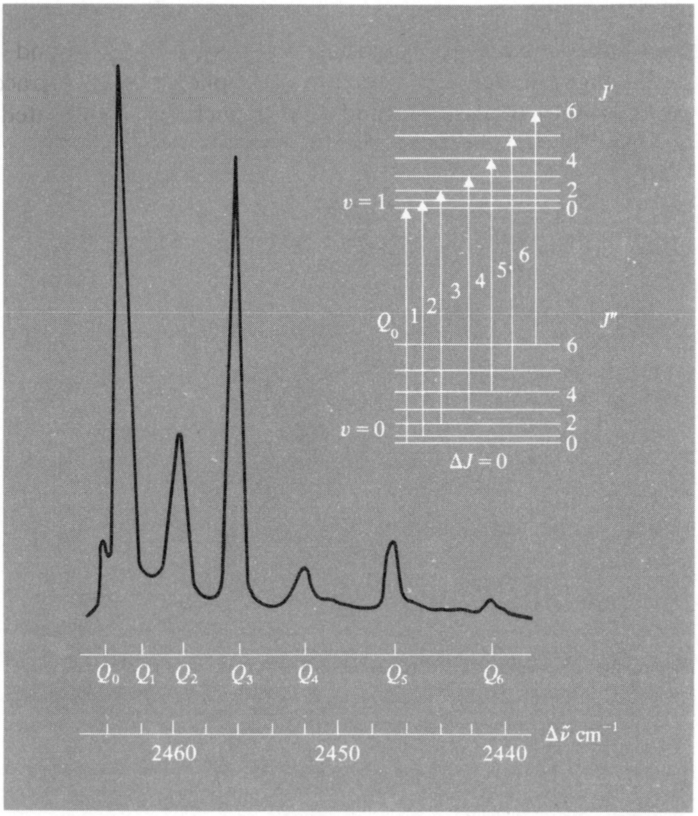

Fig. 7.10 Photoelectrically recorded vibration–rotation Raman spectrum of T_2; the Q branch of the $v=1 \leftarrow v=0$ transition; pressure 660 torr, temperature 288 K, 4880 Å (488·0 nm) excitation

is modified to take into account centrifugal stretching, the wavenumber shifts of the Q branch are given by

$$|\Delta\tilde{\nu}_Q| = \tilde{\nu}_0 + (B_1 - B_0)J(J+1) - (D_1 - D_0)J^2(J+1)^2 \tag{7.2}$$

Using B_0 and D_0 values obtained from analysis of the pure rotation Raman spectrum, the Q branch wavenumber shifts enable values of B_1, D_1, and $\tilde{\nu}_0$ to be determined. From eq. (4.109),

$$B_v = B_e - \alpha_e(v + \tfrac{1}{2}) \tag{7.3}$$

and, similarly,

$$D_v = D_0 + \beta_e(v + \tfrac{1}{2}) \tag{7.4}$$

Thus, B_e (and hence r_e), α_e, D_v, β_e, and D_0 values can be calculated. The values obtained[12] for tritium are

$$B_e = 20{\cdot}331\,4 \pm 0{\cdot}000\,6 \text{ cm}^{-1}$$

$$\alpha_e = 0{\cdot}585\,6 \pm 0{\cdot}000\,2 \text{ cm}^{-1}$$

$$10^3\beta_e = -0{\cdot}294 \pm 0{\cdot}007 \text{ cm}^{-1}$$

$$r_e = 0{\cdot}741\,51 \pm 0{\cdot}000\,02 \text{ Å}$$

The O, Q, and S rotational branches associated with the $1 \leftarrow 0$ vibrational transition in the Raman spectrum of $^{14}N_2$, $^{14}N^{15}N$, and $^{15}N_2$ have been recorded and measured. Figure 7.11 shows a microphotometer trace of the Q branch and part ($J=0$ to $J=6$) of the O and S branches associated with the $1 \leftarrow 0$ vibrational transition in $^{14}N_2$; in fact, in the O and S branches, lines out to $J=17$ were observed.

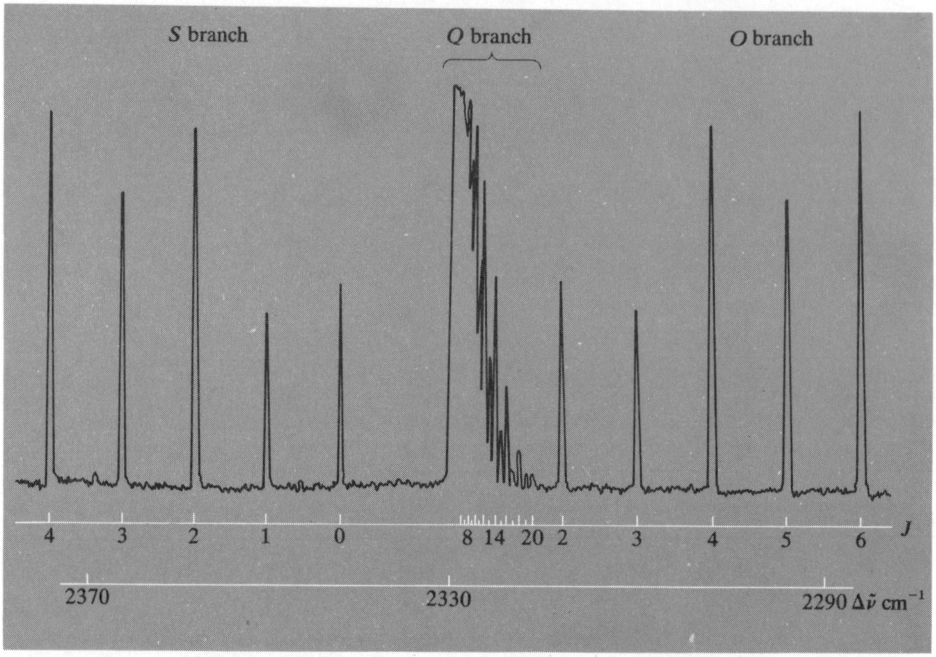

Fig. 7.11 Microdensitometer trace of part of the vibration–rotation Raman spectrum of $^{14}N_2$; *Q* branch and part of the *O* and *S* branches for the $v=1 \leftarrow v=0$ transition; pressure 760 torr, temperature 288 K, 4880 Å (488.0 nm) excitation

From a detailed analysis[3] of the vibration–rotation and pure rotation spectra of these molecules, the following values of B_0, D_0, (B_0-B_1), $\tilde{\nu}_0$, B_e, and r_e were obtained:

	$^{14}N_2$	$^{14}N^{15}N$	$^{15}N_2$
B_0/cm^{-1}	1·989 57±0·000 01	1·923 596±0·000 009	1·857 62±0·000 02
$10^6 D_0/\text{cm}^{-1}$	5·76±0·03	5·38±0·03	5·08±0·03
$10^2(B_0-B_1)/\text{cm}^{-1}$	1·738 4±0·000 3	1·650 8±0·000 2	1·566 7±0·000 2
$\tilde{\nu}_0/\text{cm}^{-1}$	2 329·916 8±0·000 3	2 291·332 4±0·000 4	2 252·124 9±0·000 3
B_e/cm^{-1}	1·998 23±0·000 01	1·931 816±0·000 009	1·865 42±0·000 02
$r_e/\text{Å}$	1·097 700±0·000 007	1·097 702±0·000 004	1·097 700±0·000 008

Within the experimental accuracy, the equilibrium bond lengths are identical in the three isotopic species as required by the Born–Oppenheimer approximation. The precision of these determinations is noteworthy.

7.2.6 *Rotational Raman spectra of asymmetric tops*

The rotational Raman spectrum of the asymmetric top molecule ethylene, C_2H_4, is shown in Fig. 7.12. The complexity of the spectrum is evident. A detailed analysis[13]

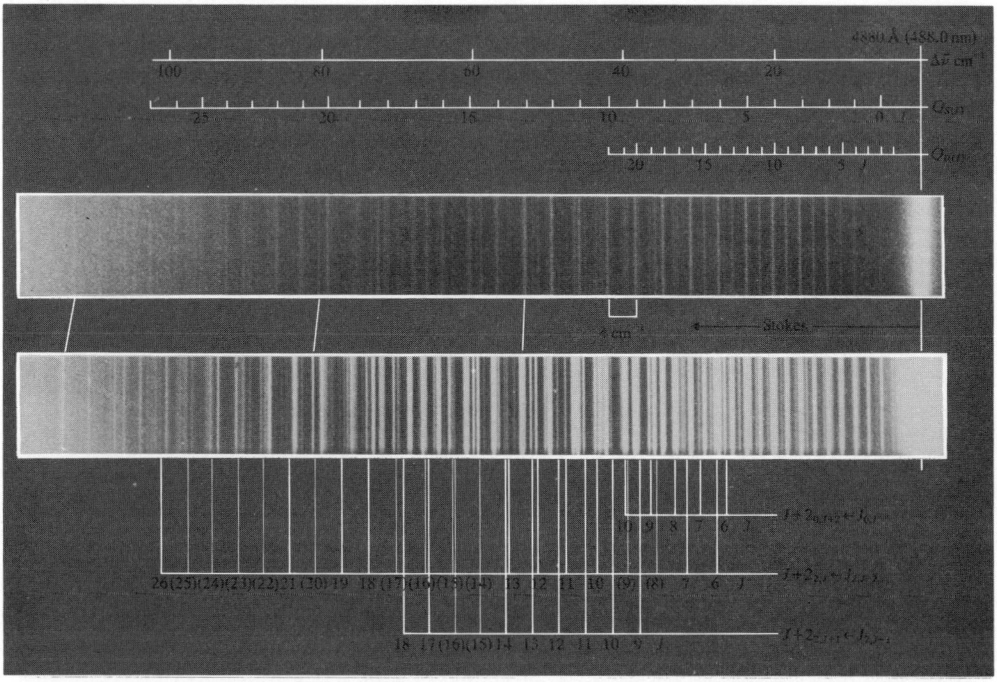

Fig. 7.12 Photographically recorded Stokes rotation Raman spectrum of C_2H_4, temperature 290 K, pressure 380 torr; upper spectrum, 2 h exposure, 4880 Å (488·0 nm) excitation; S and R branches assigned; lower spectrum, 5 h exposure, 5145 Å (514·5 nm) excitation; some single transitions assigned.

allowed the determination of all three rotational constants. The values obtained were $A_0 = 4{\cdot}866_6 \pm 0{\cdot}004_0\ \text{cm}^{-1}$, $B_0 = 1{\cdot}000\ 7_6 \pm 0{\cdot}000\ 1_3\ \text{cm}^{-1}$, $C_0 = 0{\cdot}828\ 4_8 \pm 0{\cdot}000\ 1_6\ \text{cm}^{-1}$.

7.3 Organic chemistry

The principal application of vibrational spectroscopy to organic chemistry, where, in the main, relatively large molecules are involved, is in the identification of particular structural features or characteristic groups. The use of infrared spectroscopy for this purpose has, for some time, been an almost routine procedure for the organic chemist. However, as has been stressed so often, different selection rules operate for infrared and Raman spectroscopy, and so a much more complete knowledge of the vibrational spectrum of a molecule is likely to result, if both the infrared and Raman spectra are measured. While the infrared spectrum alone may suffice to solve a particular problem, now that experimental techniques permit, it is highly desirable that the organic chemist should form the habit of measuring both infrared and Raman spectra. The additional and complementary information thus made available can often lead to the solution of structural problems that would not be amenable merely to infrared spectroscopy. Also, the pitfalls which, not infrequently, can result from deductions based on the infrared spectrum alone can be avoided. The following brief account[14] of some applications of Raman spectroscopy to organic chemistry will include specific illustrations of these points.

Table 7.1 Characteristic wavenumbers and Raman and infrared intensities of groups in organic compounds

Vibration[a]	Region (cm^{-1})	Intensity[b] Raman	Infrared
ν(O—H)	3650—3000	w	s
ν(N—H)	3500—3300	m	m
ν(≡C—H)	3300	w	s
ν(=C—H)	3100—3000	s	m
ν(—C—H)	3000—2800	s	s
ν(—S—H)	2600—2550	s	w
ν(C≡N)	2255—2220	m—s	s—0
ν(C≡C)	2250—2100	vs	w—0
ν(C=O)	1820—1680	s—w	vs
ν(C=C)	1900—1500	vs—m	0—w
ν(C=N)	1680—1610	s	m
ν(N=N), aliphatic substituent	1580—1550	m	0
ν(N=N), aromatic substituent	1440—1410	m	0
ν_a((C—)NO_2)	1590—1530	m	s
ν_s((C—)NO_2)	1380—1340	vs	m
ν_a((C—)SO_2(—C))	1350—1310	w—0	s
ν_s((C—)SO_2(—C))	1160—1120	s	s
ν((C—)SO(—C))	1070—1020	m	s
ν(C=S)	1250—1000	s	w
$\delta(CH_2)$, $\delta_a(CH_3)$	1470—1400	m	m
$\delta_s(CH_3)$	1380	m—w, s, if at C=C	s—m
ν(CC), aromatics	1600, 1580	s—m	m—s
	1500, 1450	m—w	m—s
	1000	s (in mono-; *m*-; 1,3,5-derivatives)	0—w
ν(CC), alicyclics, and aliphatic chains	1300—600	s—m	m—w
ν_a(C—O—C)	1150—1060	w	s
ν_s(C—O—C)	970—800	s—m	w—0
ν_a(Si—O—Si)	1110—1000	w—0	vs
ν_s(Si—O—Si)	550—450	vs	w—0
ν(O—O)	900—845	s	0—w
ν(S—S)	550—430	s	0—w
ν(Se—Se)	330—290	s	0—w
ν(C(aromatic)—S)	1100—1080	s	s—m
ν(C(aliphatic)—S)	790—630	s	s—m
ν(C—Cl)	800—550	s	s
ν(C—Br)	700—500	s	s
ν(C—I)	660—480	s	s
δ_s(CC), aliphatic chains			
C_n, $n = 3 \ldots 12$	400—250	s—m	w—0
$n > 12$	$2495/n$		
Lattice vibrations in molecular crystals (librations and translational vibrations)	200–20	vs—0	s—0

[a] ν stretching vibration, δ bending vibration, ν_s symmetric vibration. ν_a antisymmetric vibration.
[b] vs very strong, s strong, m medium, w weak, 0 very weak or inactive.

Table 7.1 lists, for a number of characteristic groups in organic compounds, the typical wavenumbers and intensities with which they appear in infrared and Raman spectra. It is immediately apparent from this table that a knowledge of both the infrared and the Raman spectra is, in many cases, essential, if important structural features are not to be overlooked. This is further emphasized by a comparison of the infrared and Raman spectra of, for example, 1-cystine, which are shown in Fig. 7.13. In the region of 3000 cm^{-1}, the infrared spectrum is dominated by a very broad band arising from the stretching vibrations of the NH_3^+ group, whereas the Raman spectrum shows two, strong, sharp lines associated with the stretching vibrations of the CH and CH_2 groups. In the region of 1600 cm^{-1}, the infrared spectrum shows two strong bands around 1600 cm^{-1} associated with the NH_3^+ deformation vibration and the antisymmetric stretching vibration of the carboxylate group $>CO_2^-$. Although these two vibrations also appear in the Raman spectrum, they are very much weaker. Both the infrared and Raman spectra have a strong band at 1410 cm^{-1} arising from the symmetric stretching vibration of the carboxylate group. However, the strongest Raman band at 497 cm^{-1}, which is associated with the —S—S— stretching vibration, is only just apparent in the infrared spectrum.

The Raman spectrum is particularly informative about groups like $>C=C<$, —C≡C— and —N=N—. If, in a molecule, such a group has a symmetrical situation, its vibration will be Raman active only; and in many molecules in which the symmetry is formally absent, the situation of the group may be so close to being symmetrical that its vibration will be extremely weak in the infrared. On the other hand, if such a group vibration appears strongly in both the infrared and Raman

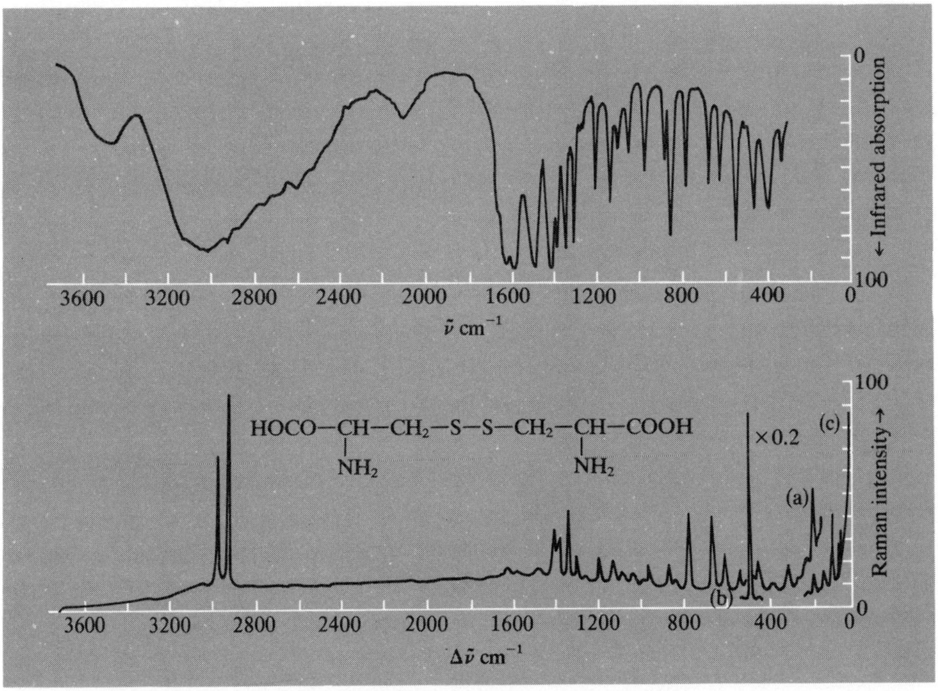

Fig. 7.13 Infrared and Raman spectra of crystalline cystine. Infrared spectrum of 0·8 mg in 400 mg KI; Raman spectrum (a), 2 mg, amplification ×1; (b), 2 mg, amplification ×0·2; and (c), disc, 140 mg, amplification ×0·3

spectra, this is evidence of a non-symmetric structure. Furthermore, since the wavenumber of these group vibrations is influenced by the extent of coupling of the stretching modes with other modes of similar wavenumber in the molecule, observed wavenumbers are not merely diagnostic of the presence of a particular group, they also provide valuable information about the structural environment of the group. Table 7.2, which lists the characteristic C=C stretching wavenumbers for a number of compounds, shows how the wavenumber associated with the C=C stretching mode varies from one type of molecule to another. A few specific examples will illustrate the points we have just made.

Table 7.2 Characteristic wavenumbers of C=C double bonds in hydrocarbons and halocarbons

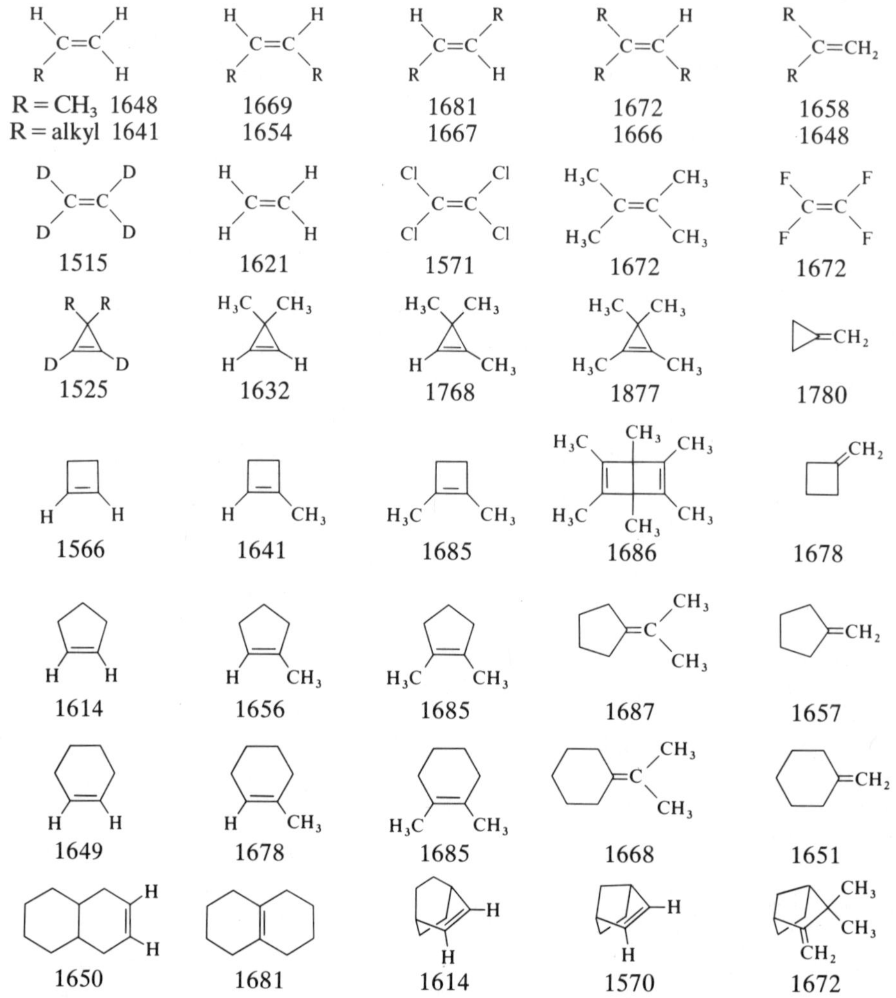

R = CH_3 1648
R = alkyl 1641
1669
1654
1681
1667
1672
1666
1658
1648
1515
1621
1571
1672
1672
1525
1632
1768
1877
1780
1566
1641
1685
1686
1678
1614
1656
1685
1687
1657
1649
1678
1685
1668
1651
1650
1681
1614
1570
1672

The Raman spectrum of the triphenylcyclopropenylium ion (Fig. 7.14) has a strong band at 1840 cm^{-1} associated with the symmetric ring stretching vibration

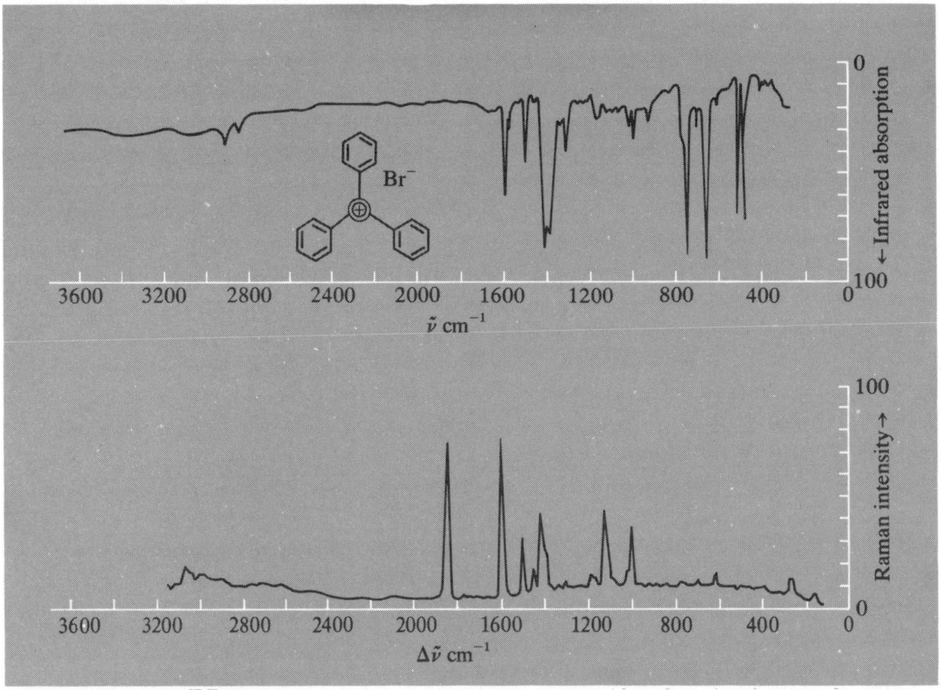

Fig. 7.14 Infrared and Raman spectra of crystalline triphenylcyclopropenylium bromide

which has substantial >C=C< character. However, because of the symmetry of this vibration, it is infrared inactive, as Fig. 7.14 shows.

The infrared and Raman spectra of a compound with the molecular formula $C_6H_6OCl_2$ are shown in Fig. 7.15. In the region 1500 to 1900 cm^{-1}, the infrared spectrum shows only one band at 1795 cm^{-1}, which is characteristic of an acid chloride group. The Raman spectrum, however, shows an additional band in this region at 1898 cm^{-1}, which is characteristic of a substituted cyclopropene derivative. On the basis of the combined evidence from the infrared and Raman spectra, the structure (I), given below, was proposed for the compound, and was subsequently confirmed by other evidence. It can be seen that in this molecule the >C=C< group is in a near symmetric situation, with the result that the vibration has such a weak infrared absorption that it is not observed.

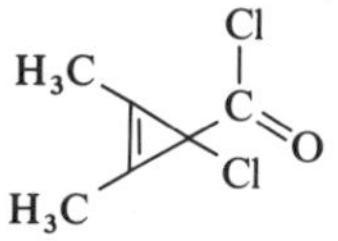

I

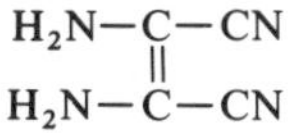

II

The Raman spectrum of $H_4C_4N_4$, the tetramer of hydrogen cyanide, has a strong band at 1621 cm^{-1} essentially coincident with a strong infrared band at 1623 cm^{-1}. The values of these wavenumbers and their observation as strong bands in both the infrared and Raman spectra provide good evidence[15] for the *cis*-ethylene structure (II). If the structure were *trans*, the >C=C< vibration would only be Raman active.

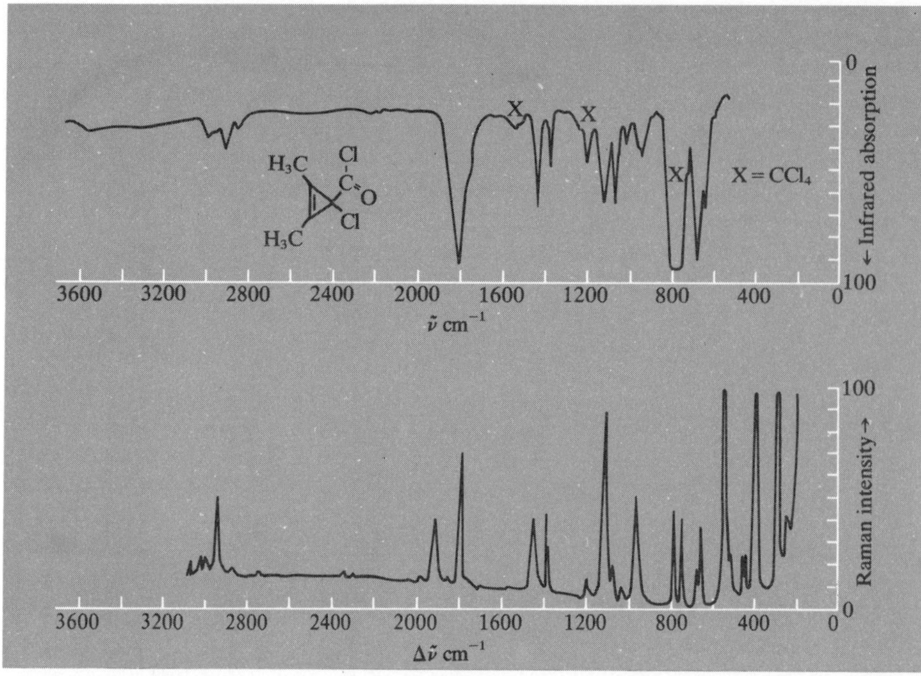

Fig. 7.15 Infrared and Raman spectra of 1-chloro-2, 3-dimethyl-2-cyclopropene-1-carbonyl chloride ($C_6H_6OCl_2$). Infrared spectrum of the solution in CCl_4, Raman spectrum of the liquid substance

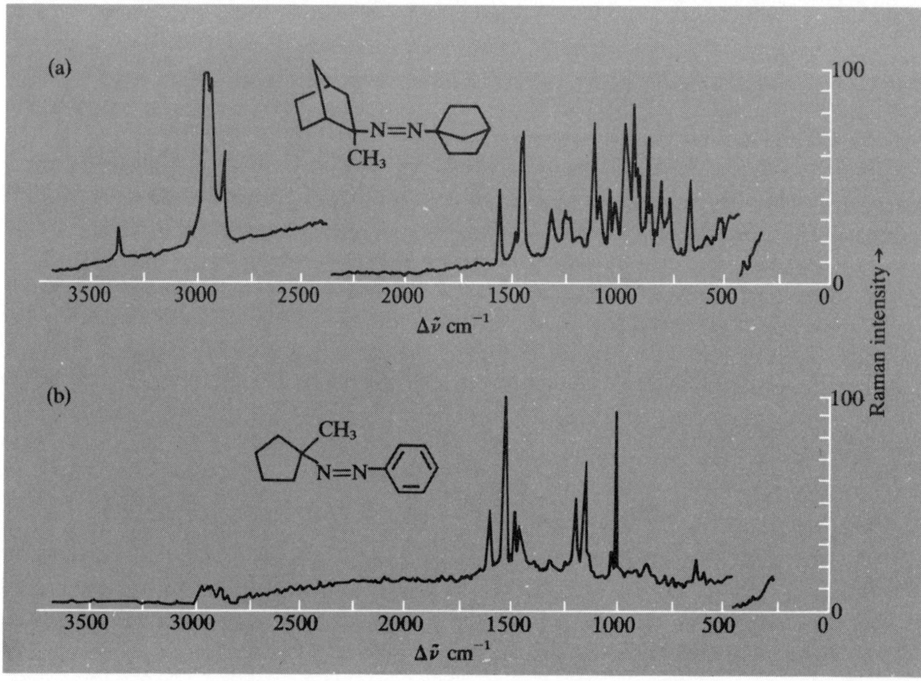

Fig. 7.16 Raman spectrum of (a) 2′-methyl-1,-*exo*-2′-azonorbornane and (b) (1-methylcyclopentylazo)benzene, crystalline samples

In azo compounds, the —N=N— vibration is forbidden in the infrared for *trans* structures and is not usually observed, because of its inherent weakness, for *cis* structures. By contrast, the —N=N— vibration is strong in the Raman spectrum for both *cis* and *trans* structures. The Raman spectra of two azo compounds are given in Fig. 7.16(a) and (b). Figure 7.16(a) shows the Raman spectrum of 2′-methyl-1,-*exo*-2′-azonorbornane, a mixed aromatic-aliphatic azo compound for which the —N=N— vibration occurs at 1520 cm^{-1}. Figure 7.16(b) shows the Raman spectrum of (1-methylcyclopentylazo)benzene, an aliphatic azo compound for which the —N=N— vibration occurs at 1570 cm^{-1}.

The infrared and Raman spectra of 2-acetoxypropionitrile (Fig. 7.17) illustrate particularly clearly the danger of relying only on infrared evidence. Although the polar —C≡N group normally has a strong characteristic absorption in the infrared in the 2100 to 2300 cm^{-1} region, the infrared spectrum of 2-acetoxypropionitrile shows no absorption characteristic of the —C≡N group. The Raman spectrum, however, has a strong —C≡N band at about 2250 cm^{-1}. In this molecule, the electronegative character of the acetoxy substituent on the carbon atom adjacent to the —C≡N group so reduces the magnitude of the dipole moment change in the —C≡N vibration that the infrared absorption is too weak to be observed.

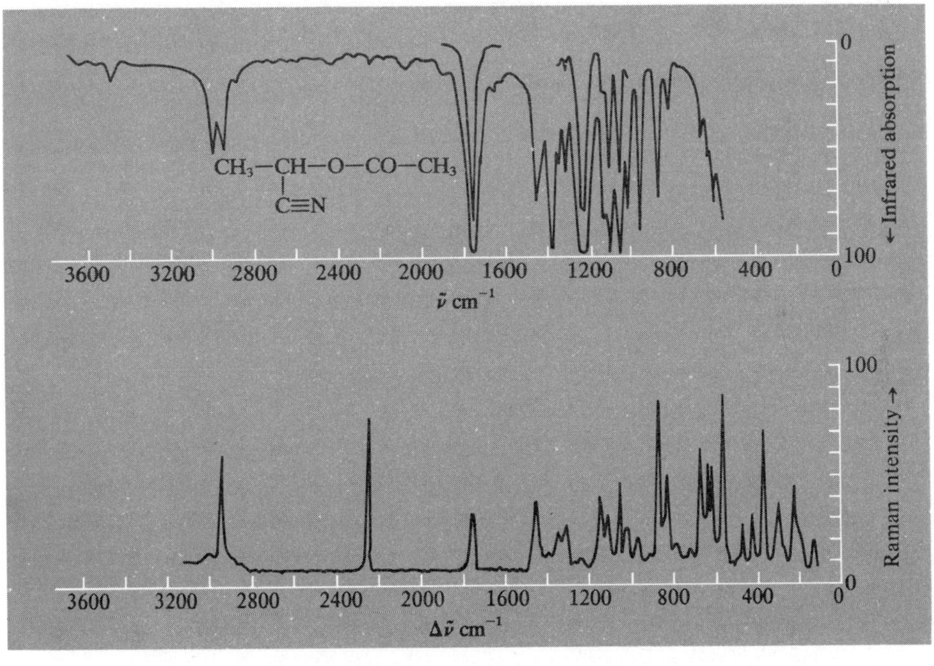

Fig. 7.17 Infrared and Raman spectra of 2-acetoxypropionitrile

Thus far, each example has been concerned mainly with one characteristic mode of vibration of the molecule in question. In relatively complex molecules, a great deal can be learned about their detailed structure if a number of characteristic bands are considered. Such analyses depend on the availability of careful comparative studies of a number of model compounds, and it is particularly important to consider all the characteristic parameters of a band, i.e., the wavenumber, intensity, band shape,

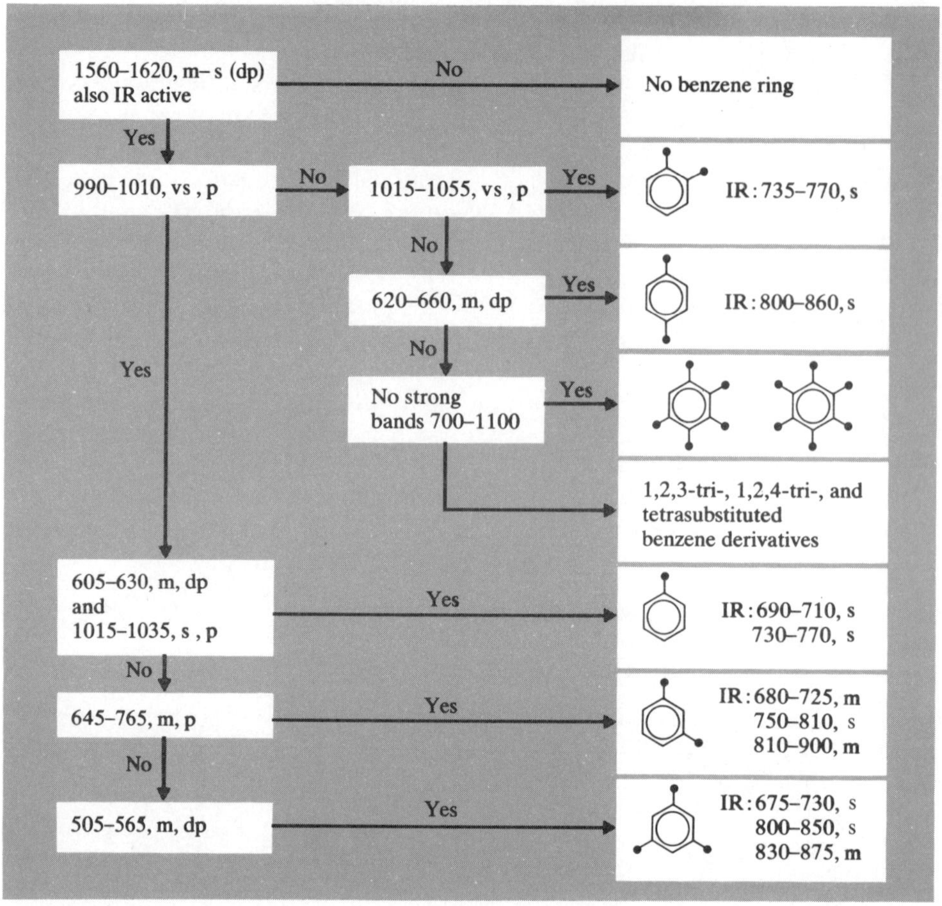

Fig. 7.18 Scheme for the determination of the substitution pattern of benzene derivatives with the aid of characteristic bands in the Raman spectrum; intensities: m medium, s strong, vs very strong, w weak, vw very weak; degree of polarization: p polarized, dp depolarized. IR denotes characteristic band in the infrared spectrum

and, in the case of a Raman band, the polarization characteristics. Studies of this kind have been made, for example, for steroids, and for substituted benzene and pyrazine derivatives. With such information available, it is possible to draw up a logical scheme for structural analysis of a particular class of compounds based on their infrared and Raman spectra. Such a scheme for substituted benzene derivatives is shown in Fig. 7.18. It can be seen that such an analytical scheme can be readily adapted for a computer, and, indeed, such computerized schemes, usually utilizing, in addition, information from ultraviolet and mass spectra, are now in use in some laboratories for the identification of unknown compounds.

A final example illustrates a situation where Raman and infrared spectroscopy can be more informative than NMR spectroscopy. The room temperature proton NMR spectrum of the norbonyl cation in "magic-acid" (FSO_3H; SbF_5; SO_2) shows three peaks with an intensity ratio of 1:6:4. This NMR spectrum does not permit a distinction to be made between the following structures for the ion: a classical structure (IIIa) undergoing a rapid Wagner–Meerwein transformation and a non-classical structure (IIIb) with delocalized electrons. This is because the lifetime of the nuclear transition (10^{-2} to 10^{-5} s) is much longer than the lifetime of a particular configuration. However, in vibrational spectroscopy, the vibrational lifetime is of the order of 10^{-11} to 10^{-13} s, which is much shorter than the lifetime of a particular configuration at room temperature. The vibrational Raman spectrum of the norbornyl cation in the wavenumber region associated with the skeletal modes is similar to that of nortricylan (IV) but quite different from that of norbornane (V). These results show that the norbonyl cation has a non-classical structure (IIIc). Low temperature ^{1}H and ^{13}C NMR spectra confirm this.

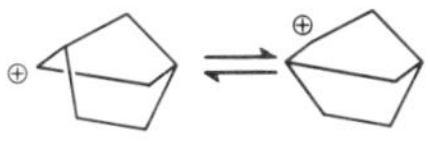

IIIa

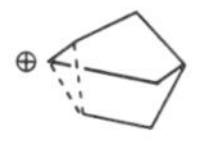

IIIb

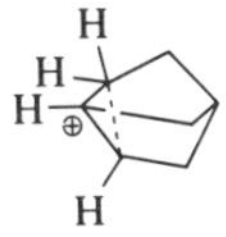

IIIc

IV

V

7.4 Inorganic chemistry

In inorganic chemistry, vibrational Raman spectroscopy, either alone or in conjunction with infrared spectroscopy, has two major applications: the identification and spectroscopic characterization of ionic or molecular species in a particular environment or environments; and the determination of the spatial configuration of such species.

An early and classical example[16] of the use of Raman spectroscopy to identify a species relates to the mercury(I) ion. The Raman spectrum of aqueous solutions of mercury(I) nitrate was found to contain, in addition to the lines known to be characteristic of the nitrate ion, a line at a wavenumber shift of 169 cm^{-1}, attributable to the stretching vibration of the diatomic mercury(I) ion, $[Hg—Hg]^{2+}$. This was the first observation of a metal–metal bond wavenumber and, as interest in such bonds has grown in recent years, Raman spectroscopy has proved a valuable technique for their study. Another important example was the identification[17] of the linear symmetric triatomic nitronium ion, $[ONO]^{+}$, in sulphuric acid–nitric acid–water systems through a strong sharp band at 1400 cm^{-1} in the Raman spectra of such acid mixtures. This wavenumber is associated with the symmetric stretching mode of this ion and its intensity may be used for quantitative estimation of the concentration of the ion in nitrating acids.

A revealing illustration of the strength and weakness of the use of vibrational spectroscopy for the elucidation of configuration relates to the investigation of the structure of disilyl ether, $(SiH_3)_2O$. The vibrations of this molecule are of two distinct kinds: those associated with the stretching of the Si—H bonds and deformation of their interbond angles; and those associated with the Si—O—Si skeleton. The vibrations and deformations involving the H atoms all occur at relatively large wavenumbers, well above the region in which the skeletal vibrations are expected to occur. In consequence, they cannot be confused with the skeletal wavenumbers; nor is there any significant coupling between the two kinds of vibration. Thus, the skeletal vibrations can be readily distinguished and regarded as those of a triatomic species $A{-}O{-}A$ in which A has a mass equal to that of the silyl unit. If this species is linear, its Raman spectrum should contain only one wavenumber associated with the symmetric stretching mode, and its infrared spectrum should contain two wavenumbers, associated with the antisymmetric stretching mode and the degenerate bending mode; neither of the infrared wavenumbers would be coincident with the Raman wavenumber. On the other hand, if the species is non-linear, the Raman and infrared spectra should both contain three wavenumbers, associated with the symmetric stretching mode, the antisymmetric stretching mode, and the angle bending mode; the wavenumbers in the Raman spectrum would be coincident with those in the infrared spectrum (see chapter 3, pages 67–71, and Figs. 3.8 and 3.9).

An early spectroscopic study[18] of disilyl ether concluded that the skeleton was linear, largely because only one line, at 606 cm^{-1}, was observed in the Raman spectrum, with no coincidence in the infrared spectrum. This deduction was at variance with electron diffraction measurements[19] which supported a non-linear structure with an Si—O—Si angle of about 141°. In a later Raman investigation,[20] the Raman spectrum of a 1:1 mixture of $(SiH_3)_2{}^{18}O$ and $(SiH_3)_2{}^{16}O$ was obtained. If the skeleton is linear, the wavenumber of the symmetric stretching mode (ν_1) would be unaffected by isotopic substitution of the central oxygen atom, but if the skeleton is bent, the wavenumber of this mode would be shifted to a lower value. Although the Raman spectra of the mixture showed no distinct resolution into two lines, a distinct broadening of the ν_1 band on the low wavenumber side was observed; and from the extent of this broadening it was estimated that there was an isotopic wavenumber shift of 3 to 4 cm^{-1}. This isotopic shift shows unambiguously that the Si—O—Si skeleton is non-linear. The extent of the isotopic shift was also shown to be compatible with an Si—O—Si angle of 141°, if reasonable values of the force constants are assumed. This example shows how dangerous it can be to base conclusions on the *absence* of lines in a spectrum. In the case of a triatomic species ABA, if the interbond angle is fairly close to 180°, the antisymmetric stretching mode and the angle bending mode, although formally Raman active, will have very low intensities in the Raman spectrum and are unlikely to be observed; similarly, the symmetric stretching mode, although formally infrared active, will have a very low intensity in the infrared spectrum and is also unlikely to be observed. On the basis of apparently absent bands, a triatomic species ABA with an interbond angle fairly close to 180° will masquerade as a linear species. In such cases, reliable inferences can only be drawn from the effect of isotopic substitution of the central atom on the wavenumber of the symmetric stretching mode, which is always strong in the Raman effect.

The ability to obtain the Raman spectra of highly coloured compounds has greatly extended its application to structural problems. Examples of the vibrational Raman spectra of highly coloured but simple inorganic compounds which have recently been obtained using laser excitation are shown in Figs. 7.19 and 7.20. Figure 7.19 shows the vibrational Raman spectrum of vanadium oxytribromide, which is dark red in colour.[21] The observation of six bands (three polarized) is consistent with a structure of C_{3v} symmetry for $VOBr_3$. Figure 7.20 gives the vibrational Raman spectrum[22] of tungsten hexachloride, which is strongly coloured. The observation of three bands (one polarized) is consistent with an octahedral structure of O_h symmetry for WCl_6.

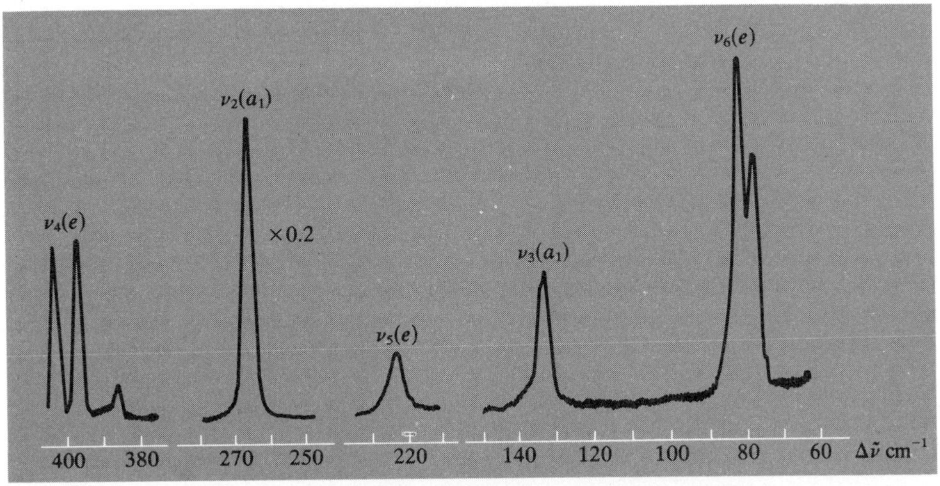

Fig. 7.19 Raman spectrum of crystalline vanadium oxytribromide (stable form A), 6471 Å (647·1 nm) excitation. Symmetry designations are given in C_{3v} nomenclature

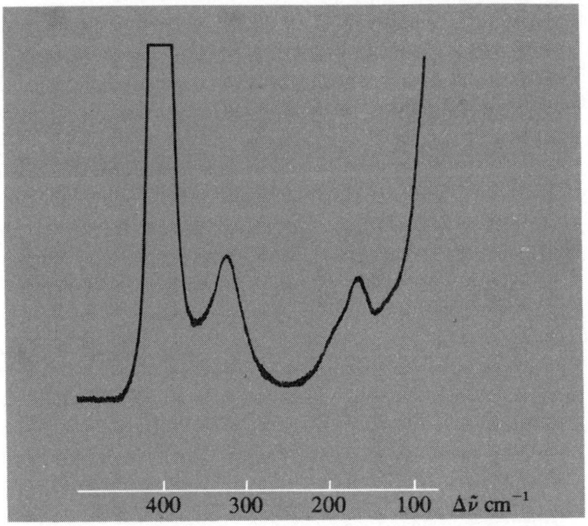

Fig. 7.20 The gas phase Raman spectrum of WCl_6

In coordination chemistry, many compounds are strongly coloured and the availability of the vibrational Raman spectra of such compounds is of particular importance because of the information that may be deduced concerning symmetry and bond strengths from the combined evidence of Raman and infrared spectra. An example is the investigation[23] of the Raman and infrared spectra of square planar palladium(II) complexes of the type *trans*-$PdX_2(SR_2)_2$, all of which are strongly coloured.

Many inorganic molecules undergo significant structural modifications with change of state. For example, in the halides of the elements, the coordination number and stereochemistry of the element in combination with the halogen are often different in different physical states. Since Raman spectra may now be quite readily obtained from samples in the solid, liquid, and vapour states, it offers a convenient and powerful method for the study of such structural changes.

Figure 7.21 shows a series of Raman spectra of aluminium bromide[24] in the vapour state at the following temperatures: (a) 880° C, (b) 620° C, (c) 480° C, and (d) 330° C.

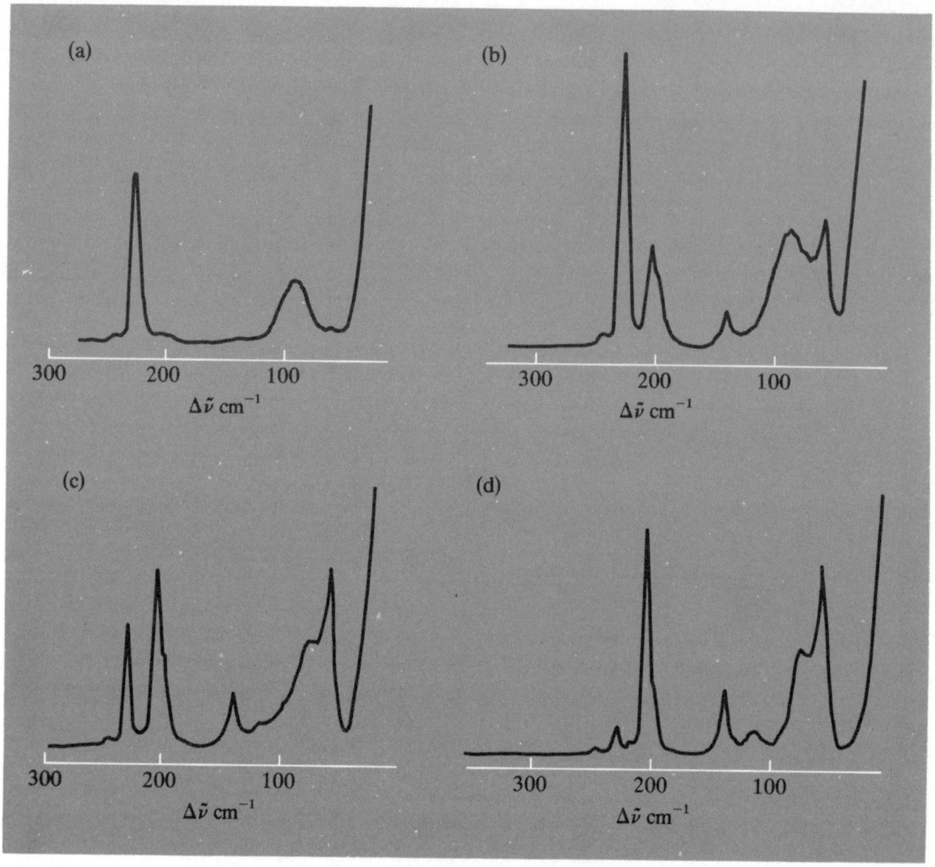

Fig. 7.21 The Raman spectrum of gaseous aluminium bromide (a) at 880° C; (b) at 620° C; (c) at 480° C; and (d) at 330° C; 5145 Å (514·5 nm) excitation. Vertical scale of (a) half that shown in (b), (c), and (d)

The Raman spectrum at 880° C is consistent with the existence of the monomer $AlBr_3$, only, at this temperature. The Raman spectra at lower temperatures show that an increasing amount of dimer, Al_2Br_6, exists as the temperature decreases. Using the intensity of the band at 228 cm^{-1} as a measure of the concentration of $AlBr_3$, and the intensity of the band at 203 cm^{-1} as a measure of the concentration of Al_2Br_6, it was possible to calculate the equilibrium constant for the dissociation reaction

$$Al_2Br_6 \rightleftharpoons 2AlBr_3$$

and hence the enthalpy of dissociation. The value obtained was $\Delta H = -96 \pm 1{\cdot}5$ kJ mol^{-1}.

Figure 7.22(a) gives the Raman spectrum of phosphorus(V) chloride in the vapour state at 430 K, and Fig. 7.22(b) the Raman spectrum of the solid formed by rapid cooling of the vapour to liquid nitrogen temperature. Comparison of the spectra[25] show that, in the vapour phase and in this metastable solid state, the halide has the simple molecular form PCl_5. In the stable solid state, it has the ionic structure $PCl_4^+PCl_6^-$, as other spectroscopic studies have shown.

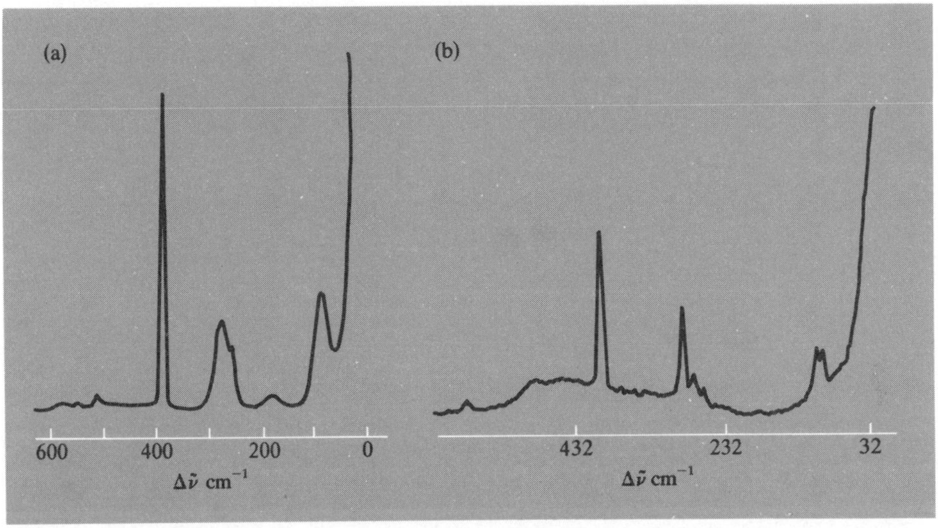

Fig. 7.22 The Raman spectrum of phosphorus(V) chloride: (a) in the vapour phase at 430 K, 4880 Å (488·0 nm) excitation, and (b) in the solid phase at 87 K formed by rapid cooling of the vapour, 6328 Å (632·8 nm) excitation

Figure 7.23 shows the Raman spectrum[26] of $HfCl_4$ in the vapour state at 500° C. The presence of four lines in the spectrum, one of which is highly polarized, is consistent with a tetrahedral structure for this species in the vapour state. This is in contrast to the solid state, where the structure consists of zig-zag chains with each metal atom octahedrally coordinated to six chlorine atoms, four of which form bridging bonds and two are terminal.

Figure 7.24 shows the Raman spectrum[27] of tin(II) chloride in the vapour state at 650° C. The spectrum is consistent with a monomeric non-linear structure $SnCl_2$, whereas in the solid and molten states polymeric structures are involved.

Now that it is possible to measure the Raman spectra of inorganic molecules in the vapour state at reasonably low pressures, band contours may often be determined. The forms of such band contours, which represent the unresolved rotational structure of the vibrational bands, are related to the symmetry of the vibration. For example,[28] in a tetrahedral molecule like SiF_4, the band contour of $\nu_1(a_1)$ consists only of an unresolved *Q* branch, whereas the band contours of $\nu_2(e)$, $\nu_3(f_2)$, and $\nu_4(f_2)$ all show unresolved *OP—Q—RS* structure, as shown in Fig. 7.25. The separation in the maxima of the unresolved *OP* and *RS* branches can be related quantitatively to the Coriolis constants, and affords a satisfactory method of determining such constants, although in cases where bands can be resolved into individual lines a more accurate value is, of course, obtainable.

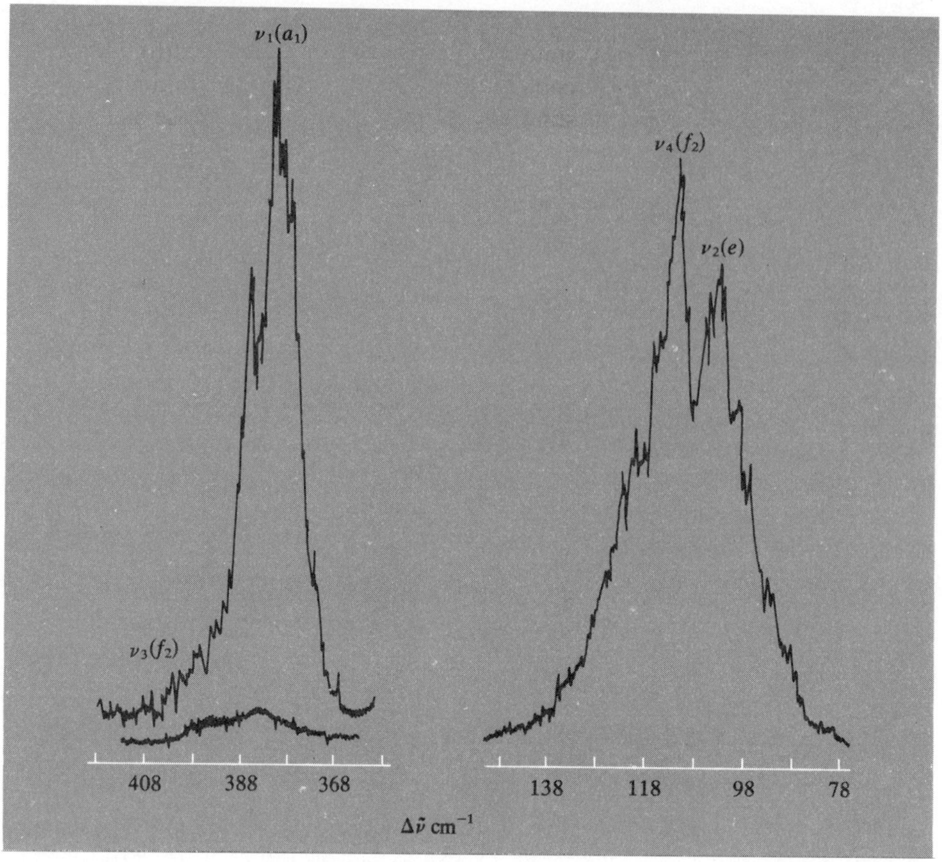

Fig. 7.23 Raman spectrum of $HfCl_4$ at 500° C, 4880 Å (488·0 nm) excitation

Raman spectroscopy is particularly suitable for the study of redistribution reactions, dissociation reactions, and equilibria in inorganic chemistry, especially in those cases where the products cannot be readily isolated. Figure 7.26 shows the Raman spectrum[29] of a cyclohexane solution of a 2:1 mixture of $VOCl_3$ and $VOBr_3$ in the region 250 to 420 cm^{-1}, where the vanadium–halogen symmetric stretching wavenumbers occur. Four bands are observed, and these are attributed to the species $VOCl_3$, $VOBr_3$, $VOCl_2Br$, and $VOClBr_2$. The redistribution reaction is virtually instantaneous at room temperature and the resulting distribution of halogen atoms is approximately statistical.

Figure 7.27 shows the Raman spectra[30] of phosphorus–arsenic mixtures in the vapour phase (660° C) with the following P_4:As_4 molar ratios: (a) 10:1, (b) 3:2, (c) 2:3, and (d) 1:10. The spectra cover the wavenumber range 150 to 650 cm^{-1}, which is the region in which the wavenumbers of the tetratomic molecules P_4 and As_4 lie. From comparison of the Raman spectra of the phosphorus–arsenic mixtures with the wavenumbers of P_4 and As_4 it can be shown that the following mixed species are present in the mixtures: P_3As, P_2As_2, and PAs_3.

The use of Raman spectroscopy to study inorganic compounds in the molten state has proved rewarding. For example, the Raman spectrum[31] of molten gallium

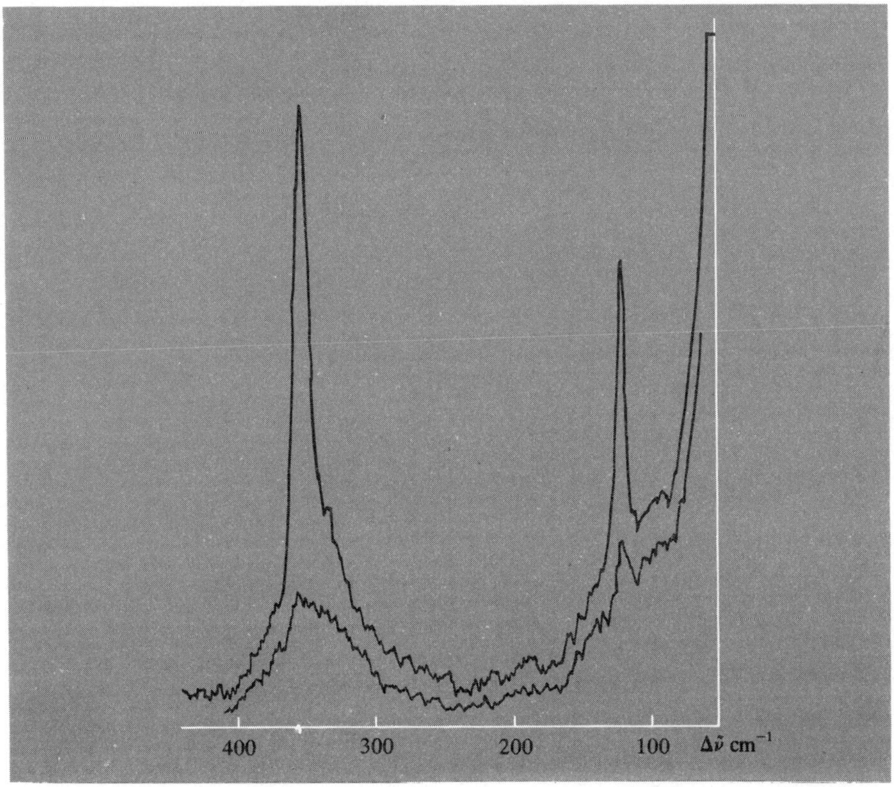

Fig. 7.24 Raman spectrum of $SnCl_2$ at 650° C, 5145 Å (514·5 nm) excitation

chloride with the empirical formula $GaCl_2$ (Fig. 7.28) leaves no doubt that it is, in fact, an ionic compound $[Ga^+][GaCl_4^-]$. By contrast the Raman spectra of the molten halides of mercury(II), $HgCl_2$ and $HgBr_2$, are generally consistent with the existence, in this state, of the linear Cl—Hg—Cl and Br—Hg—Br molecules.[32] If, however, KCl is added to molten $HgCl_2$, the Raman spectrum indicates the presence of $HgCl_3^-$ and $HgCl_4^{2-}$, as well as $HgCl_2$.[33] Detailed studies of the wavenumbers, intensities, and band widths of the vibrational Raman spectra of ionic melts, like nitrates and sulphates, have yielded valuable information on the nature of interactions in the molten state. For example, such investigations show that for molten nitrates and sulphates, the symmetric stretching wavenumber of the nitrate ion and the sulphate ion decrease approximately linearly with decrease of the ratio of Z, the cation charge, to r, the cation radius.[34]

Solvent extraction techniques have proved valuable for the investigation of the Raman spectra of complex ions. For example, in the systems $InCl_3$–HCl, $FeCl_3$–HCl,[35] $SnCl_2$–HCl, and $SnBr_2$–HBr,[36] the Raman spectra of the aqueous solutions provide evidence for the existence of *mixtures* of anionic complexes at all concentrations of the components. However, the Raman spectra of diethyl ether extracts of these systems showed that preferential extraction of the singly charged anionic complex ($M^{III}X_4^-$ or $M^{II}X_3^-$) had occurred. This method of isolating a particular complex species has been applied to a number of systems including the chloro- and bromo-complexes of mercury(II),[37] the chloro-, bromo-, and iodo-complexes of cadmium(II),[38] and the chloro- and bromo-complexes of thallium(III)[39] and arsenic(III).[40]

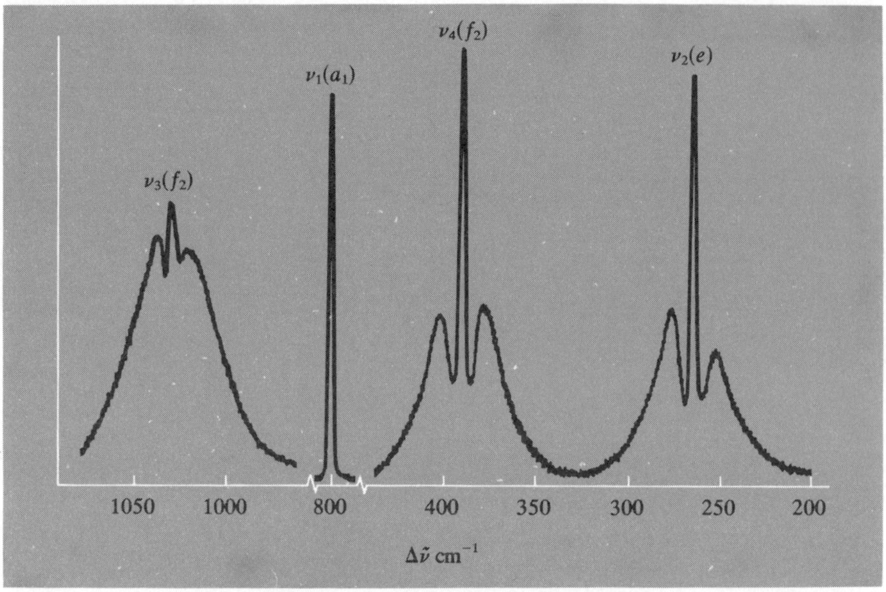

Fig. 7.25 Vapour phase Raman spectrum of silicon tetrafluoride at 22° C (various excitation wavelengths used)

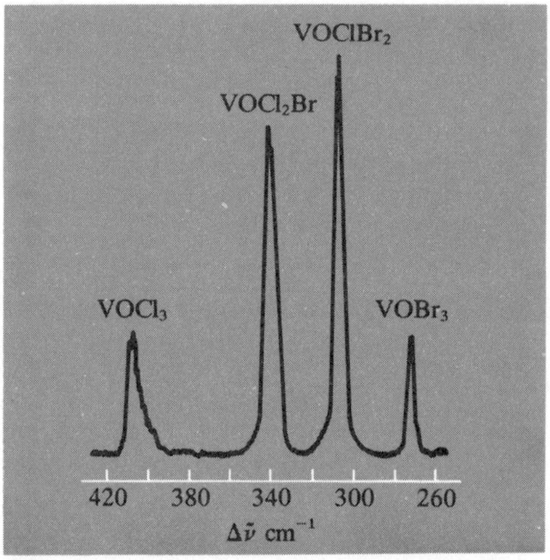

Fig. 7.2ɔ Raman spectrum of a cyclohexane solution of a 2:1 mixture of vanadium oxytrichloride and vanadium oxytribromide in the region of the ν_2 (a_1) fundamental of each molecule (6764 Å (676·4 nm), 6471 Å (647·1 nm), and 5682 Å (568·2 nm) excitation). The intensities of the a_1 bands of the bromine-containing species are greater than the abundance of such species (calculated statistically) would suggest, owing to the fact that metal–bromine bond polarizability derivatives are greater than metal–chlorine bond polarizability derivatives

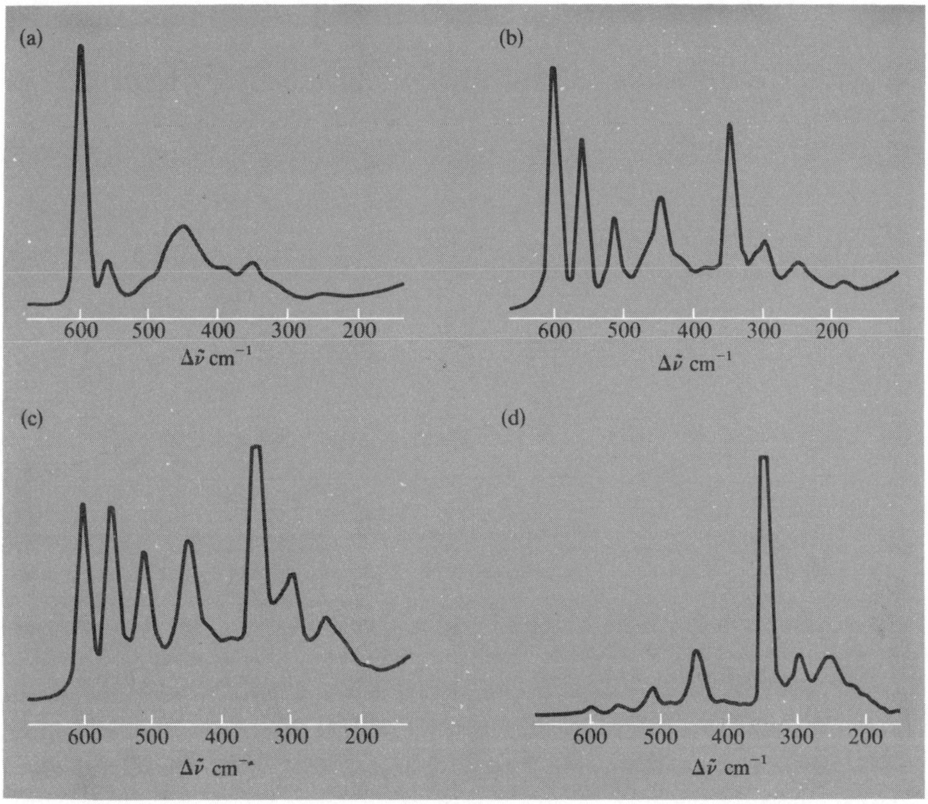

Fig. 7.27 The gas-phase Raman spectra of phosphorus–arsenic mixtures (660° C) with (a) 10:1, (b) 3:2, (c) 2:3, and (d) 1:10 molar ratios of P_4:As_4, 4880 Å (488·0 nm) and 5145 Å (514·5 nm) excitation

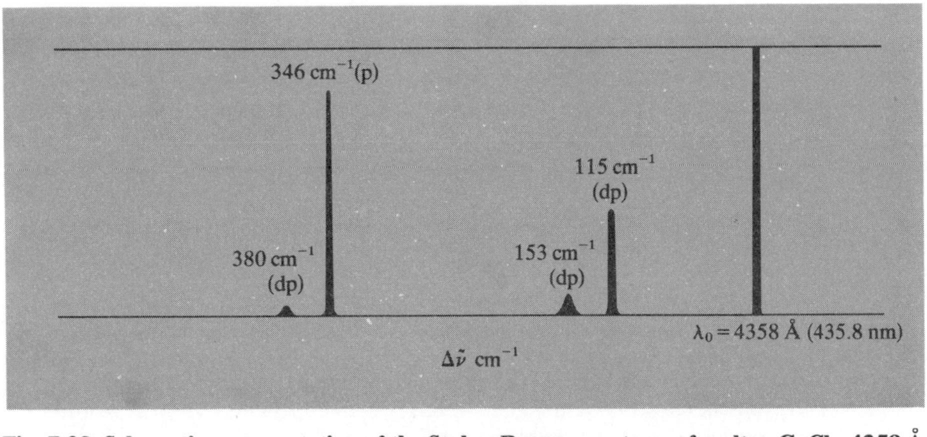

Fig. 7.28 Schematic representation of the Stokes Raman spectrum of molten $GaCl_2$, 4358 Å (435·8 nm) excitation

Raman spectroscopy can also be valuable for the study of relatively weak bonding. The Raman spectrum[41] of an aqueous solution of $[Pb_4(OH)_4](ClO_4)_4$ is illustrated in Fig. 7.29. Five of the six bands observed are attributable to $[Pb_4(OH)_4]^{4+}$, the predominant product of lead(II) hydrolysis. Of these five bands, the three strongest occur at 130 cm^{-1} (p), 87 cm^{-1} (dp), and 60 cm^{-1} (dp), and can be attributed entirely to Pb—Pb stretching in a $\{Pb(II)\}_4$ tetrahedron. Raman intensity measurements on the a_1 band at 130 cm^{-1} (see section 7.14) suggest that the Pb—Pb bond order is about 0.03 as compared with 0·4 for Hg_2^{2+}.

Further examples of the application of Raman spectroscopy in inorganic chemistry will be found in sections 7.2, 7.5, 7.7, 7.8, 7.10, 7.11, 7.12, 7.14, and 7.17.

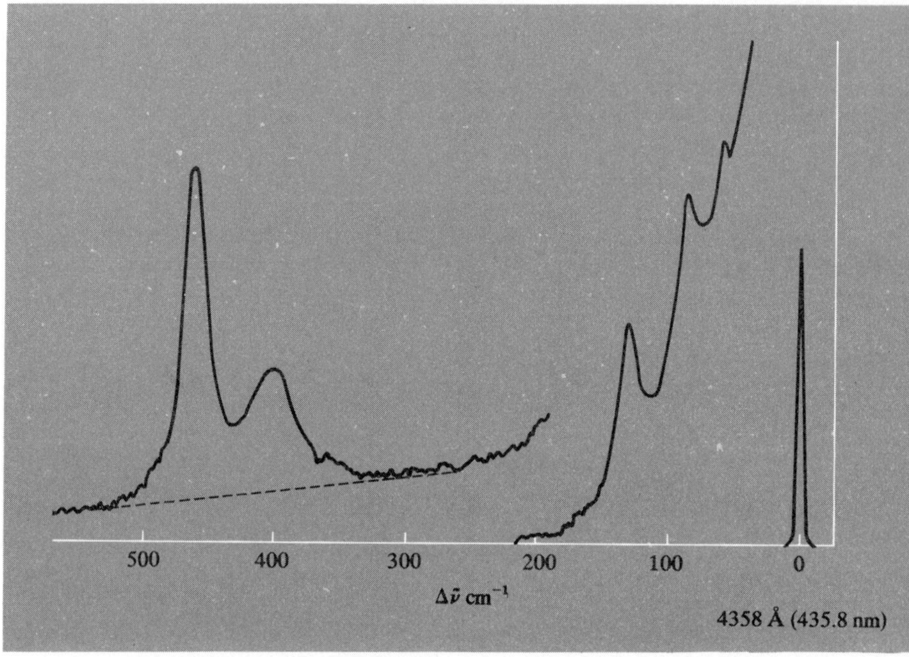

Fig. 7.29 Raman spectrum of an aqueous solution, 4·47 M in lead, 4·63 M in perchlorate, corresponding to a hydroxyl:lead ratio of 0·96; 4358 Å (435·8 nm) excitation

7.5 Matrix isolation spectroscopy

Spectroscopic studies may be made of many reactive molecules, radicals, and other normally unstable species by use of the matrix isolation technique, whereby the species of interest are preserved at low concentration in a matrix of a noble or other chemically inert gas (e.g., N_2) at low temperatures, usually in the range 4 to 20 K. Under these conditions, molecules that normally have very short lifetimes can be preserved indefinitely and investigated at leisure. The inertness of the matrix material prevents loss of reactive molecules by reaction with their environment. The rigidity of the matrix cage prevents diffusion of reactive molecules which would lead to reaction with other such species, and also rotation of all but the smallest and lightest molecules. This quenching of rotation has the effect of simplifying the vibrational spectrum and of producing sharp, line-like bands. Consequently, full advantage of isotopic substitution can be taken. The low temperature not only contributes to the rigidity of the matrix cage but also eliminates hot bands and difference combination bands, and serves to reduce the rate of possible internal rearrangements that require any activitation energy. It will be apparent that matrix isolation, in addition to being an invaluable technique for the spectroscopic investigation of reactive species, offers many advantages in the study of stable species. Although infrared studies of matrix isolated species were first made in the 'fifties and are now relatively commonplace, it has only recently become possible to obtain the Raman spectra of such species.

In concept, the experimental procedure for Raman spectroscopy of matrix isolated species is quite simple. The matrix is deposited under controlled conditions on a cold polished metal (or alkali metal halide) block. The block is placed at an angle to the laser beam (usually near grazing incidence) and the scattered radiation is collected from a solid angle which lies within that defined by the incident and reflected laser beams (see Fig. 7.30). In practice, very careful optimization of a number of factors,

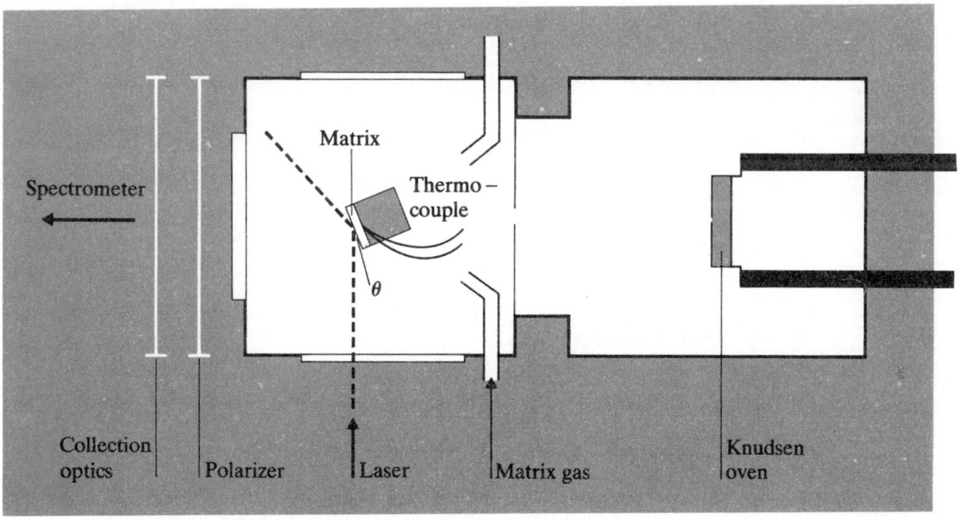

Fig. 7.30 Schematic representation of matrix isolation cell, Knudsen oven and illumination–observation geometry for Raman spectroscopy

particularly rate of deposition of the matrix, the ratio of the matrix to solute molecules, the orientation of the matrix, and the laser power are necessary, if Raman spectra of adequate quality are to be obtained. By using an inclined CsI block, it is possible to obtain the infrared and Raman spectra from the self-same matrix[42] (see Fig. 7.31).

Spectroscopic studies of the xenon–chlorine system will illustrate what can be achieved. In one study,[43] a mixture of xenon and chlorine in the ratio 25:1 was passed through a microwave discharge and condensed on to a cold tip at 20 K. The Raman spectrum was recorded at 4.2 K and contained only two strong bands in the region 60 to 650 cm^{-1}, one at 253 cm^{-1} and one around 540 cm^{-1} showing some fine structure. The Raman spectrum obtained from a matrix of xenon and chlorine (25:1) which had

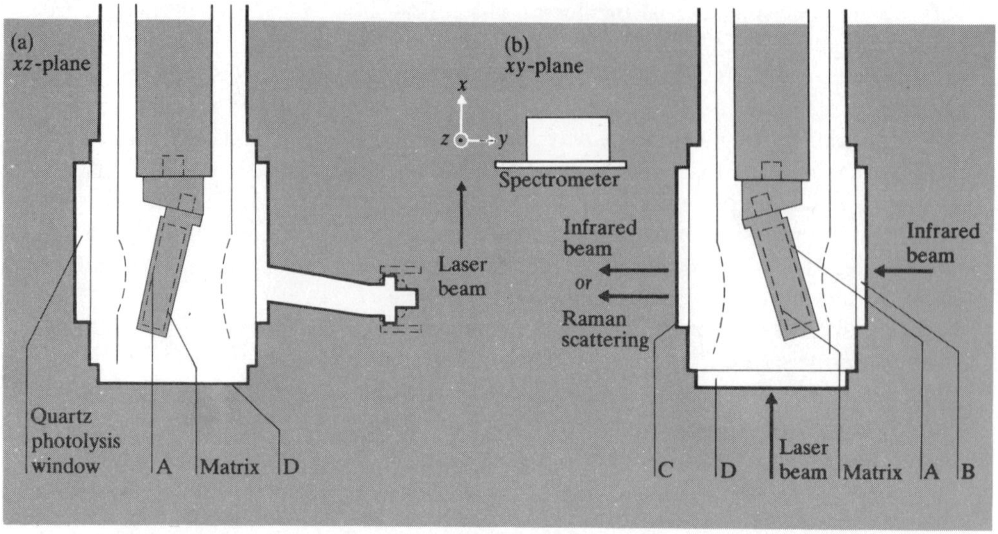

Fig. 7.31 Experimental arrangement for the acquisition of infrared absorption and Raman spectroscopic data from the self-same matrix: (a) deposition configuration and (b) spectral acquisition configuration. A, B, C, CsI windows; D pyrex window

not been subjected to a microwave discharge prior to condensation contained only one band in this region, at 540 cm^{-1} with some fine structure. From comparison of these two Raman spectra, the band at 540 cm^{-1} was assigned to unreacted Cl_2, the fine structure arising from the various isotopic molecules, $^{37}Cl_2$, $^{37}Cl^{35}Cl$, and $^{35}Cl_2$; and the band at 253 cm^{-1}, present only in the matrix formed from the gaseous mixture which had been subjected to a microwave discharge, was attributed to the $XeCl_2$ molecule.

The infrared spectrum of a similar matrix contained only one strong band at 313 cm^{-1}, which is not coincident with the band in the Raman spectrum. These spectra suggested a linear centrosymmetric structure for the $XeCl_2$ molecule. The Raman active band at 253 cm^{-1} may be assigned to the symmetric stretching mode and the infrared active band at 313 cm^{-1} to the antisymmetric stretching mode. The infrared active bending mode would be expected to occur below 200 cm^{-1}, which was the lower wavenumber limit of the infrared measurements.

Further matrix studies[44] under higher resolution have resulted in a much more conclusive spectroscopic characterization of $XeCl_2$. For the Raman studies matrix-isolated $XeCl_2$ was produced by *in-situ* photolysis of a mixture of natural xenon and natural chlorine. The photolysis was produced by the 4880 Å (488·0 nm) argon laser radiation used to excite the Raman spectrum. The band at 253 cm^{-1} was now resolved into three components in the approximate intensity ratios 9:6:1 and separated by *ca.* 3·5 cm^{-1} (Fig. 7.32). This result is unambiguous evidence for the presence of two equivalent chlorine atoms in the moiety under investigation. Further infrared studies of the antisymmetric mode using natural and isotopically enriched xenon (82 per cent ^{136}Xe) and natural and isotopically enriched chlorine (95 per cent ^{35}Cl) confirmed that the species contained xenon and two chlorine atoms. Extension of the infrared studies to 40 cm^{-1} did not reveal the bending mode.

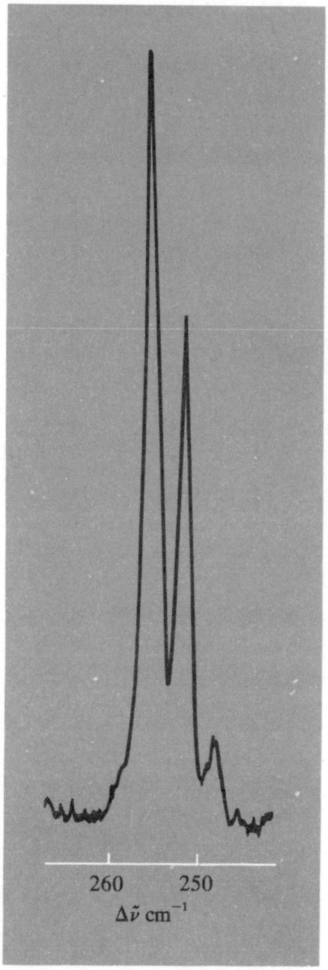

Fig. 7.32 Raman spectrum of the matrix isolated $XeCl_2$, 4880 Å (488·0 nm) excitation

7.6 Quantitative studies of ionic equilibria

It has been noted in chapter 3 that the intensity of a vibrational Raman band associated with a particular scattering species is proportional to the concentration of that species. The measurement of Raman intensities has proved a valuable method of studying certain kinds of equilibria in solution, particularly the ionization of acids in aqueous solution. We shall illustrate this application by considering some work[45] on the ionization of ethane sulphonic acid for which the following equilibrium exists in aqueous solution:

$$H_2O + C_2H_5SO_3H \rightleftharpoons C_2H_5SO_3^- + H_3O^+$$

The intensity of a Raman band assigned to the symmetric SO_3 vibration, and having a shift of 1046 cm^{-1}, was suitable for measuring the concentration of the $C_2H_5SO_3^-$ ion. The intensity of a Raman band assigned to the C—S stretching vibration in both the undissociated acid and the anion, and having a wavenumber shift of about

782 cm^{-1}, was used to measure the combined concentration of undissociated acid *and* anion, i.e., the stoichiometric amount of acid used in the preparation of the solution. Figure 7.33 shows the Stokes Raman spectrum of an aqueous solution of ethane sulphonic acid in the wavenumber shift region 200 to 1500 cm^{-1}.

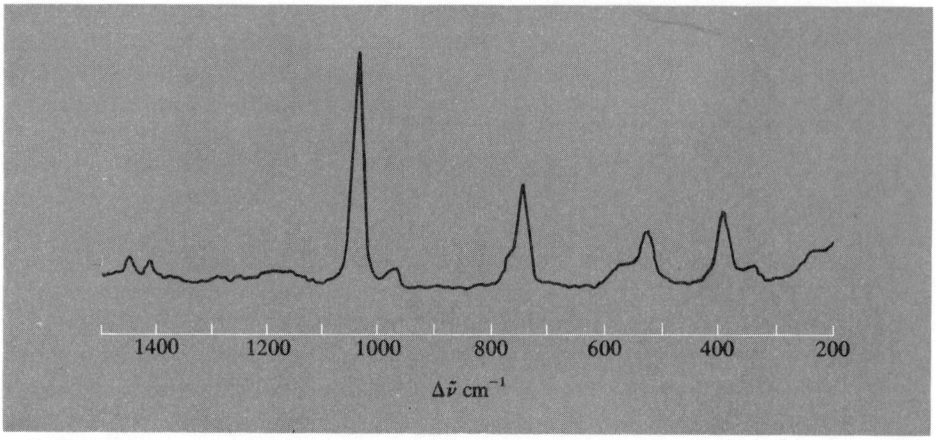

Fig. 7.33 Stokes Raman spectrum of ethane sulphonic acid in the wavenumber region $\Delta\tilde{\nu}$=200–1500 cm^{-1} (excited by 4880 Å (488·0 nm))

The relation between the intensity I of a vibrational band and the concentration C of the scattering species, may, for the present purposes, be reduced to the simple form

$$I = CJ^{-1} \tag{7.5}$$

where J^{-1} is the molar intensity. The intensity I is measured by the area under the vibrational band. It then follows that

$$\begin{aligned} \frac{I_{1046}}{I_{782}} &= \frac{C_{\text{anion}}}{C_{\text{stoich}}} \frac{J_{782}}{J_{1046}} \\ &= J'\alpha \end{aligned} \tag{7.6}$$

where α is the degree of dissociation and J' is the ratio of the molar intensities. Since $\alpha \to 1$ as $C_{\text{stoich}} \to 0$, then a plot of $J'\alpha$ against C_{stoich} should approach J' as $C_{\text{stoich}} \to 0$ with zero slope. For ethane sulphonic acid, J' was found to be 2·092. Values of α at various concentrations were then found by dividing the intensity ratio I_{1046}/I_{782} by J'. Some typical results are shown in Table 7.3. These α values lead to a value of the ionization constant of 48 ± 2 mol kg^{-2} for ethane sulphonic acid.

7.7 Ionic interactions

Ionic interactions in aqueous solution may be investigated using Raman spectroscopy. An interesting, and revealing, example concerns the nitrite ion.[46] The free nitrite ion $(ONO)^-$ has a non-linear structure and so has three Raman active vibrations: $\nu_1(a_1)$ ca. 1330 cm^{-1}, $\nu_2(a_1)$ ca. 815 cm^{-1}, and $\nu_3(b_1)$ ca. 1240 cm^{-1}. The Raman spectra of aqueous solutions of $LiNO_2$, $NaNO_2$, KNO_2, and $CsNO_2$ show

Table 7.3 Ethane sulphonic acid: degrees of dissociation

$C/\text{mol l}^{-1}$	α
0·710	0·992
1·290	0·984
1·310	0·984
1·382	0·982
1·915	0·970
2·675	0·947
3·000	0·936
3·046	0·935

only these three lines, even at high concentrations. Furthermore, the integrated intensities of each band are proportional to concentration and show no specific dependence on the cation. These results suggest that the residence time of a nitrite ion adjacent to alkali metal ions, with appropriate orientation for binding, is very short, and that no contact–ion–pairs are formed. By contrast, the Raman spectra of aqueous solutions of $Zn(NO_2)_2$ show lines additional to those characteristic of the free nitrite ion. As can be seen from Fig. 7.34, which shows the Raman spectrum of a 1·0 M $Zn(NO_2)_2$ solution in the region 750 to 1500 cm^{-1}, two additional bands are observed at 867 cm^{-1} and 1400 cm^{-1}. These are attributed to the deformation and antisymmetric stretching modes, respectively, of bound NO_2^-. The relatively high wavenumber of the antisymmetric stretching mode probably indicates a nitrito

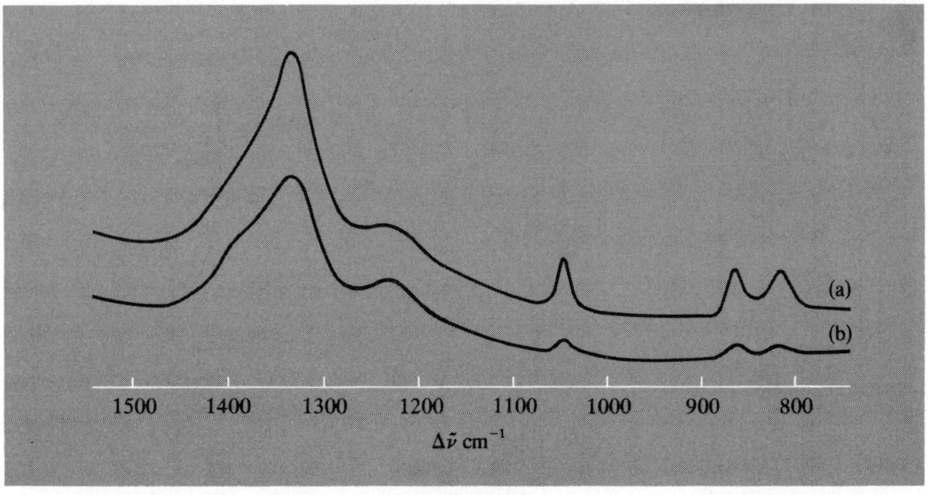

Fig. 7.34 Stokes Raman spectra of 1·0 M $Zn(NO_2)_2$ excited by 4358 Å (435·8 nm): (a) cross-polarized, $I_\perp$, and (b) axially polarized, $I_\parallel$. The band at 1050 cm^{-1} is due to the nitrate ion (0·067 M)

linkage Zn—(ONO). However, in the low wavenumber region of the Raman spectrum (not shown in Fig. 7.34), a weak band is found at 360 cm^{-1}, which may be a metal–nitrogen stretch and so indicate a nitro structure Zn—(NO_2). Quantitative studies of the intensities of the 815 cm^{-1} band (characteristic of the free nitrite) and the 867 cm^{-1} band (characteristic of the bound nitrite), as the ratio of zinc(II) ions to nitrite ions was varied, indicate that in aqueous zinc(II)–nitrite systems, stepwise formation of the 1 : 1 ($Zn(NO_2^+)$) and 1 : 2 ($Zn(NO_2)_2$) complexes occur as follows:

$$Zn^{2+} + NO_2^- \rightleftharpoons ZnNO_2^+, (K_1)$$

$$ZnNO_2^+ + NO_2^- \rightleftharpoons Zn(NO_2)_2, (K_2)$$

Analysis of the intensity data gave the following values for the equilibrium constants: $K_1 = 1{\cdot}4$ and $K_2 = 0{\cdot}9$.

7.8 Resonance Raman spectroscopy

If a scattering system has an absorption band close to the excitation frequency, resonance Raman scattering results. The intensity of resonance Raman scattering is usually many orders of magnitude greater than normal Raman scattering. The form of the spectrum may also be different; for example, overtones may be observed with appreciable intensity, and new bands may appear. The availability of a range of

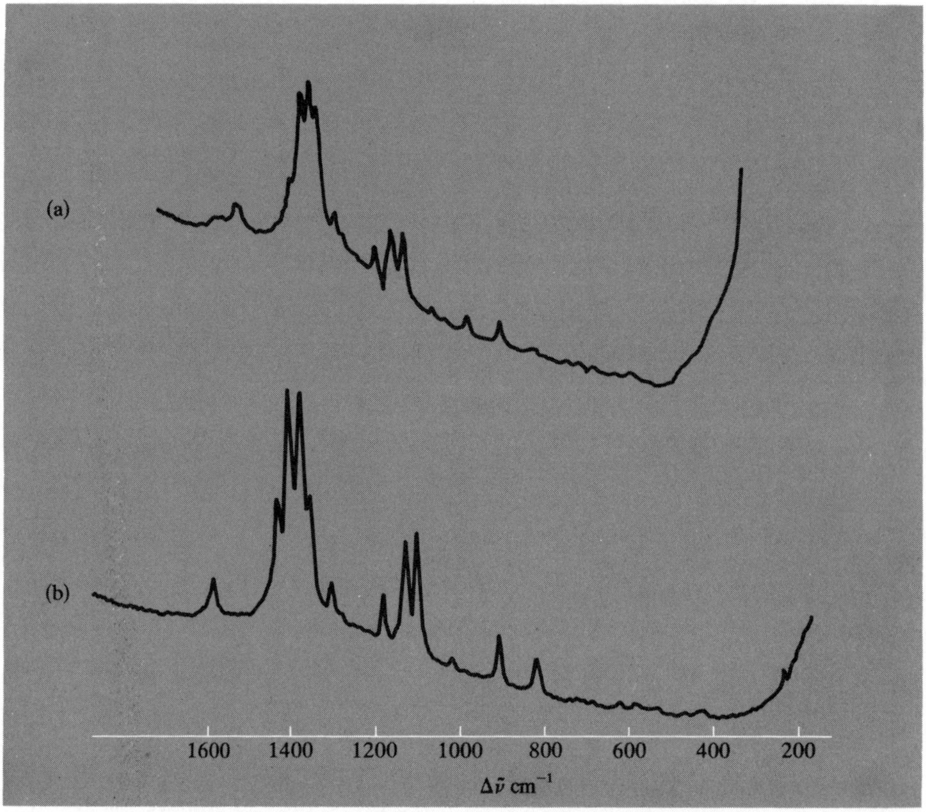

Fig. 7.35 Stokes resonance Raman spectra of methyl orange, 4880 Å (488·0 nm) excitation, rotating sample cell: (a) 10^{-4} M aqueous solution and (b) powder

excitation frequencies from laser sources and the development of tunable laser sources has resulted in a growing exploitation of this form of Raman scattering.

An obvious, but important, application of resonance Raman scattering is to use the intensity enhancement to enable spectra to be obtained from low concentrations of material. For example, resonance Raman spectra[47] of methyl orange are shown in Figs. 7.35a and 7.35b; Fig. 7.36 shows the resonance Raman spectrum[48] of some chlorophylls; and Fig. 7.37 shows the resonance Raman spectrum[49] of a solution of halophile bacteria which has a typical carotenoid-like spectrum.

A good illustration of how the form of a resonance Raman spectrum can differ from a normal Raman spectrum is provided by Fig. 7.38. Figure 7.38(a) shows the resonance Raman spectrum[50] of solid K_2CrO_4 obtained with 3638 Å excitation and a rotating sample cell; Fig. 7.38(b) shows the normal Raman spectrum[51] of K_2CrO_4 in aqueous solution obtained with 6328 Å excitation; and Fig. 7.38(c) shows an absorption spectrum of K_2CrO_4 in aqueous solution. The normal Raman spectrum shows only the four bands characteristic of a tetrahedral CrO_4^{2-} ion. By contrast, the resonance Raman spectrum shows a progression of *ten* harmonics of the symmetric stretching mode ($\tilde{\nu}_M \approx 853\ \mathrm{cm^{-1}}$). The intensity decreases but the bandwidth increases as the progression moves to higher overtones. The observation of such a progression enables the anharmonicity to be determined.

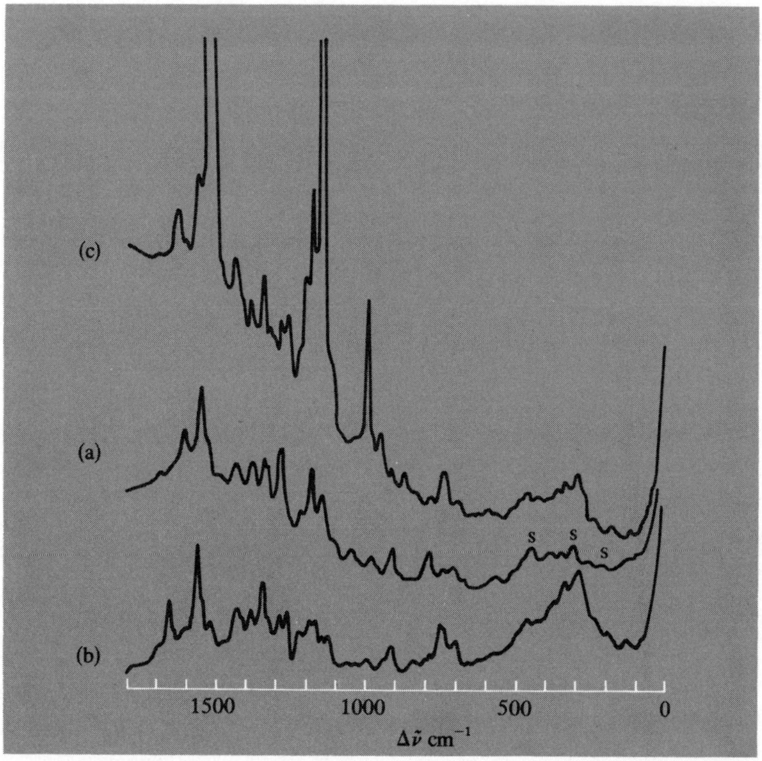

Fig. 7.36 Stokes resonance Raman spectra of chlorophylls: (a) dehydrated chlorophyll-a, 10^{-2} M in CCl_4, 4579 Å (457·9 nm) excitation (solvent bands marked s); (b) non-dehydrated chlorophyll-b, 10^{-1} M in acetone, 4579 Å (457·9 nm) excitation; and (c) chloroplast fragments, equivalent to 10^{-3} g chlorophyll-a per 10^{-3} dm^{-3} water, 4765 Å (476·5 nm) excitation

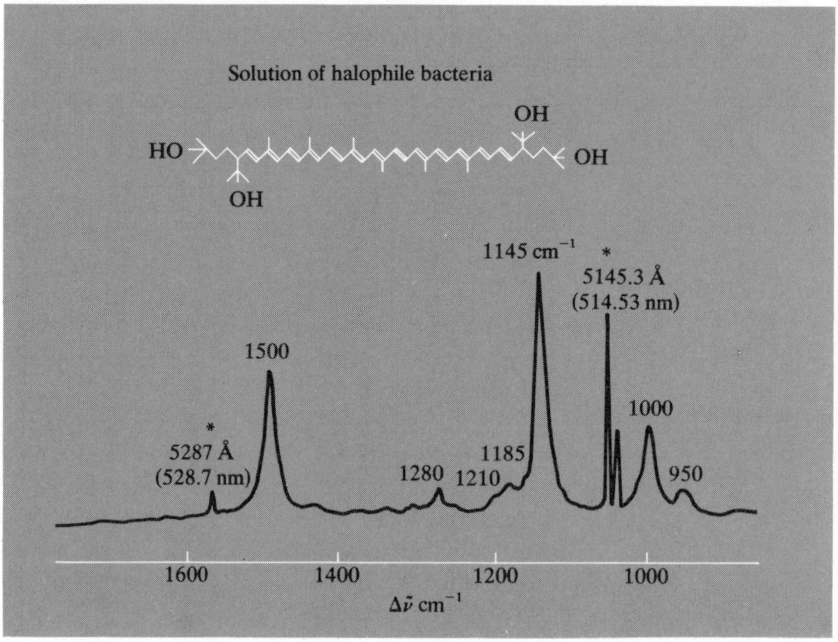

Fig. 7.37 Stokes Raman spectrum of a solution of halophile bacteria, 4880 Å (488·0 nm) excitation

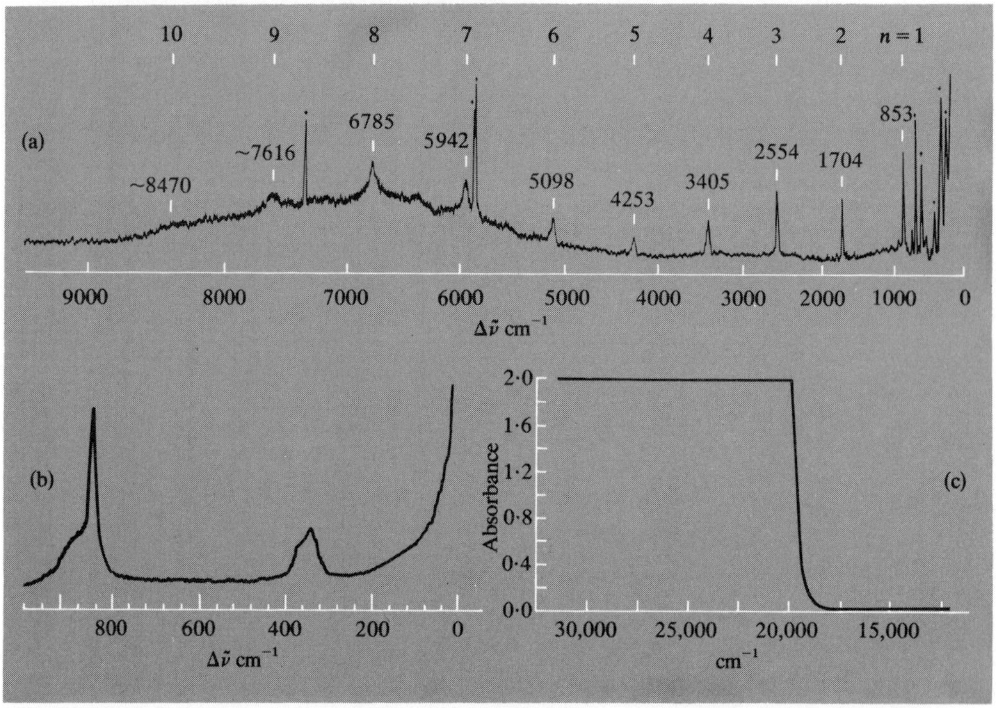

Fig. 7.38(a) Stokes Raman spectrum of solid K_2CrO_4, 3638 Å (363·8 nm) excitation, rotating sample cell; lines marked + are laser lines. (b) Normal Raman spectrum, 6328 Å (632·8 nm) excitation and (c) Absorption spectrum of K_2CrO_4 aqueous solution 1·0 M

Figure 7.39(a) shows the Stokes Raman spectra of quinoxaline[52] in aqueous solution in the wavenumber shift range $\Delta\tilde{\nu} = 800\text{–}1100\ cm^{-1}$ obtained with six different excitation wavelengths. The solutions also contained 0·5 M $(NH_4)_2SO_4$. Figure 7.39(b) compares the dependence on excitation wavenumber of the experimentally determined relative intensities of the 877 cm^{-1} band of quinoxaline (individual points) with that predicted on the basis of eq. (5.87) (continuous curve) using $\tilde{\nu}_{e^r e^0} = 29\,500\ cm^{-1}$ and $\tilde{\nu}_{e^{r'} e^0} = 31\,650\ cm^{-1}$, which are associated with $^1B_1$ and $^1A_1$ electronic states respectively of quinoxaline. The intensities of the 877 cm^{-1} band were measured relative to the intensity of the symmetric stretching vibration of the SO_4^{2-} ion, which serves as an internal standard since it exhibits no significant resonance effects at these excitation wavenumbers. There is good general agreement between theory and experiment.

The resonance Raman effect can also be helpful in studies of matrix-isolated species. When sodium atoms and ozone molecules at high dilution in argon are

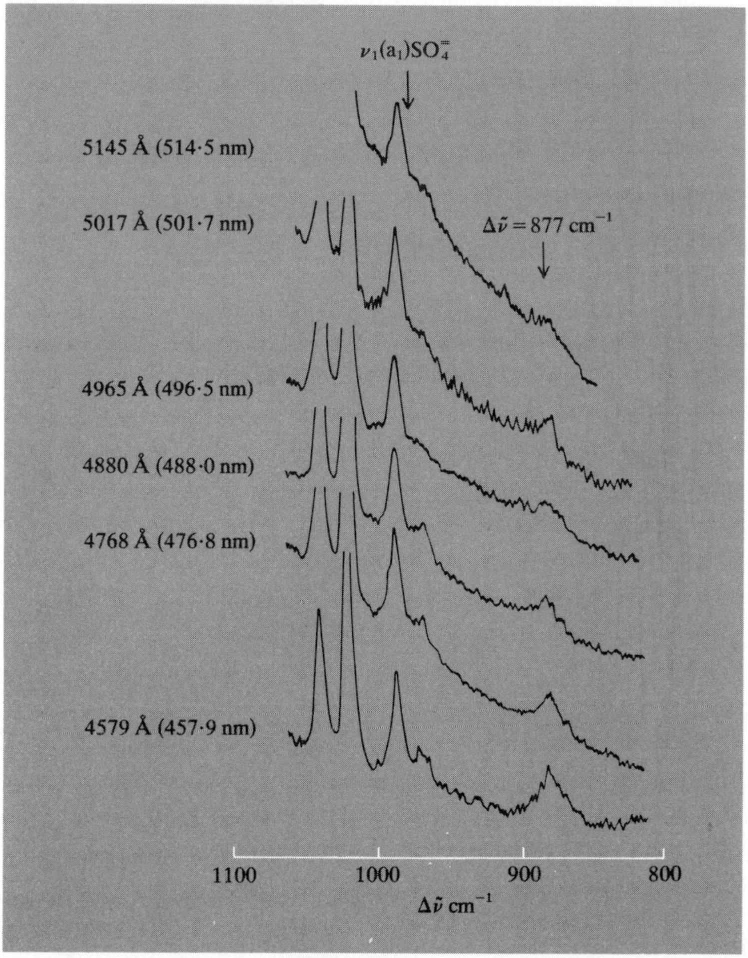

Fig. 7.39(a) Stokes Raman spectra of quinoxaline in aqueous solution in the wavenumber shift range $\Delta\tilde{\nu} = 800\text{–}1100\ cm^{-1}$ obtained with six different excitation wavelengths. The solutions also contained 0·5 M $(NH_4)_2SO_4$

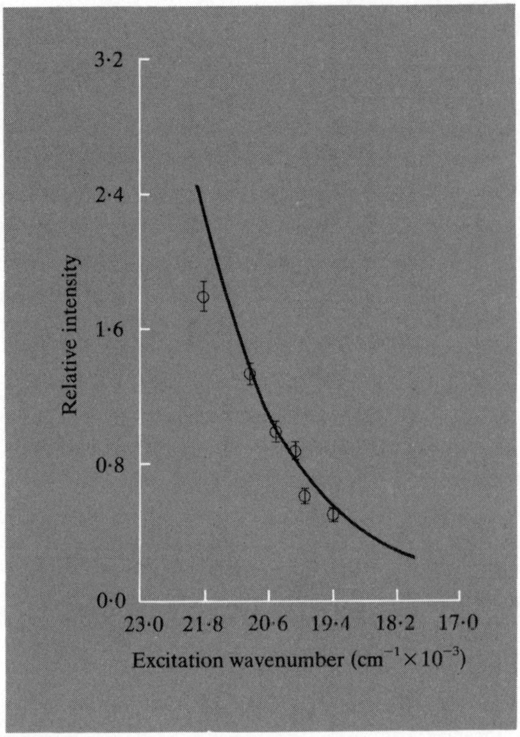

Fig. 7.39(b) Comparison of the dependence on excitation wavenumber of the experimentally determined relative intensities of the 877 cm^{-1} band of quinoxaline (individual points) with that predicted by eq. (5.87) (continuous curve)

condensed on to a tilted copper wedge at 16 K, the resulting matrix samples are deep orange in colour. The Raman spectrum[53] of such a matrix excited with 6471 Å (647.1 nm) radiation contained a band at 1011 cm^{-1} attributed to the symmetric oxygen–oxygen vibration of the ozonide ion in the species $Na^+O_3^-$. When radiation of shorter wavelength was used to excite the Raman spectra, bands at 2013 cm^{-1} and 3001 cm^{-1} were observed with appreciable intensity, in addition to the band at 1011 cm^{-1} (see Fig. 7.40). These additional bands are assigned to the first and second overtones of the 1011 cm^{-1} fundamental, their high intensity resulting from resonance enhancement. From an analysis of these bands the anharmonicity was determined.

The resonance Raman effect can be used as a subtle and sensitive probe of environmental and structural factors. For example, the nucleic acids and polynucleotides, although colourless, have absorption bands in the ultraviolet region which are very sensitive to the polynucleotide conformation. It is thought that changes in the stacking of the purine and pyrimidine bases are responsible for the changes in the ultraviolet absorption spectrum. The ultraviolet spectrum itself shows only broad features and a detailed analysis is not possible. However, the changes in the ultraviolet absorption spectrum produce changes in the intensity of bands in the Raman spectra excited in the visible region.[54] A detailed quantitative understanding

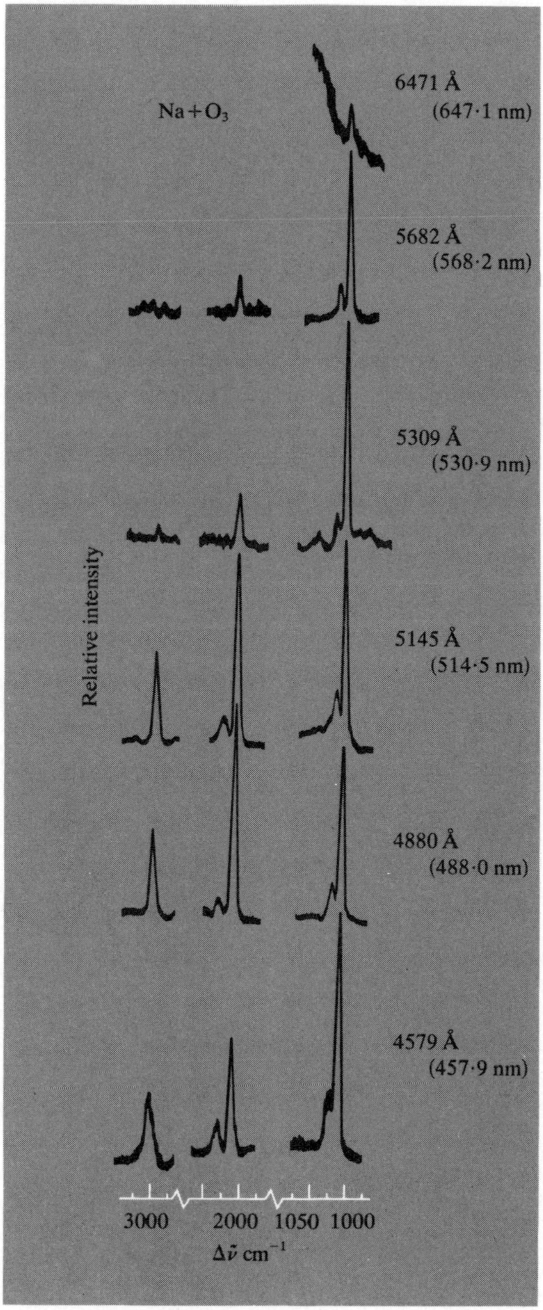

Fig. 7.40 Raman spectra of sodium–ozone matrix reaction products with $Ar/O_3 = 100$ and $Ar/Na > 200$

of the dependence of intensities in the Raman spectrum on the positions of the absorption bands is not yet possible. However, on a semi-empirical basis, useful correlations can be made. For example, Fig. 7.41 shows the variation with temperature of the intensity of a uracil ring vibration occurring at 1236 cm^{-1} in the Raman spectrum of a poly AU complex (=polyriboadenylic and polyribouridylic acids) in buffered aqueous solution at pH 7·0. The sharp rise in intensity at 59° C is thought to be due to a cooperative transition from the helix to random-coil conformation. The gradual increase in intensity from 15° C to 58° C is thought to be indicative of a weakening of the poly U (=polyribouridylic acid) structure as the transition temperature is approached. The structural changes produce changes in the ultraviolet absorption spectra which manifest themselves, through resonance effects, in the observed intensity changes in the Raman spectrum.

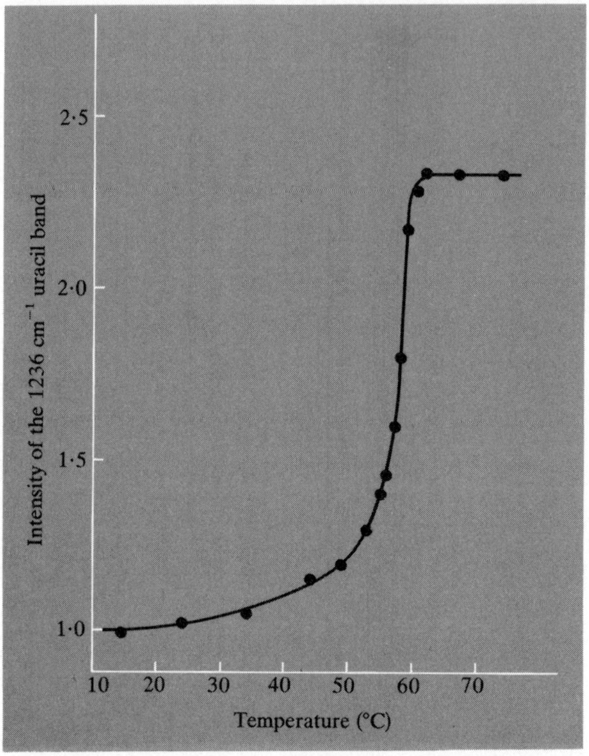

Fig. 7.41 Plot of the intensity of the uracil band at 1236 cm^{-1} in poly AU versus temperature

The investigation of the interaction of oxygen with haemoglobin provides another example of the valuable role played by Raman spectroscopy in the study of complex systems. The resonance Raman spectra[55] (parallel and perpendicular components) of oxyhaemoglobin and ferrocytochrome-C are shown in Fig. 7.42(a) and (b), respectively. These spectra were obtained from aqueous solutions in which the concentration of the haem protein was $0{\cdot}5\times10^{-3}$ M. Several features of these spectra are noteworthy. The spectra are relatively simple because the bands associated with the chromophoric centres are resonance-enhanced, but the vibrations

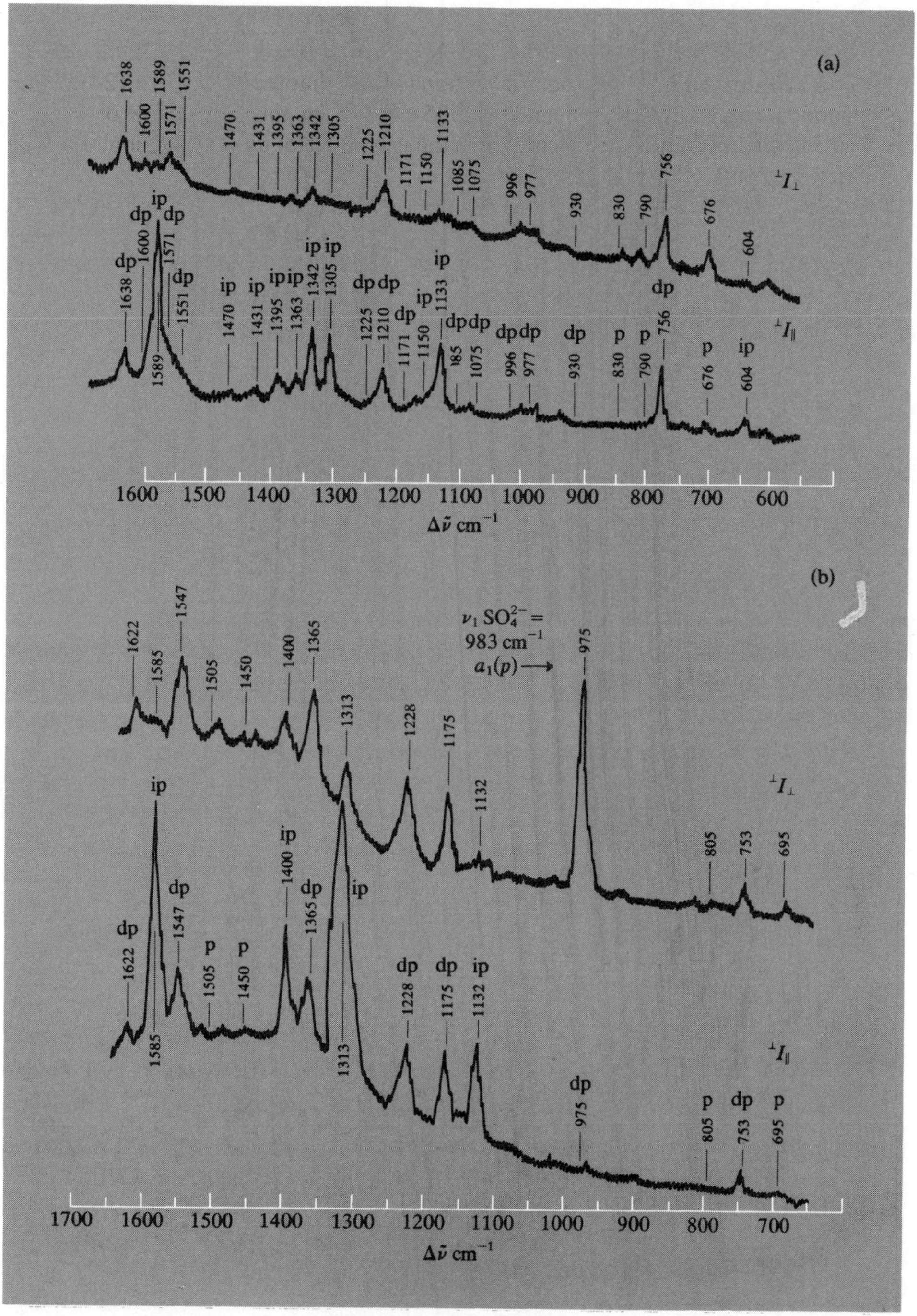

Fig. 7.42 Resonance Raman spectra of (a) oxyhaemoglobin, concentration 0·5 mM, 5682 Å (568·2 nm) excitation, and (b) ferrocytochrome-C, concentration 0·5 mM, 5145 Å (514·5 nm) excitation

of the surrounding large protein molecules are not. This selective enhancement is a valuable aspect of resonance Raman scattering. A number of bands in these spectra also exhibit inverse polarization (see chapter 5, page 124). Striking examples are the bands at 1313 cm^{-1} and 1585 cm^{-1} in the spectra of ferrocytochrome-C and at 1133 cm^{-1} and 1589 cm^{-1} in the spectra of oxyhaemoglobin; $\rho_{\perp}(\pi/2)$ is estimated to be >100 for these bands. If the haem centre possessed D_{4h} symmetry, these bands would have $\rho_{\perp}(\pi/2)=\infty$; the lower values observed indicate that the effective symmetry of the haem centre is lower than D_{4h} in these molecules. Most of the bands which do not exhibit inverse polarization have $\rho_{\perp}(\pi/2)\approx\frac{3}{4}$; the few bands with $\rho_{\perp}(\pi/2)\ll\frac{3}{4}$ are very weak. This means that bands of a_{1g} symmetry are either absent or very weak. This is in accord with the theory of resonance enhancement in this molecule.

7.9 Macromolecules including those of biological interest

Normal Raman spectroscopy is also finding increasing application to studies of macromolecules of biological interest, largely because such molecules can be investigated under conditions approximating to those *in vivo*, namely, in aqueous solution at near-neutral pH. As an example of what can now be achieved, Fig. 7.43(a) shows

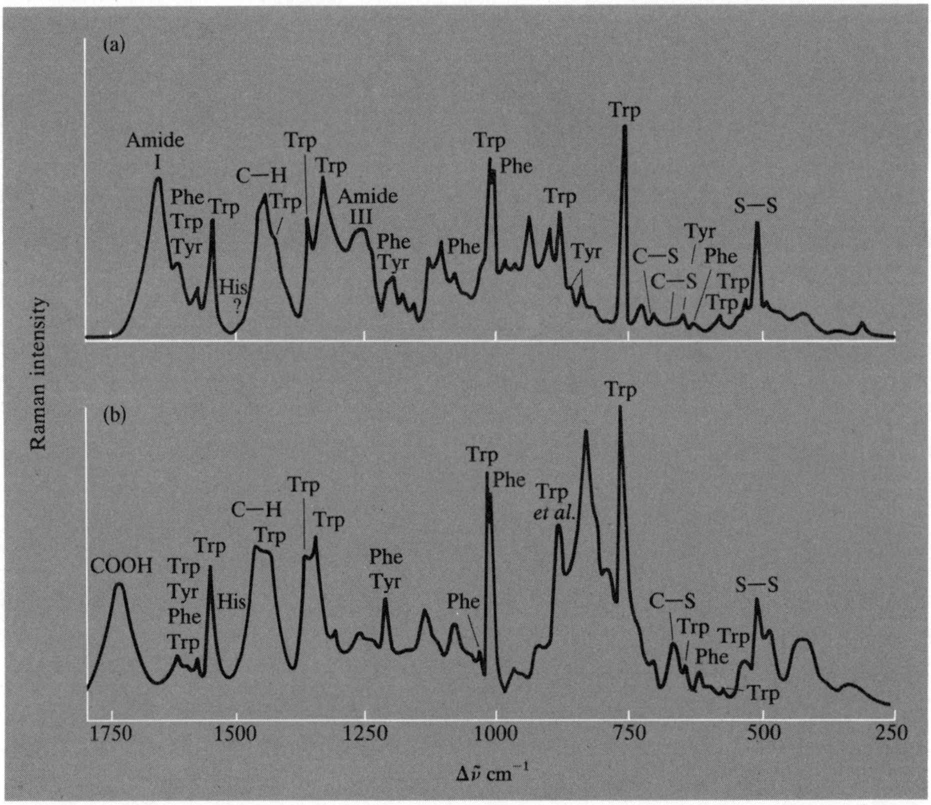

Fig. 7.43 The Raman spectrum of (a) chicken egg-white lysozyme in water pH 5·2 and (b) superposition of the Raman spectra of constituent amino acids at pH 1·0, 6328 Å (632·8 nm) excitation

the Raman spectrum[56] of chicken egg-white lysozyme in water at pH 5·2. Figure 7.43(b) shows the Raman spectrum of an aqueous solution containing the constituent amino acids of lysozyme in approximately the same proportions as in the protein. The general similarity of these two spectra is quite striking. A more detailed analysis of the lysozyme spectrum enables deductions concerning conformation and the environment of the —S—S— bridging group to be made.

Another example relates to bilayers of egg lecithin–water, which provide a good model of a biological membrane.[57] It can be seen from Fig. 7.44 that there are changes in the relative intensities of a number of bands in the Raman spectra of egg lecithin–water systems as the temperature and water content are varied. These intensity changes can be correlated with conformation changes in the hydrocarbon

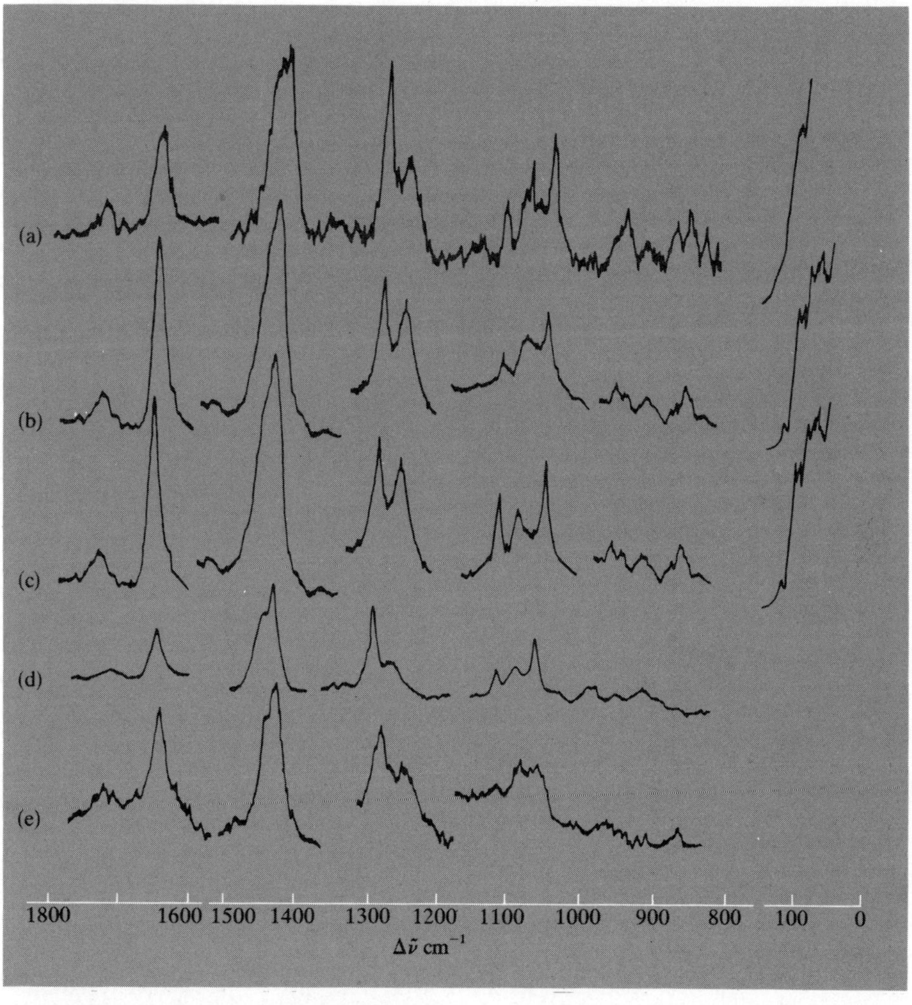

Fig. 7.44 Raman spectra of ex-hen egg lecithin-water systems in several regions in the range 100–1800 cm^{-1}: (a) 3° C, water content 1·0–2·5%, (b) 3° C, water content 5–8%, (c) −20° C, water content 5–8%, (d) −10° C, water content 30%, and (e) 15° C, water content 30%; 4880 Å (488·0 nm) excitation

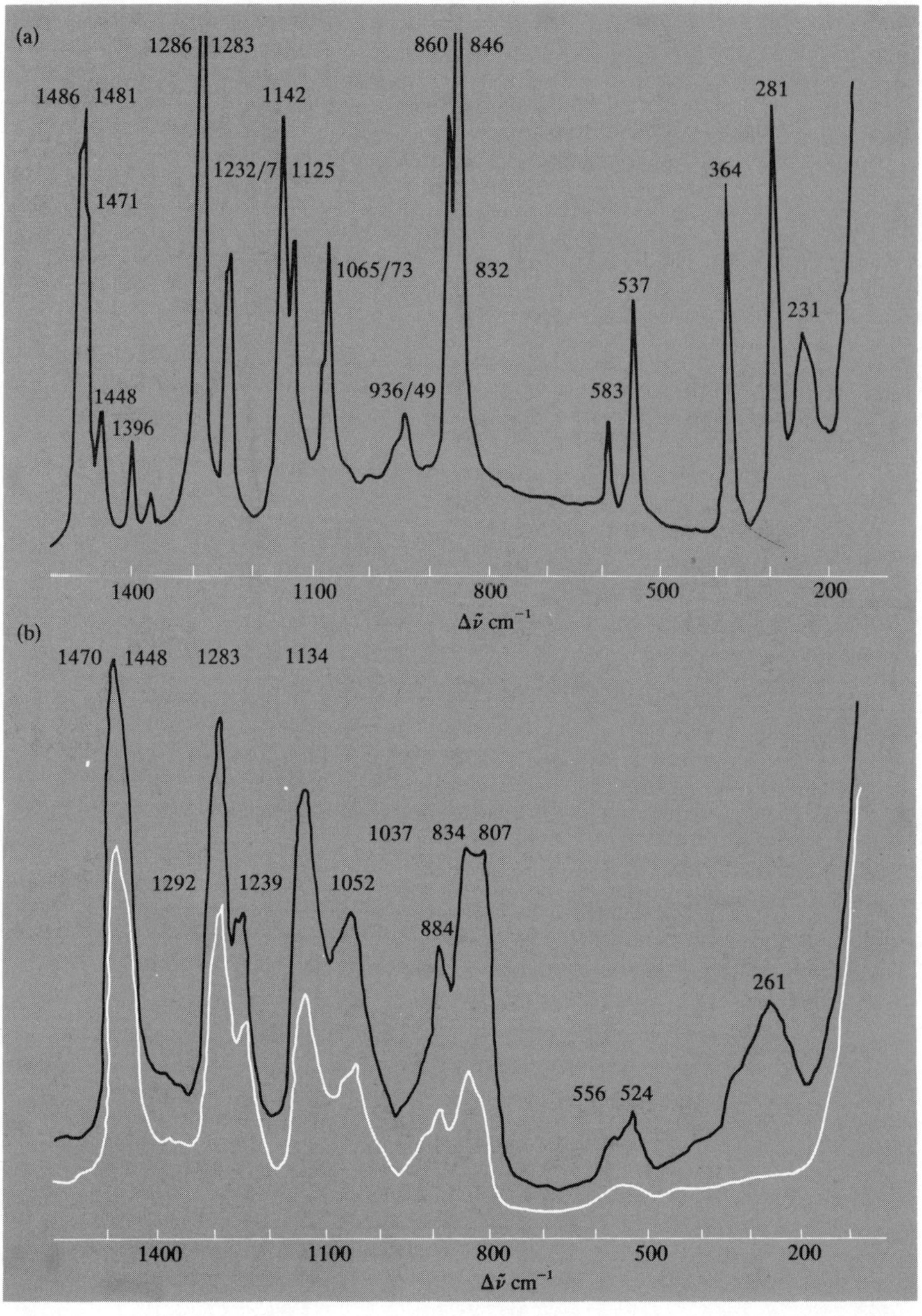

Fig. 7.45 (a) Raman spectrum of crystalline polyethylene glycol. (b) Raman spectrum of molten polyethylene glycol: (——) electric vector of incident beam perpendicular to scattering plane; (▭) electric vector of incident beam parallel to scattering plane. 4880 Å (488·0 nm) excitation

chains in the lecithin. The interaction of an antibiotic with the egg lecithin–water has also been investigated[58] by comparing the Raman spectra of the antibiotic–lecithin system with the Raman spectra of the components.

Spectra of excellent quality may also be obtained from synthetic polymeric materials which had previously proved intractable. In particular, orientation studies akin to those for single crystals may be made and the low-lying wavenumbers characteristic of accordion-like vibrations of the whole skeleton may be observed. Figure 7.45(a) shows a typical Raman spectrum[59] of the crystalline polymer polyethylene glycol. On melting, the spectrum shown in Fig. 7.45(b) is obtained. In this spectrum, the bands are much broader because of the disorder in the molten polymer. Single fibres may also be studied by Raman spectroscopy; and dyestuffs bound to the fibres may be investigated by resonance Raman spectroscopy.

7.10 Qualitative and quantitative analysis

In many situations Raman spectroscopy may be used very effectively for qualitative and quantitative analysis, usually without harming the sample in any way. As far as qualitative analysis is concerned, it can be convenient to record Raman spectra in digital form and to build up a memory bank of characteristic spectra. Computer search techniques may then be used for the identification of spectra from samples of unknown nature. As an illustration of quantitative analysis,[60] Fig. 7.46 shows that it is possible to detect and estimate benzene in carbon tetrachloride at concentration levels down to at least one part in 10^4. A further example of quantitative analysis is provided by Fig. 7.47, which shows part of the vibrational Raman spectrum[61] of a sample of ^{13}C-enriched $CHCl_3$. From the intensities of the two bands in the region of $670\ \mathrm{cm}^{-1}$, characteristic of $^{12}CHCl_3$ and $^{13}CHCl_3$, it was found that the sample contained 58·7 per cent $^{13}CHCl_3$. This analysis was carried out without unsealing the manufacturer's ampoule—which was, of course, transparent. The application of Raman spectroscopy to measurement of pollution in the atmosphere is now receiving attention.

A Raman microprobe and microscope has recently been developed.[62] This enables individual components in a sample to be identified by illuminating the sample with radiation of wavenumber $\tilde{\nu}_0$ but 'viewing' the sample via radiation at $\tilde{\nu}_0 - \tilde{\nu}_M$ produced by Raman scattering, where $\tilde{\nu}_M$ is a characteristic wavenumber of the component. Figure 7.48(a) to (c) shows micrographs obtained in this way from a sample which was a mixture of potassium chromate(VI) and hexacyanoferrate(II) microcrystals with dust particles. The micrograph in Fig. 7.48(a) was obtained by 'viewing' with $\tilde{\nu}_0$; the micrographs in Fig. 7.48(b) and (c) were obtained by 'viewing' with $(\tilde{\nu}_0 - 850)\ \mathrm{cm}^{-1}$ and $(\tilde{\nu}_0 - 2056)\ \mathrm{cm}^{-1}$, respectively, where $850\ \mathrm{cm}^{-1}$ and $2056\ \mathrm{cm}^{-1}$ are characteristic vibrational wavenumbers of the chromate(VI) and hexacyanoferrate(II) ions, respectively. It can be seen that particles of potassium chromate(VI) and hexacyanoferrate(II) are readily distinguished.

Raman spectroscopy is also beginning to find application to the measurement of temperature in flames. One method is based on the relative intensities of the unresolved Stokes Q branches associated with the fundamental ($v = 1 \leftarrow v = 0$) and 'hot-band' ($v = 2 \leftarrow v = 1$, $v = 3 \leftarrow v = 2$, etc.) transitions. Because of anharmonicity, these transitions have slightly different wavenumbers. The intensity of the fundamental transition decreases and the intensities of the 'hot-band' transitions increase

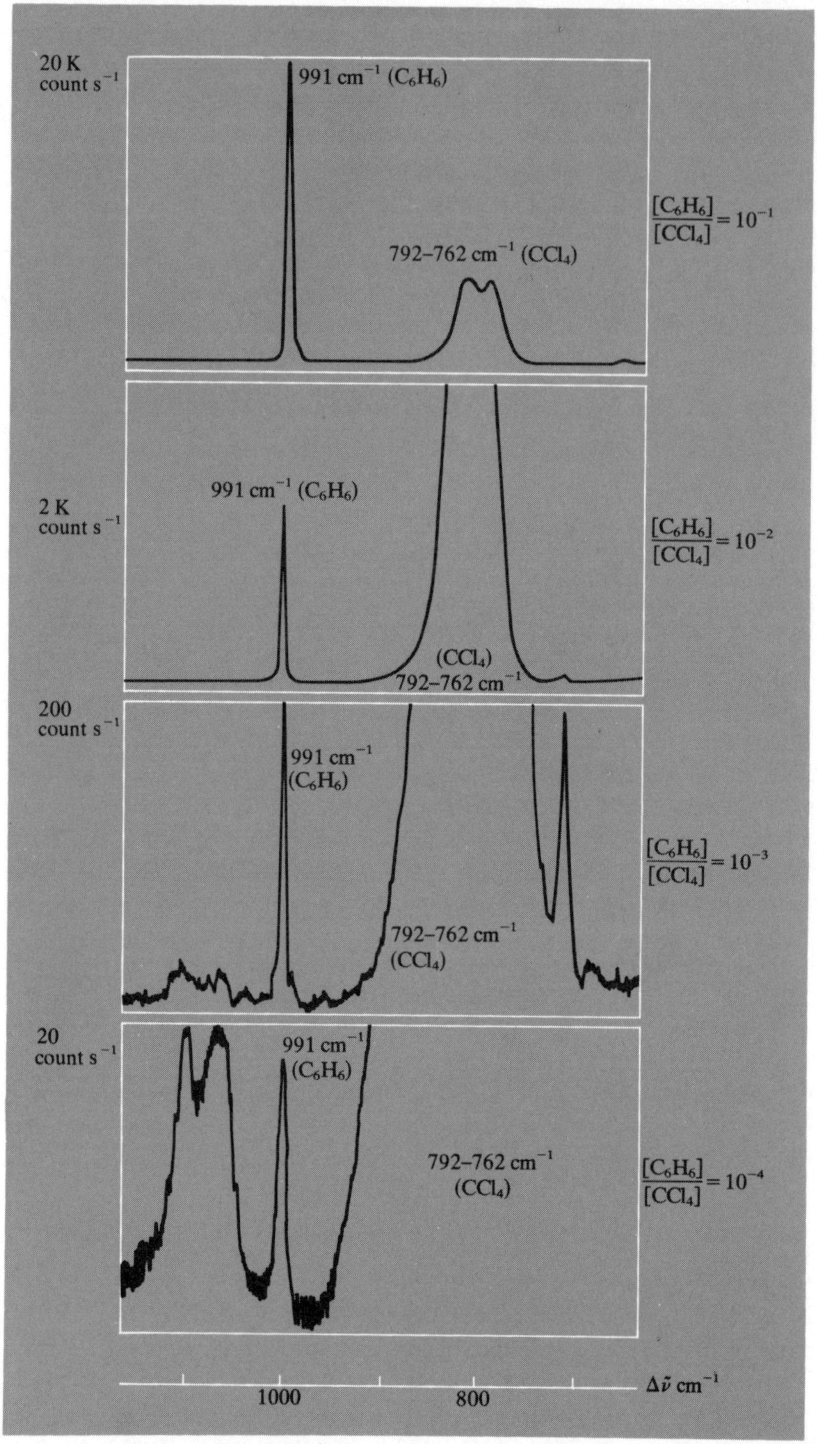

Fig. 7.46 Quantitative analysis of benzene in carbon tetrachloride using Raman spectroscopy: 4880 Å (488·0 nm) excitation

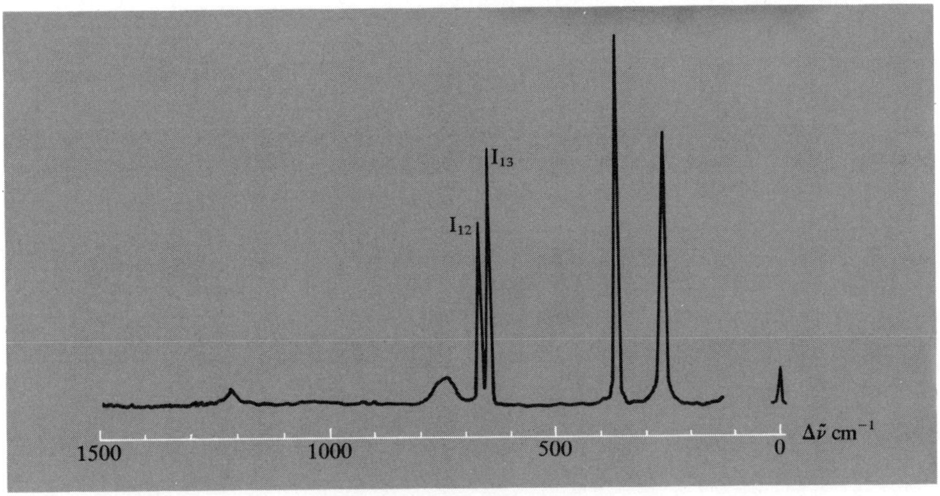

Fig. 7.47 Raman spectrum of chloroform 57·6% labelled with ^{13}C, *in the original ampoule*. The intensities of the two bands near 670 cm^{-1} yield $I_{13}/(I_{12}+I_{13}) = 58{\cdot}7\%$ ^{13}C

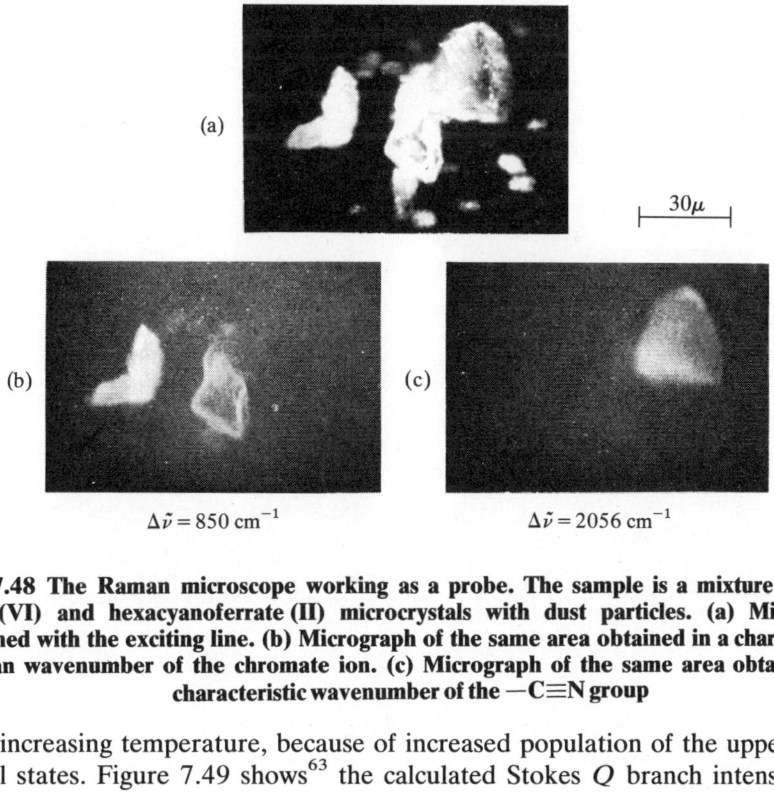

Fig. 7.48 The Raman microscope working as a probe. The sample is a mixture of chromate (VI) and hexacyanoferrate (II) microcrystals with dust particles. (a) Micrograph obtained with the exciting line. (b) Micrograph of the same area obtained in a characteristic Raman wavenumber of the chromate ion. (c) Micrograph of the same area obtained in a characteristic wavenumber of the —C≡N group

with increasing temperature, because of increased population of the upper vibrational states. Figure 7.49 shows[63] the calculated Stokes Q branch intensities for nitrogen over the temperature range 300 to 3500 K. A typical experimental result[64] obtained for nitrogen in a methane–air flame at a point 2·5 cm from the burner plug of a laminar-flow burner is shown in Fig. 7.50. Computer simulation of the observed spectral profile achieves a best-fit for a temperature of 1650 ± 180 K.

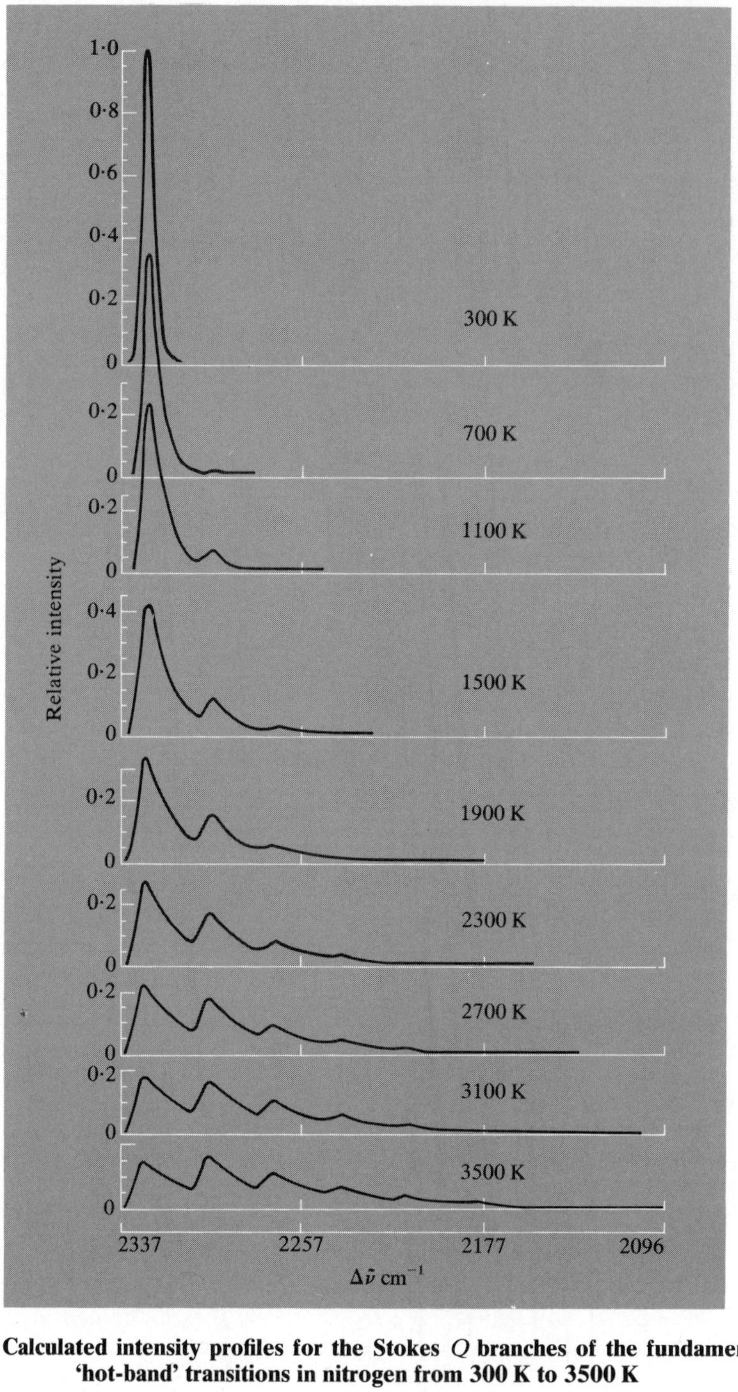

Fig. 7.49 Calculated intensity profiles for the Stokes Q branches of the fundamental and 'hot-band' transitions in nitrogen from 300 K to 3500 K

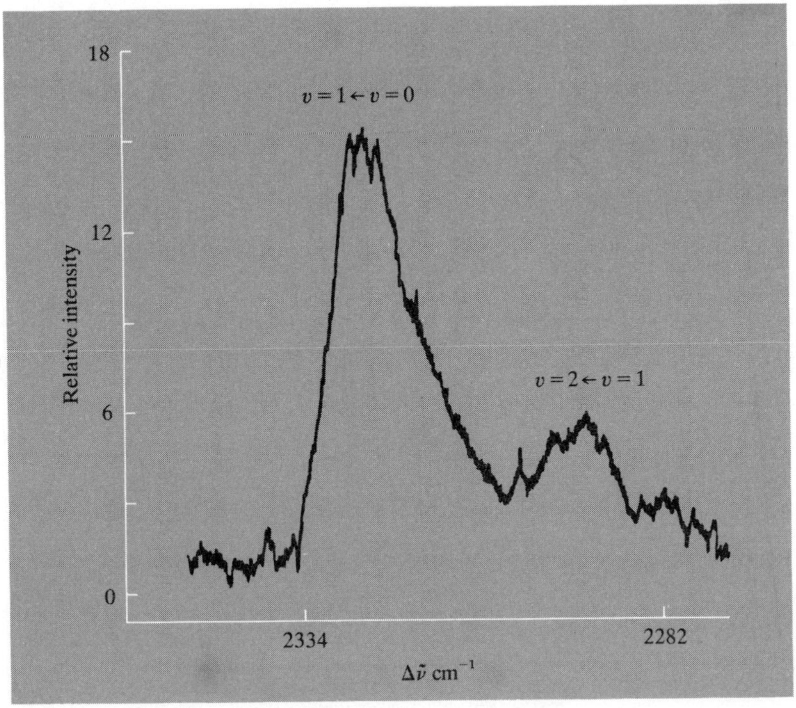

Fig. 7.50 Directly recorded Stokes Q branch fundamental (v=1←v=0) and first hot-band (v=2←v=1) of N_2 in a methane–air flame from a laminar-flow burner. Exciting radiation: 4880 Å (488·0 nm). Burner gases: 7 l m^{-1} air; 0·3 l m^{-1} methane. 'Best-fit' temperature at a position 2·5 cm from the burner plug, 1650 ± 180 K

7.11 Rapid-scan Raman spectroscopy

In many cases, the sensitivity of Raman spectroscopy is such that spectra (or relevant parts thereof) may be recorded in very short periods of time. The scan time for, say, 100 cm^{-1} can range from the order of minutes to 10^{-2} minutes or less, depending on the techniques employed. The detection of short-lived species and the monitoring of chemical reactions is therefore possible. Figure 7.51 shows the evolution with time of the Raman spectra in the region 50 to 350 cm^{-1} of a mixture of acetic acid and bromine.[65] The progressive disappearance of molecular bromine Br_2 and the growth and decay of the associated ions Br_5^- and Br_3^- can be clearly seen. From the time dependence of the intensities of bands in the spectra in Fig. 7.51 the concentration–time plots for the species Br_2, Br_5^-, and Br_3^-, shown in Fig. 7.52, can be obtained. The scan times for the spectra in Fig. 7.51 were about two minutes. Acetyl bromide catalyses the reaction of bromine and acetic acid, but it was possible to obtain the concentration–time plots, shown in Fig. 7.53, from the Raman spectra of the reaction mixture using scan speeds of the order of 4 seconds per spectrum.

An interesting new development[66] in high-speed Raman spectroscopy involves scanning the spectrum by rapid variation of the exciting frequency and observation at a fixed frequency. The rapid variation of the exciting frequency can be achieved using a pulsed dye laser (see Fig. 7.54). If such a 'scanning' dye laser is used in conjunction with an image intensifier and vidicon television camera detection system, ultra-fast

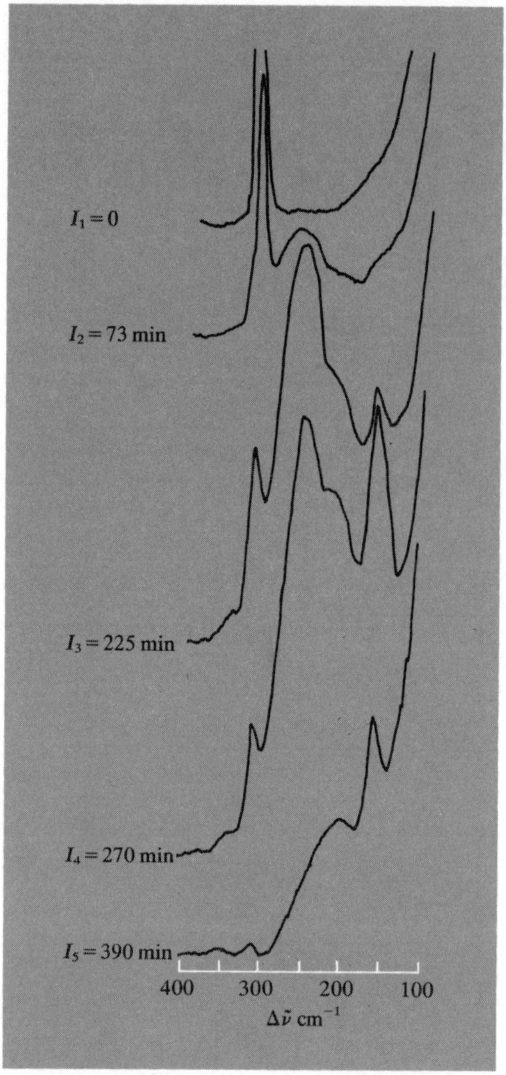

Fig. 7.51 Time evolution of the Raman spectra of a mixture of acetic acid and bromine at 373 K showing the disappearance of bromine ($\Delta\tilde{\nu} = 310$ cm^{-1}) and the rise and fall in concentration of Br_3^- ($\Delta\tilde{\nu} = 170$ and 210 cm^{-1}) and Br_5^- ($\Delta\tilde{\nu} = 250$ cm^{-1})

scanning of small regions of Raman spectra can be achieved. Figure 7.55 shows some typical spectra obtained using this technique.

7.12 Electronic transitions

The study of Raman scattering associated with electronic transitions in appropriate ions has recently become possible. Figure 7.56 shows the $x(zx)z$ Raman spectrum[67] in the region 0 to 1000 cm^{-1} of a single crystal of CaF_2 doped with Eu^{3+} ions such that

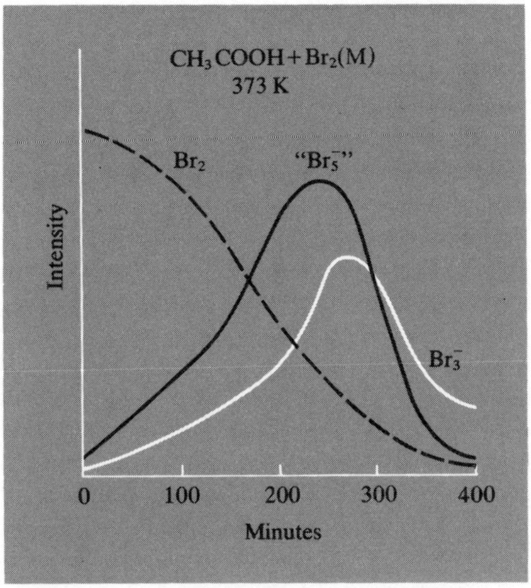

Fig. 7.52 Raman intensity as a function of time for the species Br_2, Br_3^-, and Br_5^- in a mixture of $CH_3COOH+Br_2$ (0·5 M) at 373 K

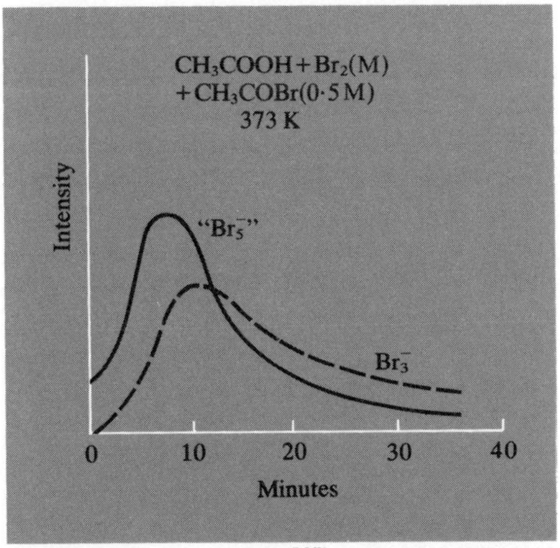

Fig. 7.53 Raman intensity as a function of time for the species Br_3^- and Br_5^- in a mixture of CH_3COOH and Br_2 (1·0 M) and CH_3COBr (0·5 M)

$Eu^{3+} : Ca^{2+} = 0{\cdot}005$. In this experiment the radiation was incident along x and the direction of observation was z which was coincident with the four-fold crystal axis. In addition to a number of lattice vibrations, several bands associated with electronic transitions in the Eu^{3+} ion are observed. Some lie in the region 260 to 500 cm^{-1} and are associated with the transition ${}^7F_1 \leftarrow {}^7F_0$; others, including the strongest band at 797 cm^{-1}, lie in the region 700 to 1000 cm^{-1} and are associated with the transition ${}^7F_2 \leftarrow {}^7F_0$;

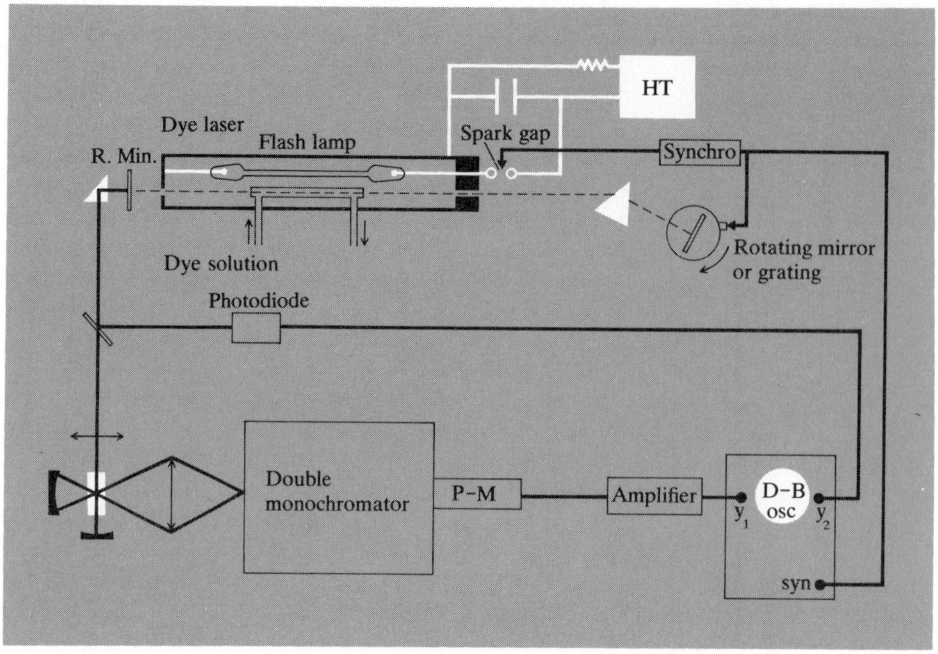

Fig. 7.54 Schematic diagram of a 'scanning laser' Raman spectrometer for fast recording of Raman spectra

7.13 Bandwidths in liquids and molecular motions

The width and profile of vibrational bands in liquids are produced by both vibrational and reorientational relaxation mechanisms. Whereas infrared spectroscopy cannot distinguish between these two contributions, Raman spectroscopy can, in certain cases at least, provided polarization studies are made of the band contours. This kind of investigation is not particularly easy: good band contours are difficult and tedious to obtain experimentally and the mathematical analysis is complicated. Figure 7.57 shows the polarized and depolarized components of the Stokes Raman band associated with the $\nu_3(a_1)$ C—I stretching mode of CH_3I centred at 526 cm^{-1}. The depolarized band is much broader than the polarized band. From an analysis[68] of the profiles of these bands, it is found that the vibrational half-width is about 2 cm^{-1}. The rotational half-width is about 4 cm^{-1} corresponding to a 'tumbling' rotation diffusion constant of the order of 10^{11} s^{-1}, which is in good agreement with values obtained from other measurements.

7.14 Vibrational Raman intensities and bond parameters

It has been shown in chapter 3 that a combination of intensity and polarization measurements on vibrational bands in liquids can lead to values of the squares of the invariants of the derived polarizability tensor, $(a')^2$ and $(\gamma')^2$, and hence to their magnitudes. The values of such invariants can be related to polarizability changes in individual bonds in the molecule.[69,70] For example, the value of a'_{Q_1} for the spherically symmetric stretching mode in the tetrahedral group IV halides, MX_4, is

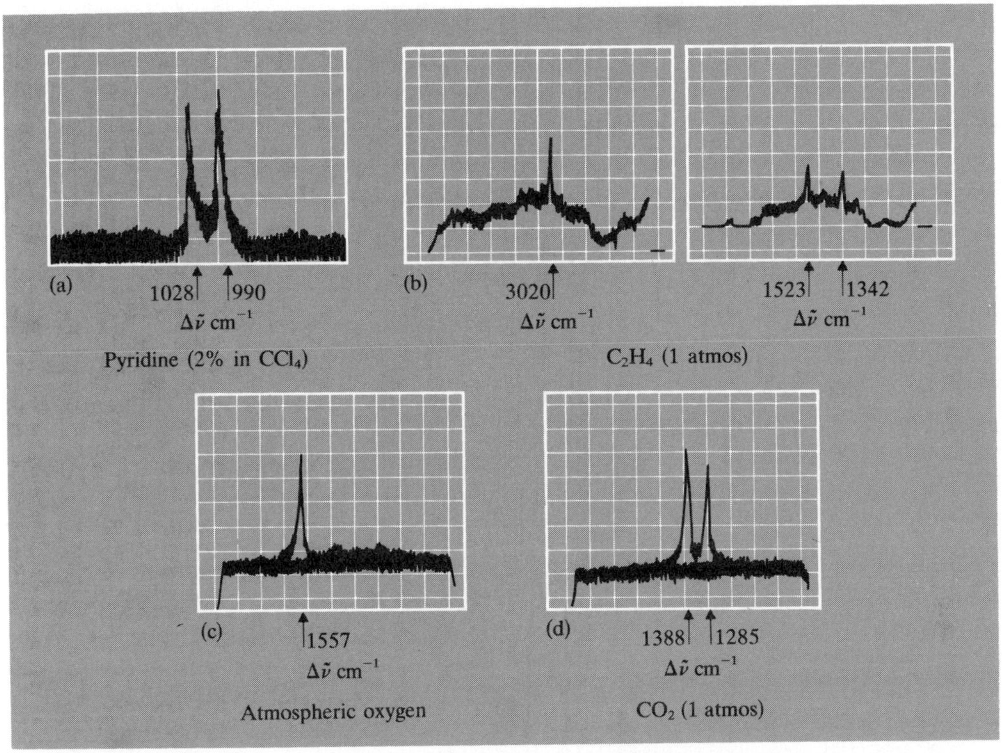

Fig. 7.55 Examples of portions of Raman spectra recorded with an ultra-fast Raman spectrometer system using a 'scanning' laser in conjunction with an image intensifier and vidicon television camera; the scan time in each case is 2×10^{-3} s: (a) pyridine (2% in CCl_4), 5471 Å (547·1 nm) excitation, (b) ethylene (760 torr), 4880 Å (488·0 nm) excitation, (c) oxygen in the air (150 torr), 4880 Å (488·0 nm) excitation, and (d) carbon dioxide (760 torr), 4880 Å (488·0 nm) excitation

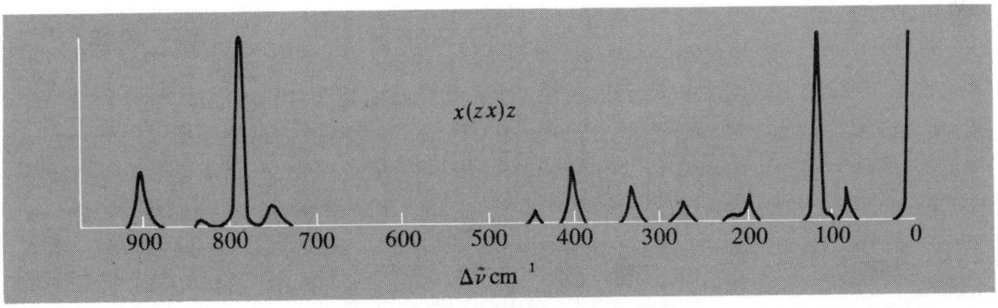

Fig. 7.56 ***x(zx)z*** **polarized Raman spectrum of a Eu^{3+}:CaF_2 single crystal at 3000 K. Various excitation wavelengths were used**

related to a'_{M-X}, the derived mean polarizability of the $M—X$ bond, by the equation

$$a'_{Q_1} = \pm 2 m_X^{-\frac{1}{2}} a'_{M-X}$$

where m_X is the relative atomic mass of the atom X. The normal coordinate Q_1 is 'mass-adjusted' and so has the dimension of $m^{\frac{1}{2}}$ whereas the internal coordinate $\Delta(M—X)$ is not.

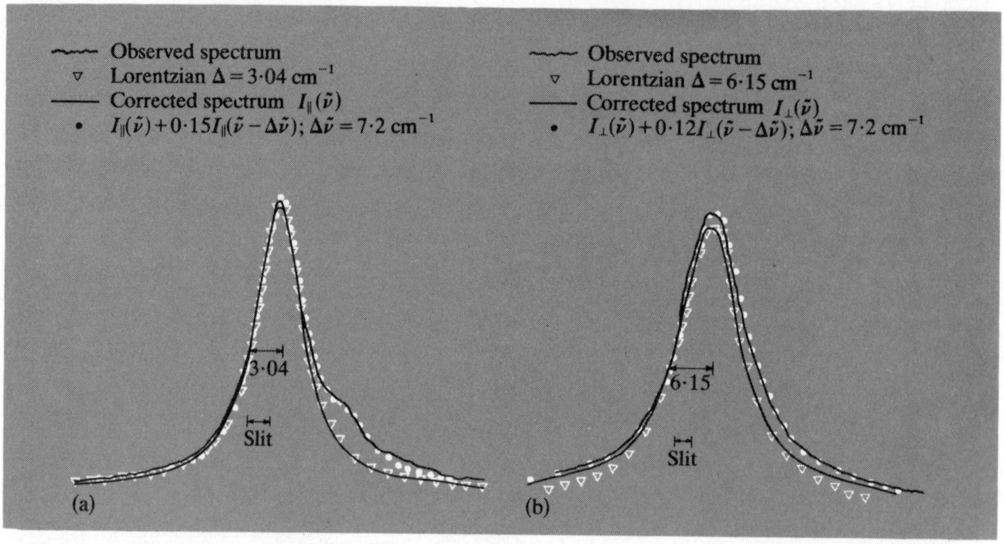

Fig. 7.57 The Stokes Raman band associated with the $\nu_3(a_1)$ C—I stretching mode of CH_3I, centred at 526 cm^{-1}: (a) the polarized component and (b) the depolarized component. In the corrected spectrum, a partially overlapping hot band has been removed

Bond polarizability derivatives determined in this way may be correlated with the nature of the M—X bond, particularly the degree of bond order, although this involves the assumption of a particular model. Bond orders for some polyatomic ions calculated on the basis of two different models[71,72] are given in Table 7.4. Bond polarizabilities may also be used to interpret vibrational Raman intensities.[73]

Table 7.4 Bond orders for some polyatomic ions

Ion	a'_{M-X} experimental	Bond order Model I (ref. 71)	Bond order Model II (ref. 72)
SO_4^{2-}	1·37	1·47	1·08
ClO_4^-	1·73	1·76	1·26
IO_4^-	2·74	1·80	1·29
$ZnCl_4^{2-}$	0·87	0·25	0·35
$CdCl_4^{2-}$	1·04	0·21	0·33
$HgCl_4^{2-}$	2·10	0·35	0·56
$GaCl_4^-$	1·12	0·43	0·46
$ZnBr_4^{2-}$	1·80	0·38	0·61
$CdBr_4^{2-}$	2·46	0·40	0·67
$HgBr_4^{2-}$	5·11	0·66	1·14
$GaBr_4^-$	3·08	0·91	1·07

a'_{M-X} in units of 10^{-16} cm (10^{-2} nm^2).

7.15 Raman optical activity

As has been shown in chapter 5, sections 5.7.2—5.7.4, the intensity of Raman scattering from chiral molecules is slightly different according as the incident radiation is right or left circularly polarized. Part of the depolarized Raman and Raman CID spectra[74] of (+)α-pinene are shown in Fig. 7.58(a) to (c). Figure 7.58(a) shows $^{®}I_{\|}+{}^{©}I_{\|}$, Fig. 7.58(b) shows $^{®}I_{\|}-{}^{©}I_{\|}$, and Fig. 7.58(c) shows a derived CID spectrum[74] in which the background has been subtracted in estimating $\Delta_{\|}=({}^{®}I_{\|}-{}^{©}I_{\|})/({}^{®}I_{\|}+{}^{©}I_{\|})$. It can be seen that six bands exhibit Raman CID. The development of molecular models which will enable the relation between Raman CIDs and molecular vibration to be more fully understood is awaited.

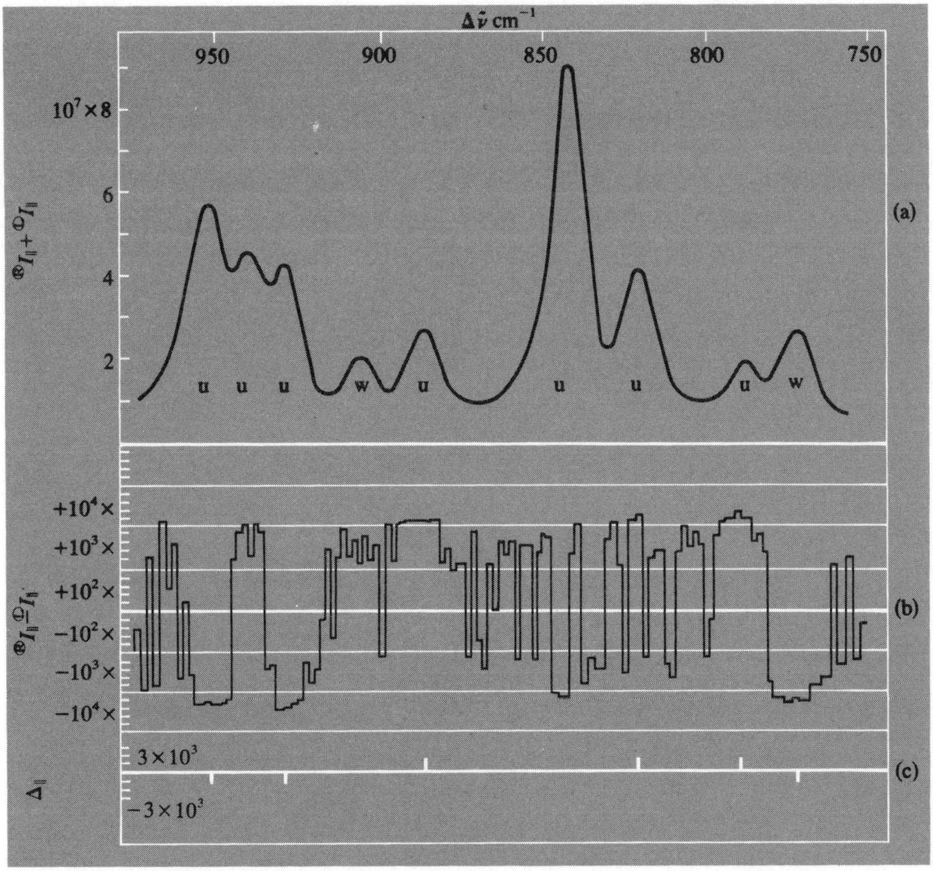

Fig. 7.58 Part of the depolarized Raman and Raman CID spectra of (+)-pinene; (a) $^{®}I_{\|}+{}^{©}I_{\|}$, (b) $^{®}I_{\|}-{}^{©}I_{\|}$ and (c) the derived CID spectrum in which the background has been subtracted in estimating $\Delta_{\|}$. u and w indicate unpolarized and weakly polarized bands. Instrumental conditions: laser wavelength 4880 Å (480·0 nm), laser power 1 W, slit width 10 cm^{-1}, scan speed 1 cm^{-1}/mm

7.16 Magnetic Raman optical activity

Very recently, the first observation of magnetic Raman optical activity has been reported.[75] Using a magnetic field of about 0·7 T, Raman CIDs have been induced in resonance Raman bands of ferrocytochrome-C. The spectra are shown in Fig. 7.59(a) to (c), and it can be seen that magnetic CIDs occur in most of the strong

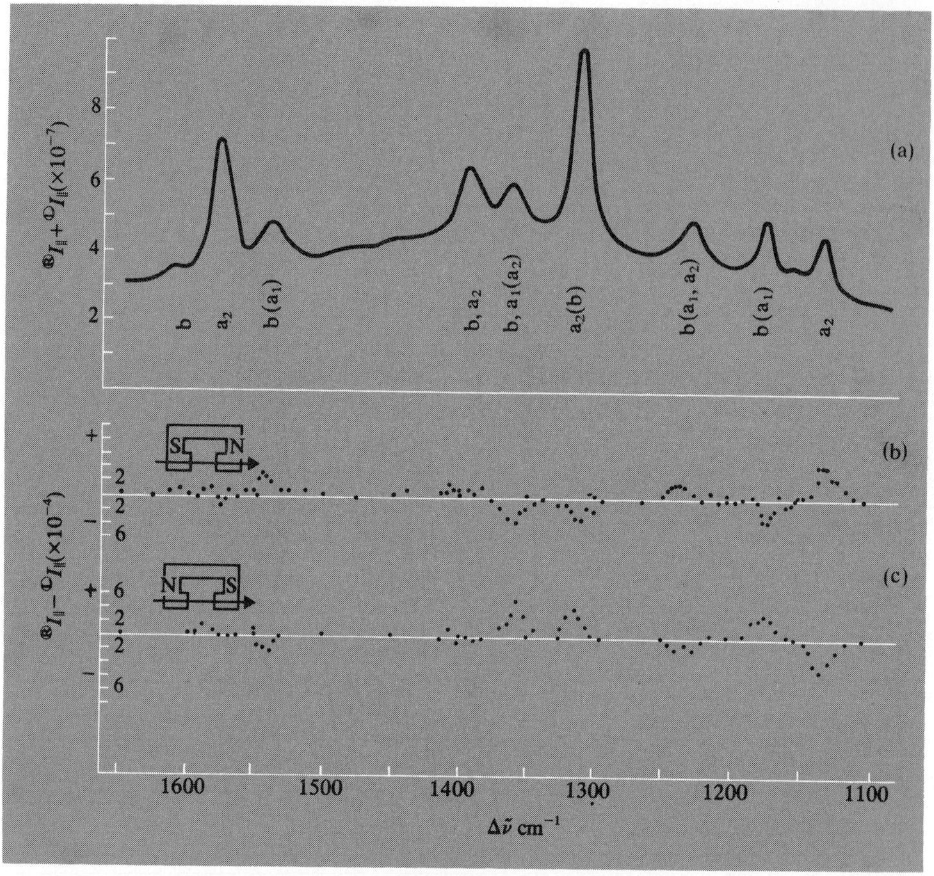

Fig. 7.59 The depolarized magnetic resonance Raman spectra of ferrocytochrome-C (a) ${}^{\circledR}I_{\parallel}+{}^{\copyright}I_{\parallel}$ and (b) and (c) ${}^{\circledR}I_{\parallel}-{}^{\circledR}I_{\parallel}$. The magnetic field directions are opposite in (b) and (c). Experimental conditions: 1×10^{-4} M aqueous solution of cytochrome-C with excess sodium dithionite; magnetic field strength about 0·7 T; 5145 Å (514·4 nm) excitation. The ${}^{\circledR}I_{\parallel}-{}^{\copyright}I_{\parallel}$ spectra were recorded manually at fixed points with a counting period of about 8 min. To estimate the dimensionless CIDs, the background should first be subtracted from the ${}^{\circledR}I_{\parallel}+{}^{\copyright}I_{\parallel}$ spectrum

resonance-enhanced Raman bands between 1100 and 1600 cm^{-1}. There is good reflection symmetry on reversing the magnetic field direction, and the CIDs disappeared when the field was removed. The interpretation of these interesting results awaits further development of the theory.

7.17 The solid state

7.17.1 *Introduction*

The theoretical development in this book has been related to an assembly of non-interacting randomly oriented molecules in the gas or liquid states. Before considering some examples of Raman spectra of solids, we shall examine briefly to what extent this theory can be carried over to the solid state.

Consider first a molecular crystal. A simple picture of such a crystal is the *oriented gas model,* in which the molecular entity has a relatively fixed orientation with respect to the crystal axes. In such a model, the molecule retains its internal vibrations but the rotation and translation degrees of freedom of the free molecule are replaced by external vibrations, torsional motions of the molecule about its axes on the lattice site (librations), and restricted translational displacements within the lattice. According to this picture, the Raman spectra of molecular crystals might be expected to be attractively simple, since the absence of rotational fine structure should lead to sharper vibrational lines than in the liquid state.

However, this is not the case. Although there is a broad correlation between the internal vibrations of molecules in the gas phase and in the crystal, the Raman spectrum of a molecular crystal shows many differences from that of the gas phase. For example, in the spectrum of the crystal there are many new bands with low wavenumber shifts ($\tilde{\nu}_M < 800\,\text{cm}^{-1}$) arising from external vibrations; there are substantial changes in the shape and intensity of the internal vibrational bands; and gas phase fundamental vibrations can be split into additional bands. This splitting can be due to the symmetry of the site (site group splitting) which, if lower than that of the molecule, may remove some or all of the degeneracy of internal vibrations; to interactions with internal vibrations of *other* molecules in the same unit cell (correlation field splitting); to coupling between internal and external vibrations; and in certain cases to long-range electrostatic forces in the crystal.

The nature and extent of such splitting depends partly on the symmetry of the molecule and the crystal, and partly on the strength of the coupling between different types of motion. Often, the intermolecular coupling is much weaker than the intramolecular forces, and so the effects of the crystal field can be treated as a perturbation of the molecular field. Thus, in certain circumstances, the perturbed oriented gas phase model is likely to give a reasonable account of the internal vibrations of molecules in crystals.

There are, however, many systems to which this model cannot be applied. Such systems include those which have no internal modes, e.g., covalent crystals like diamond, and ionic solids like sodium chloride. The analysis of the Raman spectra of such systems demands an entirely different approach, in which the dynamics of the crystal lattice as a whole are considered.

It will be clear from this discussion that a general treatment of Raman scattering from solids would require an extensive development of group theory, intermolecular force fields, and lattice dynamics. This cannot be achieved within the confines of this book. We must therefore content ourselves with presenting a few examples of Raman spectra of solids which lend themselves to discussion, at least in qualitative terms, without the need for further theoretical development.

7.17.2 *Raman spectrum of calcite*

The crystal structure of calcite is well established: it has a trigonal structure, with two molecules per unit cell. As can be seen from Fig. 7.60, the calcium ions and the carbon atoms of the carbonate ions all lie on the trigonal axis (z-axis), and the orientations of the two carbonate ions are staggered relative to each other so that there is a centre of symmetry in the unit cell. Since the unit cell contains ten atoms,

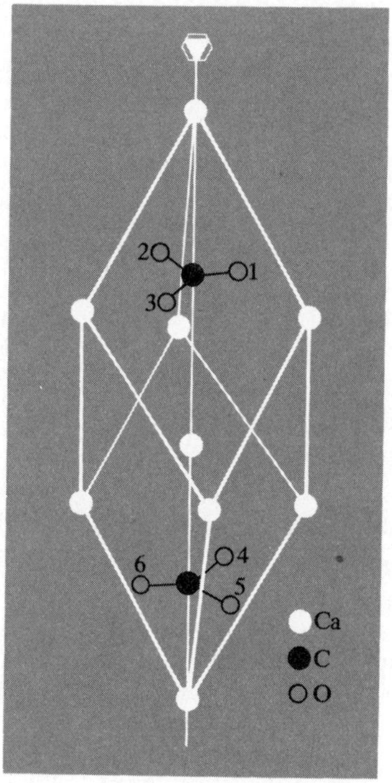

Fig. 7.60 Unit cell of calcite

there are $\{(3\times 10)-3\}=27$ vibrational degrees of freedom, but we shall concern ourselves only with those which are Raman active. The point group symmetry is D_{3d}, and symmetry arguments enable the vibrational Raman active modes to be classified as one vibration of a_{1g} symmetry and four doubly degenerate pairs of vibrations of e_g symmetry. We note that all these vibrations have *gerade* symmetry, i.e., they conserve the symmetry with respect to inversion at the centre of symmetry. The forms of some of the Raman active vibrations in the crystal may be correlated with the vibrations of the free CO_3^{2-} ion (point group symmetry D_{3h}), and are termed internal modes; the forms of the others are related to overall motions (translations and restricted rotations or librations) of the two CO_3^{2-} ions relative to each other, and are termed external modes. Figure 7.61(a) shows how the Raman active vibrations in the crystal are correlated with the internal vibrations and relative overall motions of the free ions. The a_{1g} vibration is an internal mode and involves two carbonate ions performing *in phase* vibrations each of a_1' symmetry in the free ion. Similarly, two of the degenerate pairs of vibrations of e_g symmetry in the crystal are internal modes, and involve the two carbonate ions performing *in phase* vibrations each of e' symmetry in the free ion. The remaining two degenerate pairs of vibrations of e_g symmetry in the crystal are external modes. One pair involves librations of the two ions and the other pair involves translational motions of the two ions. In both cases, the motion of one CO_3^{2-} ion relative to the other must be such that the inversion symmetry is preserved.

From symmetry considerations alone, the pattern of entries in the scattering tensors associated with the a_{1g} and e_g vibrations may be deduced. These are shown in Fig. 7.61(b). The tensor components are referred to a coordinate system xyz fixed in the crystal with z, the trigonal (optical) axis, and y, a binary axis. As we have already seen in chapter 3, planned selection of the direction and state of polarization

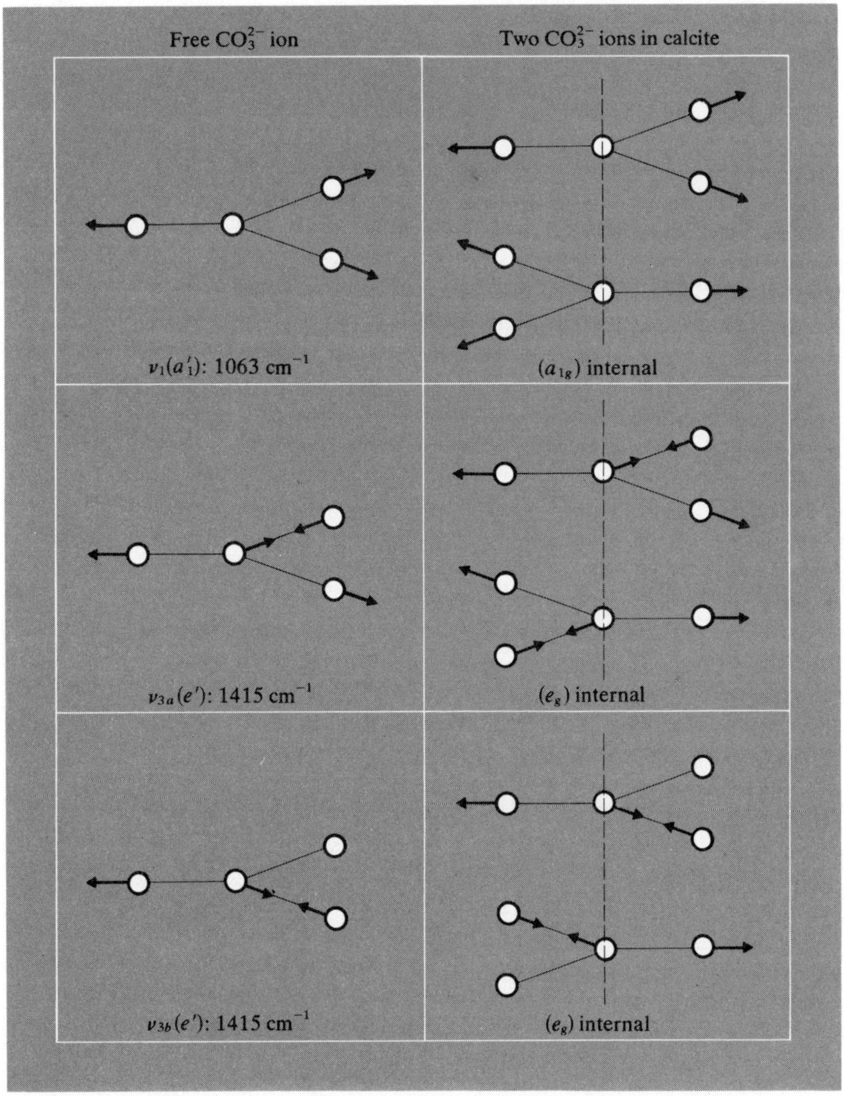

Fig. 7.61(a) The forms of the Raman active vibrations in calcite

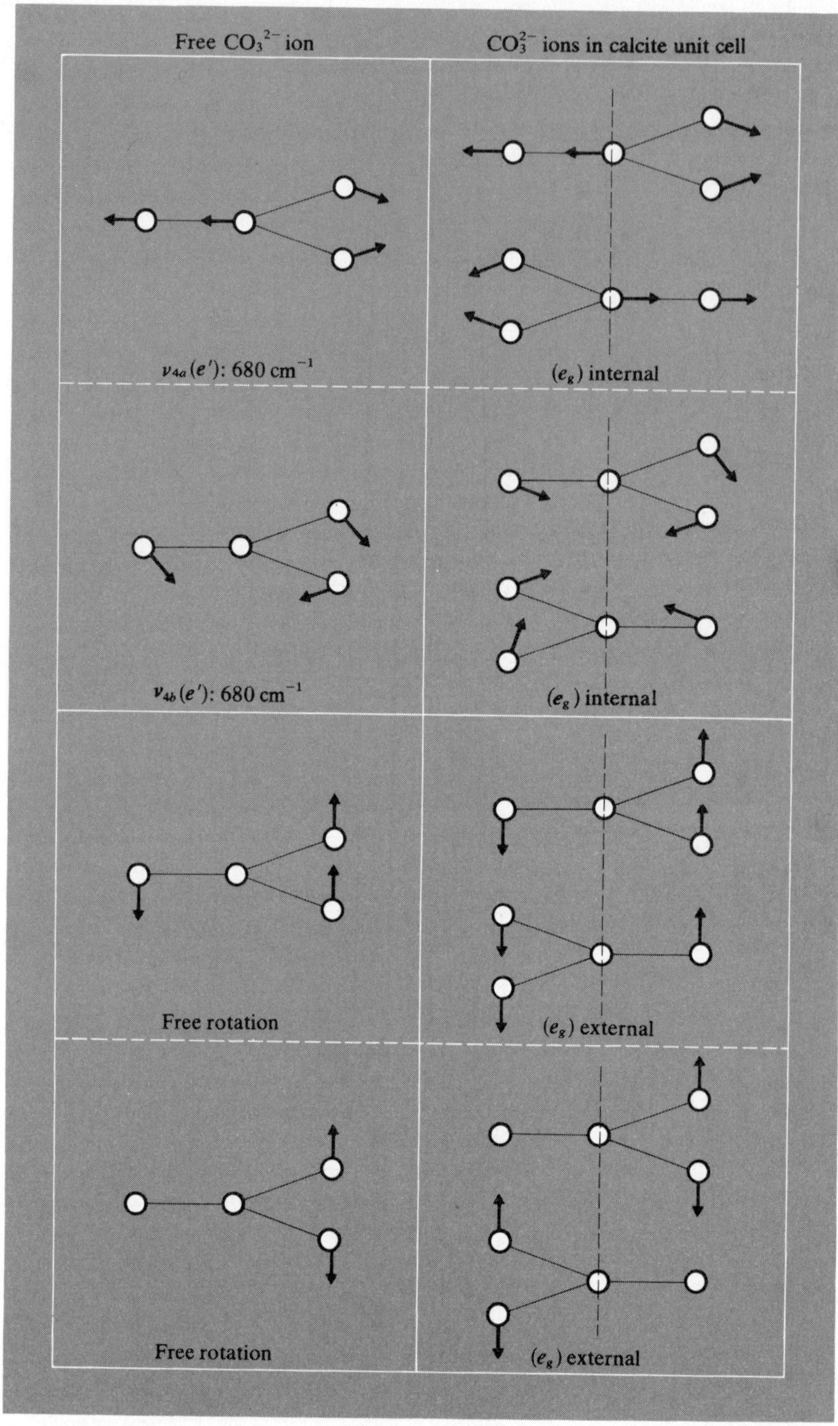

Fig. 7.61(a) continued

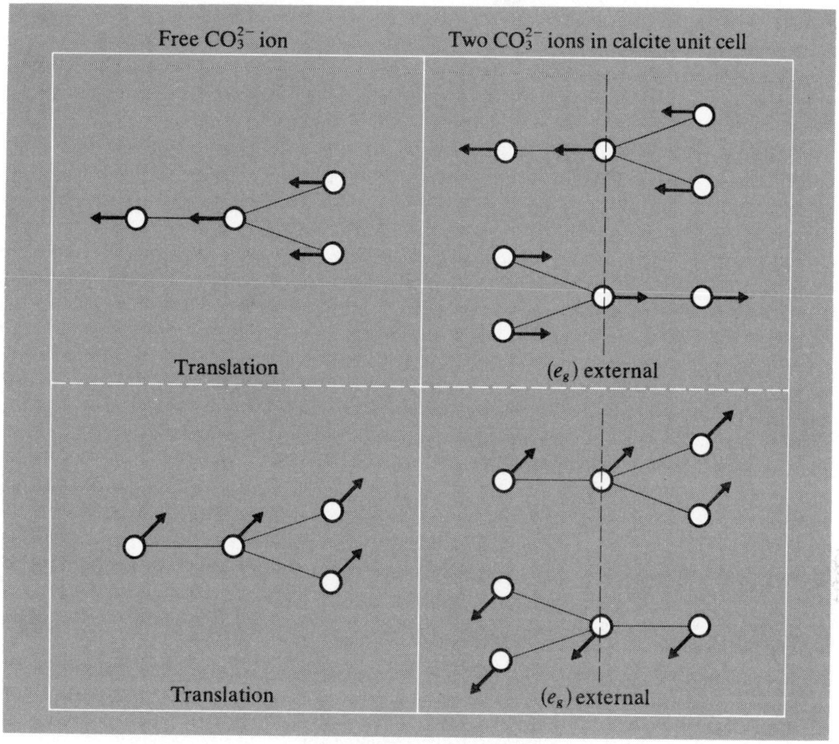

Fig. 7.61(a) continued

Scattering tensors

$$a_{1g}: \begin{matrix} a & 0 & 0 \\ 0 & a & 0 \\ 0 & 0 & b \end{matrix}$$

$$e_g: \begin{matrix} c & 0 & 0 \\ 0 & -c & d \\ 0 & d & 0 \end{matrix} \quad \text{and} \quad \begin{matrix} 0 & -c & -d \\ -c & 0 & 0 \\ -d & 0 & 0 \end{matrix}$$

Fig. 7.61(b) The vibrational scattering tensors

of both the incident radiation and the observed Raman radiation enables the magnitudes of the tensor components to be determined in turn. Hence, the symmetry classes of the Raman active vibrations may be identified. Thus, in the case of calcite, for illumination along the z-axis of the crystal with light polarized along the x-axis, and observation along the y-axis of the Raman scattering which is polarized along the x-axis, the scattered power is proportional to α_{xx}^2. Since α_{xx} is non-zero for the a_{1g} mode and one component of each of the e_g modes, all the Raman active wavenumbers will be observed with this illumination–observation configuration, which is described in the Porto notation (page 64) by $z(xx)y$. For the illumination–configuration $y(xy)x$, the scattered power is proportional to α_{xy}^2, and since this is non-zero only for one component of each of the e_g modes, only the e_g vibrations will

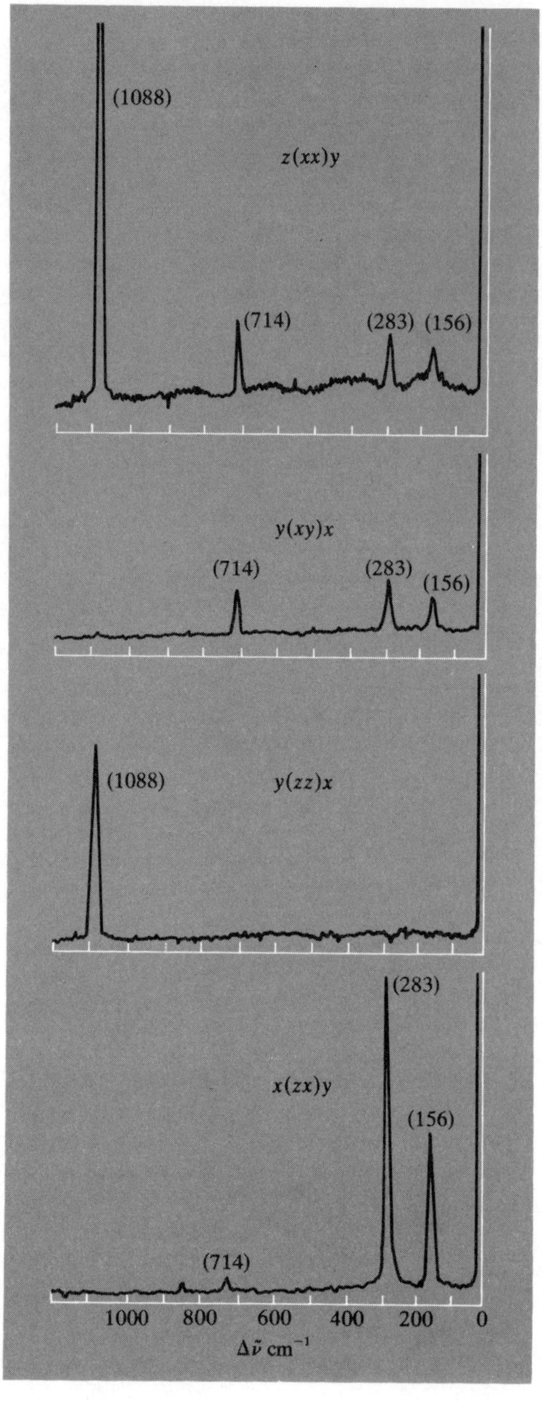

Fig. 7.62 Polarized Stokes Raman spectra of calcite, $\Delta\tilde{\nu} = 0\text{–}1200\ cm^{-1}$; 6328 Å, (632·8 nm) excitation

be observed for this experimental arrangement. For the illumination–observation configuration $y(zz)x$, the scattered power is proportional to α_{zz}^2, and since this is only non-zero for the a_{1g} mode, only this vibration will be observed for this experimental arrangement. Similarly, in the $x(zx)y$ illumination–observation configuration, only one component of each of the e_g modes will be observed, since α_{zx} is involved. Raman spectra[76] obtained from a single crystal of calcite for the $z(xx)y$, $y(xy)x$, $y(zz)x$, and $x(zx)y$ illumination–observation configurations are shown in Fig. 7.62 for the wavenumber shift range 0–1200 cm^{-1}. From such spectra an unambiguous identification of the symmetry classifications of the five Raman active wavenumbers in calcite can be made as follows:

$$a_{1g} \text{ (internal)} = 1088 \text{ cm}^{-1}$$

$$e_g \text{ (internal)} = 714 \text{ and } 1432 \text{ cm}^{-1}$$

$$e_g \text{ (external)} = 156 \text{ and } 283 \text{ cm}^{-1}$$

It is interesting to note that a number of experimental studies of Raman scattering from calcite were made, before the discovery of lasers, using mercury arc excitation. All these earlier studies found that the a_{1g} mode at 1088 cm^{-1} had a non-zero value of the α_{xy} tensor component, in apparent contradiction of the theory. A number of possible explanations for this anomaly were proposed, but the results of the recent study using laser excitation show that the apparent anomaly found when mercury arc excitation was used was merely an artifact of the experimental technique. In the experiments with a mercury arc as a source, the crystal was illuminated by a highly converging beam, and the scattered radiation collected over an appreciable solid angle. In calcite, as a result of the birefringence properties of the crystal, radiation which is initially linearly polarized can become elliptically polarized as it travels through the crystal, particularly when the propagation direction departs from the optic axis. In consequence, the crystal appears to have a non-zero α_{xy} tensor component associated with the a_{1g} mode. This anomaly disappears when the crystal is illuminated by a laser beam with low convergence (about 1°) and the scattered radiation collected over a small angle (about 6°).

7.17.3 *Low-wavenumber Raman spectrum of naphthalene*

Naphthalene crystallizes in a monoclinic unit cell of space group C_{2h}^5, and the unit cell contains two naphthalene molecules. Group theory arguments show that the point group appropriate to the description of the symmetry of the unit cell is C_{2h}, and there are six Raman active intermolecular modes, all librational in nature: three of a_g symmetry and three of b_g symmetry. The scattering tensors have the following forms:

$$a_g: \begin{matrix} \alpha_{aa} & 0 & \alpha_{ac} \\ 0 & \alpha_{bb} & 0 \\ \alpha_{ac} & 0 & \alpha_{cc} \end{matrix}$$

$$b_g: \begin{matrix} 0 & \alpha_{ab} & 0 \\ \alpha_{ab} & 0 & \alpha_{bc} \\ 0 & \alpha_{bc} & 0 \end{matrix}$$

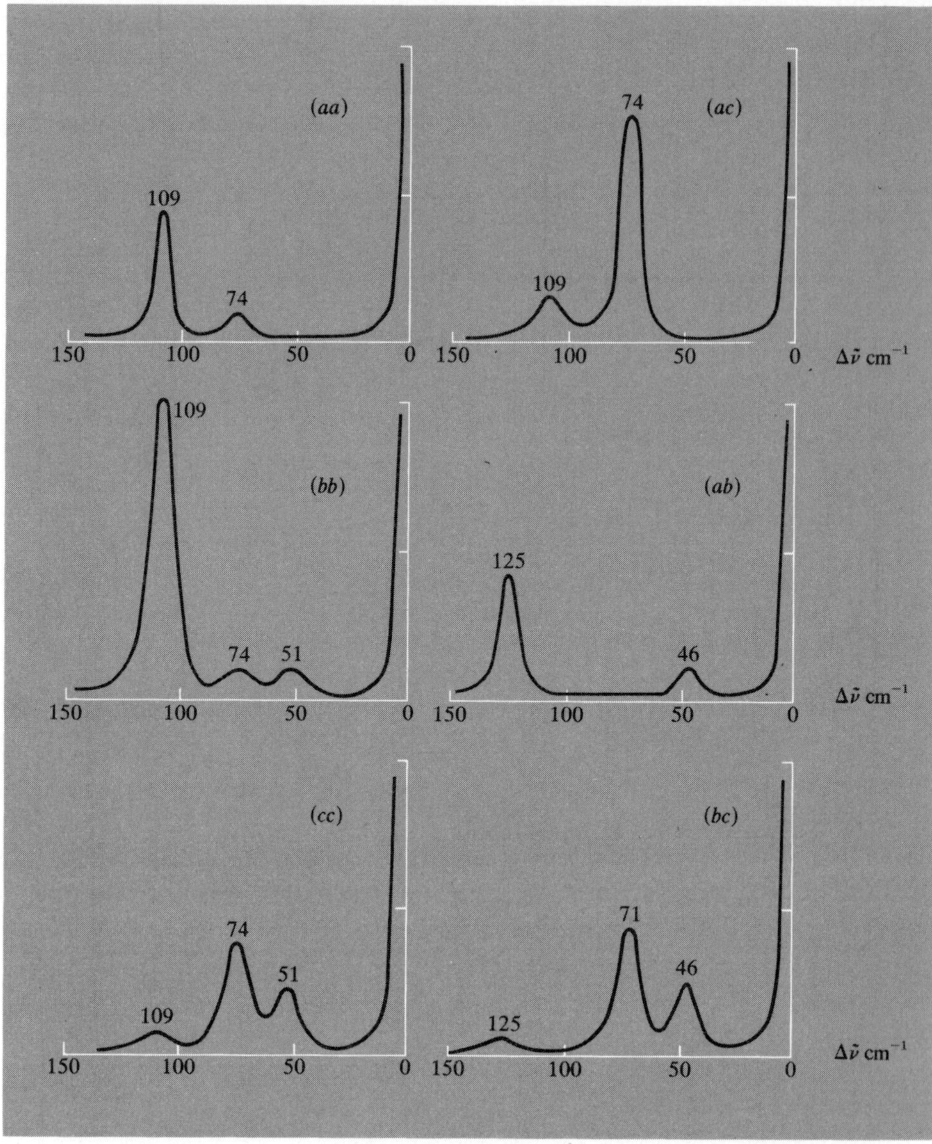

Fig. 7.63 Microphotometer tracings of the polarized Raman spectra of naphthalene crystal in the lattice vibrational region at 20° C; 4880 Å (488·0 nm) excitation

where *a*, *b*, *c* refer to the crystal axes. Thus, a_g modes will appear only in those spectra for which the illumination–observation geometry gives the *aa*, *bb*, *cc*, and *ac* polarizability components, and the b_g modes only in those spectra associated with the *ab* and *bc* polarizability components.

The observed Raman spectra[77] for the Stokes shift region 0 to 150 cm^{-1} are shown in Fig. 7.63. Six Raman lines are indeed observed in this region, and the spectra for the various orientations enable the a_g and b_g modes to be unambiguously identi-

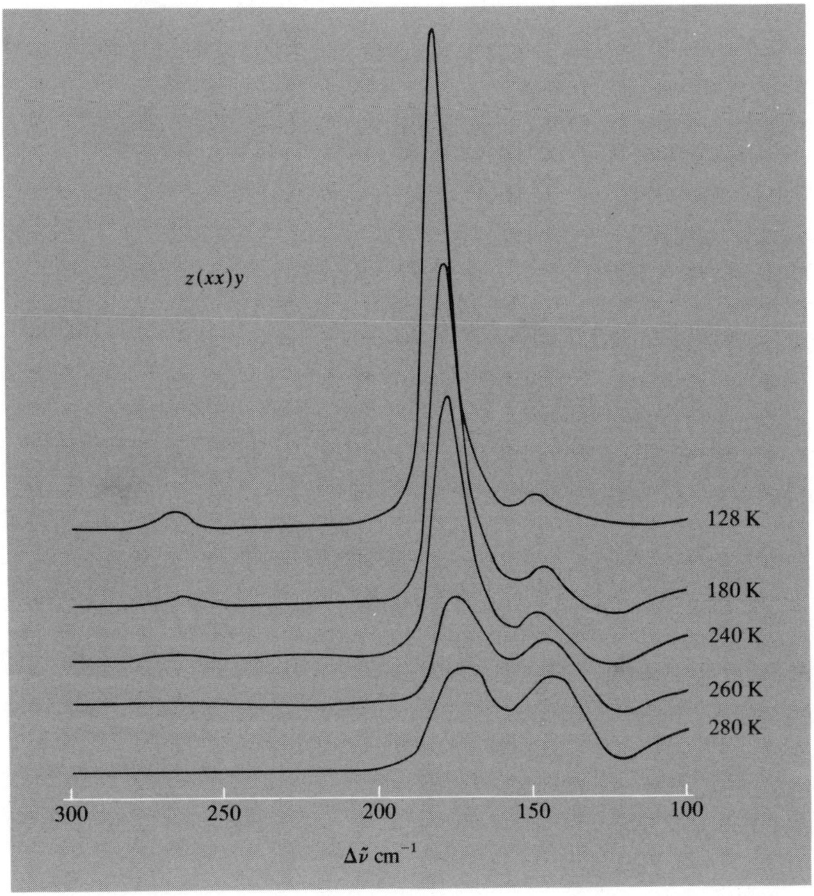

Fig. 7.64 Polarized Raman spectrum of a single crystal of NH_4Cl, $\Delta\tilde{\nu} = 100\text{–}300\ cm^{-1}$ over temperature range 128–280 K; 4880 Å (488·0 nm) excitation

fied as follows: a_g symmetry 51, 74, and 109 cm^{-1}; b_g symmetry 46, 71, and 125 cm^{-1}.

The unambiguous determination of the symmetry of vibrations as illustrated here for calcite and naphthalene, is a valuable aspect of the study of Raman spectra of oriented single crystals. A further example is considered in Appendix III.

7.17.4 *Phase transitions*

The study of the Raman spectra of oriented crystals as a function of temperature and pressure can yield valuable information concerning the structural changes associated with phase transitions. Figure 7.64 shows $z(xx)y$ Raman spectra[78] of a single crystal of NH_4Cl over the temperature range 128 to 280 K. Ammonium chloride undergoes an order–disorder phase transition at 243 K, and the changes in the wavenumbers, intensities, and bandwidths in the Raman spectra of both the lattice and internal modes as a function of temperature may be correlated with the

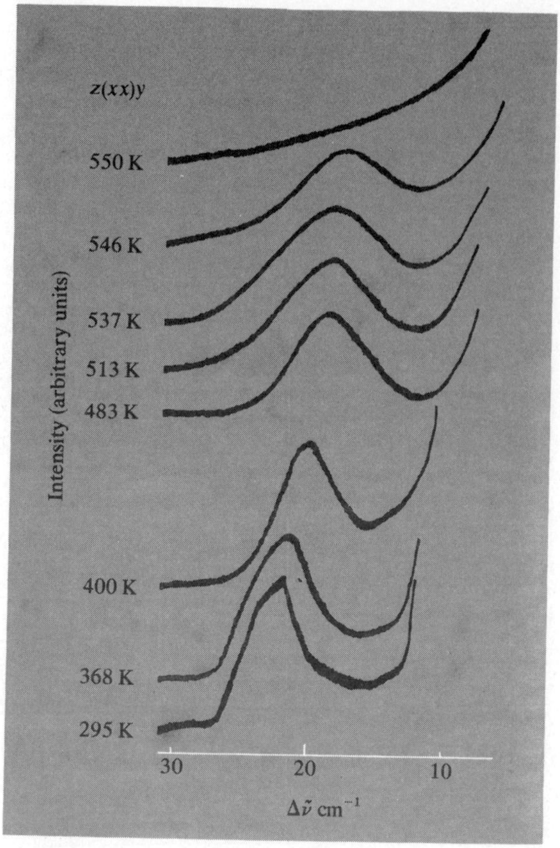

Fig. 7.65 The intensity of the Stokes Raman band ($\Delta\tilde{\nu} \approx 22\ cm^{-1}$) of $NaNO_3$, as a function of temperature over the range 295–550 K. The transition temperature is 548 K: 4880 Å (488·0 nm) and 5145 Å (514·5 nm) excitation

structural changes resulting from the phase transition. Figure 7.65 shows the $z(xx)y$ low wavenumber Raman spectra[79] of a single crystal of $NaNO_3$ in the temperature range 295–550 K. Sodium nitrate undergoes a phase transition at 548 K and the Raman spectra in Fig. 7.65 show that a low wavenumber band disappears above the transition temperature. Two interesting deductions may be made from these observations. Firstly, the occurrence of this low wavenumber mode which is totally symmetric is not compatible with a planar structure for the nitrate ion in the low temperature phase of the crystal. Its occurrence can be explained only by assuming that the nitrogen atoms are pushed out of the NO_3^- plane along the z-axis. Secondly, the disappearance of this wavenumber above the transition temperature indicates that there is free rotation of the nitrate ion about the z-axis in the high temperature phase.

Figure 7.66 shows the Raman spectra[80] of a potassium selenate crystal in the Stokes and anti-Stokes wavenumber shift range 0 to 50 cm^{-1}, in the temperature range 54 to 123 K. It can be seen that a low wavenumber lattice vibration moves from

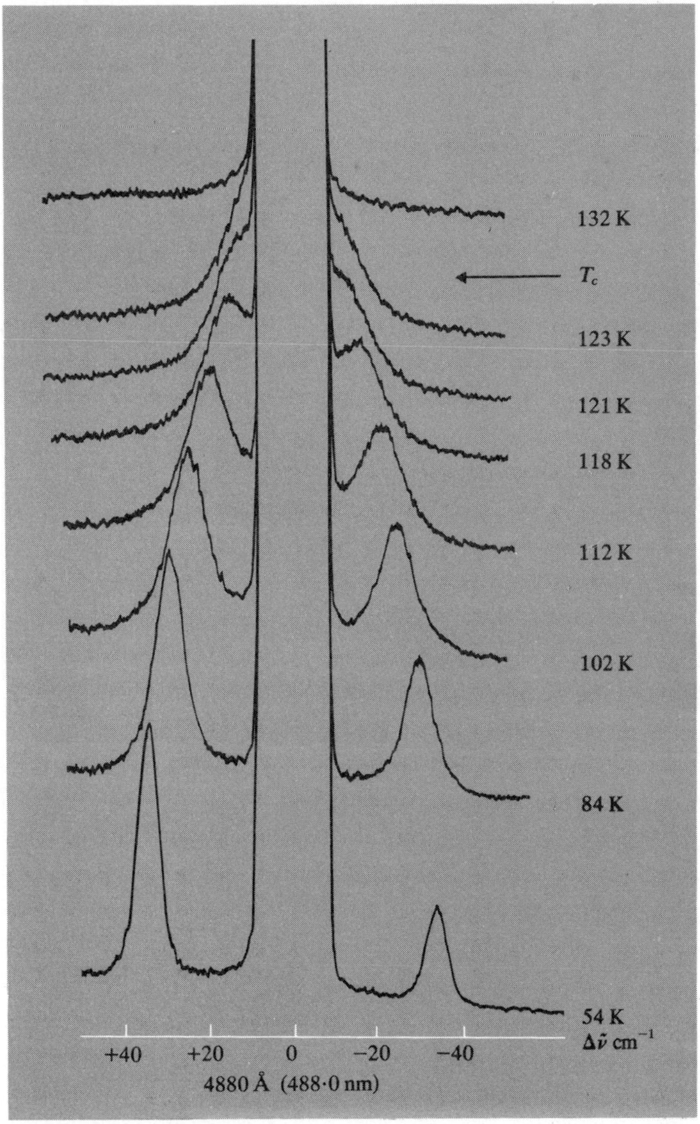

Fig. 7.66 Stokes and anti-Stokes Raman spectra of a K_2SeO_4 single crystal, wavenumber shifts 0 ± 50 cm^{-1}, over the temperature range 54–123 K. The 'soft' mode moves from 35 cm^{-1} at low temperature towards zero frequency at the phase transition temperature of 129·5 K. 4880 Å (488·0 nm) excitation

35 cm^{-1} at low temperature towards zero wavenumber at 129·5 K. Potassium selenate undergoes a transition at this temperature from space group $Pna2_1$ to *Pnam*. Because of its behaviour as the phase transition temperature is approached, this mode is described as 'soft'. This 'soft' mode behaviour is characteristic of a displacement of ions in the lattice at the phase transition.

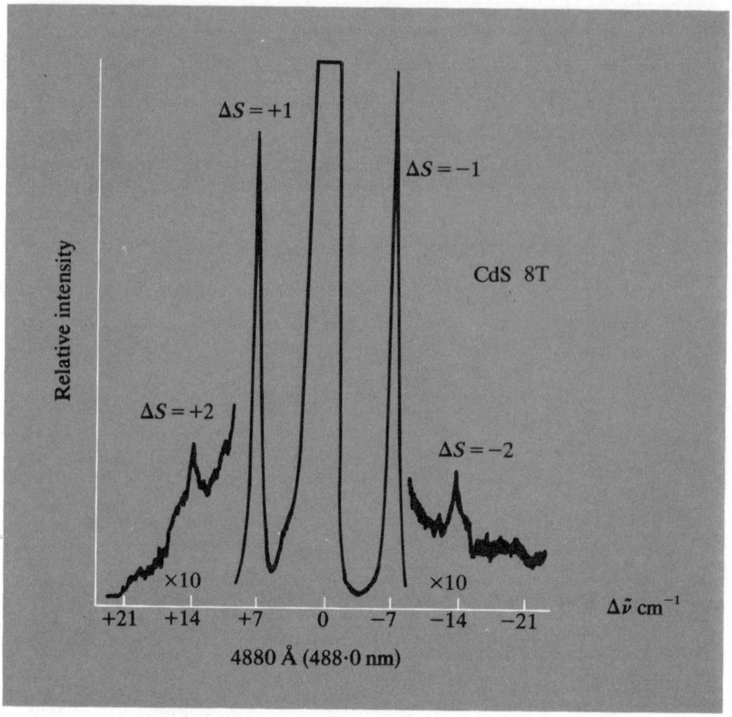

Fig. 7.67 Single and double spin–flip Raman spectra of CdS at 10 K; 4880 Å (488·0 nm) excitation

7.17.5 *Spin–flip transitions*

Raman scattering has recently been observed from an unusual type of transition in the wide band gap II–VI semiconductors, CdS and ZnSe, when subjected to an intense magnetic field at low temperatures. An electron in a valence band is excited to a $\downarrow$ state above the Fermi surface and another electron in the $\uparrow$ state of the same Landau level drops down to fill the hole in the valence band. The overall process is simply one of spin–flip, i.e., change in the resultant spin quantum number. The spin–flip energy is proportional to the magnetic field intensity and has a value of about 7 cm^{-1} when the magnetic field is 10 T. Figure 7.67 shows the Stokes and anti-Stokes Raman spectra[81] arising from spin–flip transitions in CdS at 10 K with a magnetic field of 8 T. The spectra were excited with 4880 Å radiation. The strong band at 7 cm^{-1} arises from a single spin–flip ($\Delta S = \pm 1$) and the much weaker band at 14 cm^{-1} from a double spin–flip ($\Delta S = \pm 2$). The variation of the observed spin–flip wavenumbers with magnetic field is shown in Fig. 7.68. This variation of the wavenumber of the spin–flip transition with magnetic field intensity may be used to construct a tunable spin–flip Raman laser. For example, if a pulsed CO_2 laser (output 943 cm^{-1}) is used for excitation of InSb the resulting spin–flip Raman laser can be tuned over the region 700–1100 cm^{-1}. Such lasers have very narrow line width, and are becoming important for high resolution studies.

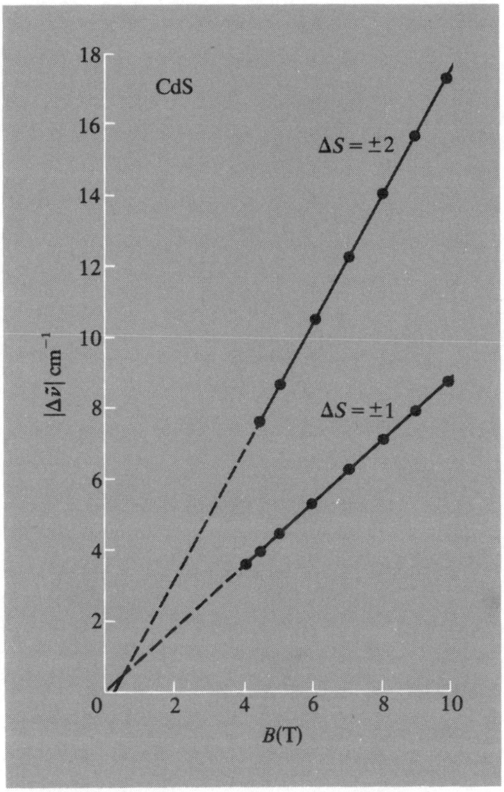

Fig. 7.68 Wavenumber shift in single and double spin–flip Raman spectra of CdS at 10 K as a function of magnetic field

7.17.6 *Raman spectra of glasses*

Raman spectra characteristic of glasses[82] are shown in Fig. 7.69. The spectrum in Fig. 7.69(a) is of boric oxide glass and shows features characteristic of a borate ring structure. The spectrum in Fig. 7.69(j) is of sodium metaphosphate glass and has features characteristic of the metaphosphate chain structure. The spectra in Fig. 7.69(b) to (h) are of mixed boric oxide–sodium metaphosphate glasses, the percentage of sodium metaphosphate increasing from spectrum 7.69(b) to 7.69(h). These spectra may be interpreted in terms of the gradual breakdown of the borate ring structure and its replacement by the metaphosphate chain structure. The spectra of the glasses of intermediate composition show bands characteristic of P—O—P, B—O—P, and B—O—B linkages. In these spectra, the bands are generally broad as a result of the increased disorder of the lattice.

7.17.7 *Concluding remarks*

Raman scattering has been observed from many other types of solids: these include metals, covalent systems like diamond, ionic crystals like NaCl, ferroelectric crystals, mixed crystals, semi-conductors, *U*-centres, V_k-centres, and doped crystals. Raman

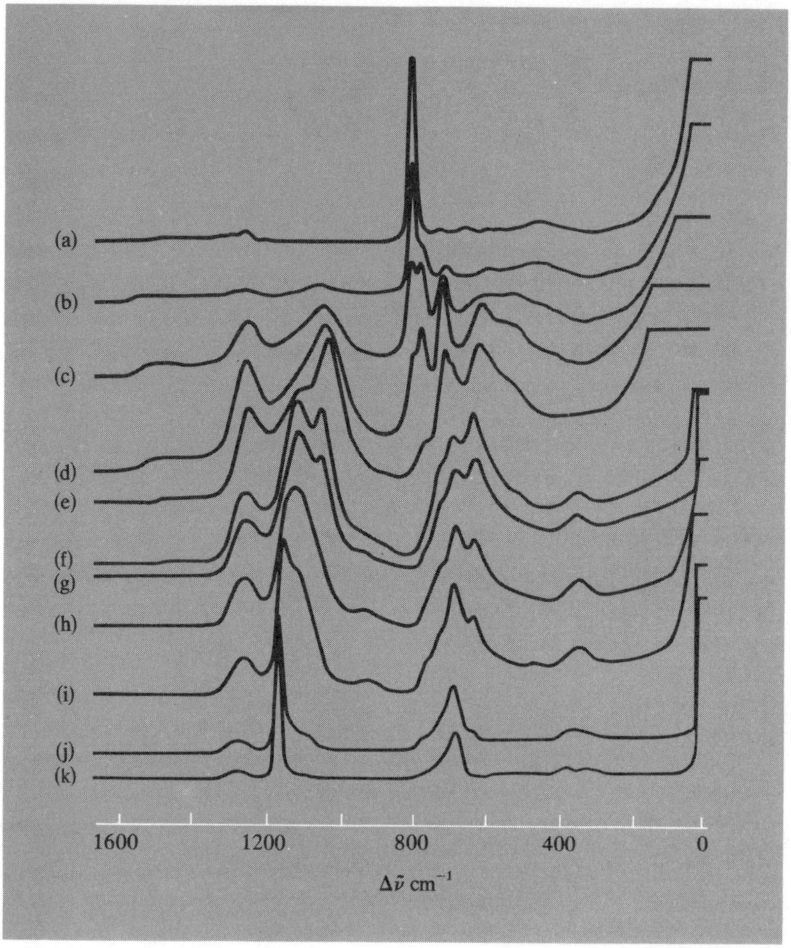

Fig. 7.69 Raman spectra of boric oxide/sodium metaphosphate glasses; 4880 Å (488·0 nm) excitation

spectroscopy is now established as a powerful and important technique for the study of the solid state. It has yielded much useful information, but a complete theoretical treatment of many systems has not yet been achieved.

References

1. L. A. Woodward, *Ann. Reports Chem. Soc.*, **31**, 21, 1934.
2. H. G. M. Edwards, E. A. M. Good, and D. A. Long, *J. Chem. Soc.* (*Faraday II*), **72**, 865, 1976.
3. J. Bendtsen, *J. Raman Spectroscopy*, **2**, 133, 1974.
4. H. G. M. Edwards, E. A. M. Good, and D. A. Long, *J. Chem. Soc.* (*Faraday II*), **72**, 984, 1976.
5. H. G. M. Edwards, E. A. M. Good, and D. A. Long, *J. Chem. Soc.* (*Faraday II*), **72**, 927, 1976.

6. R. J. Butcher, D. V. Willetts, and W. J. Jones, *Proc. Roy. Soc. London*, **A.324**, 231, 1971.
7. P. A. Freedman, W. J. Jones, and A. Rogstad, *J. Chem. Soc.* (*Faraday II*), **71**, 286, 1975.
8. S. Brodersen and J. Bendtsen, *J. Raman Spectroscopy*, **1**, 97, 1973.
9. H. G. M. Edwards, D. A. Long and H. R. Mansour, in E. D. Schmid (Ed.), *Proceedings Fifth International Conference on Raman Spectroscopy*, p. 420, Schulz Verlag, Freiburg, 1976.
10. B. P. Stoicheff, *Canad. J. Phys.*, **32**, 339, 1954.
11. H. G. M. Edwards, E. A. M. Good, and D. A. Long, to be published.
12. H. G. M. Edwards, D. A. Long, and R. Love, in J. P. Mathieu (Ed.), *Advances in Raman Spectroscopy*, vol. 1, p. 504, Heyden, London, 1972.
13. G. W. Hills and W. J. Jones, *J. Chem. Soc.* (*Faraday II*), **71**, 812, 1975.
14. B. Schrader, *Angewandte chemie*, **12**, 884, 1973.
15. D. A. Long, W. O. George, and A. E. Williams, *Proc. Chem. Soc.*, 285, 1960.
16. L. A. Woodward, *Phil. Mag.*, **18**, 823, 1934.
17. R. J. Gillespie and D. J. Millen, *Q. Rev. Chem. Soc.*, **2**, 277, 1948.
18. R. C. Lord, D. W. Robinson, and W. C. Schumb, *J. Amer. Chem. Soc.*, **78**, 1327, 1956.
19. L. E. Sutton, personal communication, see ref. 20.
20. D. C. McKean, R. Taylor, and L. A. Woodward, *Proc. Chem. Soc.*, 321, 1959.
21. R. J. H. Clark and P. D. Mitchell, *J. Chem. Soc.* (*Dalton*), 2429, 1972.
22. G. A. Ozin, *Prog. Inorg. Chem.*, **14**, 173, 1971.
23. J. R. Allkins and P. J. Hendra, *J. Chem. Soc.* (*A*), 1325, 1967; R. J. H. Clark, G. Natile, V. Belluco, L. Cattalini, and C. Fillipin, *J. Chem. Soc.* (*A*), 659, 1970.
24. I. R. Beattie and J. R. Horder, *J. Chem. Soc.* (*A*), 2656, 1969.
25. G. A. Ozin, *Prog. Inorg. Chem.*, **14**, 173, 1971; I. R. Beattie, *Chem. in Britain*, **3**, 347, 1967.
26. D. A. Long, Plenary Lecture, *Colloquium Spectroscopicum Internationale XVI*, Hilger, London, 1971.
27. R. O. Perry, *J. Chem. Soc.* (*D*), 886, 1969.
28. R. J. H. Clark and D. M. Rippon, *J. Chem. Soc.* (*D*), 1295, 1971.
29. R. J. H. Clark and P. D. Mitchell, *J. Chem. Soc.* (*Dalton*), 2432, 1972.
30. G. A. Ozin, *J. Chem. Soc.* (*A*), 2308, 1970.
31. L. A. Woodward, G. Garton, and H. L. Roberts, *J. Chem. Soc.*, 3723, 1956.
32. G. J. Janz and D. W. James, *J. Chem. Phys.*, **38**, 902, 1963.
33. G. J. Janz and D. W. James, *J. Chem. Phys.*, **38**, 905, 1963.
34. G. E. Walrafen, *J. Chem. Phys.*, **43**, 479, 1965.
35. L. A. Woodward and M. J. Taylor, *J. Chem. Soc.*, 4473, 1960.
36. L. A. Woodward and M. J. Taylor, *J. Chem. Soc.*, 407, 1962.
37. J. E. D. Davies and D. A. Long, *J. Chem. Soc.* (*A*), 2564, 1968.
38. J. E. D. Davies and D. A. Long, *J. Chem. Soc.* (*A*), 2054, 1968.
39. J. E. D. Davies and D. A. Long, *J. Chem. Soc.* (*A*), 2050, 1968.
40. J. E. D. Davies and D. A. Long, *J. Chem. Soc.* (*A*), 1761, 1968.
41. V. A. Maroni and T. G. Spiro, *J. Amer. Chem. Soc.*, **89**, 45, 1967.
42. J. M. Grzybowski, B. R. Carr, B. M. Chadwick, D. G. Cobbold, and D. A. Long, *J. Raman Spectroscopy*, **4**, 421, 1976.
43. D. Boal and G. A. Ozin, *Spectroscopy Letters*, **4**, 43, 1971.
44. I. R. Beattie, A. German, H. E. Blayden, and S. B. Brumbach, *J. Chem. Soc.* (*Dalton*), 1659, 1975.
45. A. K. Covington and R. Thompson, *J. Solution Chemistry*, **3**, 603, 1974.
46. M. H. Brooker and D. E. Irish, *Trans. Faraday Soc.*, **67**, 1927, 1971.
47. H. J. Bernstein, in J. P. Mathieu (Ed.), *Advances in Raman Spectroscopy*, p. 314, Heyden, London, 1973.
48. M. Lutz, *Compt. rend.*, **275B**, 497, 1972.
49. H. J. Bernstein, in J. P. Mathieu (Ed.), *Advances in Raman Spectroscopy*, p. 315, Heyden, London, 1973.
50. H. J. Bernstein, in J. P. Mathieu (Ed.), *Advances in Raman Spectroscopy*, p. 312, Heyden, London, 1973.
51. V. Fawcett, private communication.

52. A. Aminzadeh, V. Fawcett, and D. A. Long, in E. D. Schmid (Ed.), *Proceedings Fifth International Conference on Raman Spectroscopy*, p. 304, Schulz Verlag, Freiburg, 1976.
53. L. Andrews and R. C. Spiker, *J. Chem. Phys.*, **59**, 1863, 1973.
54. E. W. Small and W. L. Peticolas, *Biopolymers*, **10**, 1377, 1971.
55. T. G. Spiro and T. C. Strekas, *Proc. Nat. Acad. Sci. U.S.A.*, **69**, 2622, 1972.
56. R. C. Lord and N. T. Yu, *J. Mol. Biol.*, **50**, 509, 1970.
57. R. Faiman and D. A. Long, *J. Raman Spectroscopy*, **3**, 379, 1975.
58. R. Faiman and D. A. Long, *J. Raman Spectroscopy*, **5**, 3, 1976.
59. J. L. Koenig and A. C. Angood, *J. Polymer Sci.*, *A*-2, *Polymer Phys.*, **8**, 1787, 1970.
60. D. A. Long, in L. A. K. Staveley (Ed.), *The Characterization of Chemical Purity, Organic Compounds*, p. 149, Butterworths, London, 1971.
61. B. Schrader, *Angewandte chemie*, **12**, 891, 1973.
62. M. Delhaye and P. Dhamelincourt, *J. Raman Spectroscopy*, **3**, 33, 1975.
63. M. Lapp, in J. P. Mathieu (Ed.), *Advances in Raman Spectroscopy*, p. 258, Heyden, London, 1973.
64. L. Beardmore, H. G. M. Edwards, D. A. Long and T. K. Tan, *Project BELT Report*, 1975.
65. F. Wallart, These, Docteur es Sciences, Universite de Lille, 1970.
66. M. Bridoux, A. Chapput, M. Crunelle, and M. Delhaye, in J. P. Mathieu (Ed.), *Advances in Raman Spectroscopy*, p. 65, Heyden, London, 1973.
67. G. Lucazeau and J. A. Koningstein, in J. P. Mathieu (Ed.), *Advances in Raman Spectroscopy*, p. 385, Heyden, London, 1973.
68. H. Goldberg and P. S. Pershan, in J. P. Mathieu (Ed.), *Advances in Raman Spectroscopy*, p. 437, Heyden, London, 1973.
69. M. W. Wolkenstein, *Compt. rend. Acad. Sci. U.S.S.R.*, **32**, 185, 1941.
70. D. A. Long, *Proc. Roy. Soc.*, **A217**, 20, 1953.
71. T. V. Long, Jr., and R. A. Plane, *J. Chem. Phys.*, **43**, 457, 1965.
72. E. R. Lippincott and G. Nagarajan, *Bull. Soc. chim. belges*, **74**, 551, 1965.
73. D. A. Long, A. H. S. Matterson, and L. A. Woodward, *Proc. Roy. Soc.*, **A224**, 33, 1954.
74. L. D. Barron and A. D. Buckingham, *Ann. Rev. Phys. Chem.*, **26**, 381, 1975.
75. L. D. Barron, *Nature*, **257**, 372, 1975.
76. S. P. S. Porto, J. A. Giordmaine, and T. C. Damen, *Phys. Rev.*, **147**, 608, 1966.
77. M. Suzuki, T. Yokoyama, and M. Ito, *Spectrochim. Acta*, **24A**, 1091, 1968.
78. D. A. Long, Plenary Lecture, *Colloquium Spectroscopicum Internationale XVI*, Hilger, London, 1971.
79. A. D. Prasad Rao, R. S. Katiyar, and S. P. S. Porto, in J. P. Mathieu (Ed.), *Advances in Raman Spectroscopy*, p. 174, Heyden, London, 1973.
80. C. Caville, V. Fawcett, and D. A. Long, in E. D. Schmid (Ed.), *Proceedings Fifth International Conference on Raman Spectroscopy*, p. 626, Schulz Verlag, Freiburg, 1976.
81. J. F. Scott, in J. P. Mathieu (Ed.), *Advances in Raman Spectroscopy*, p. 353, Heyden, London, 1973.
82. V. Fawcett, D. A. Long, and L. Taylor, in E. D. Schmid (Ed.), *Proceedings Fifth International Conference on Raman Spectroscopy*, p. 112, Schulz Verlag, Freiburg, 1976.

8 Non-linear Raman effects

"Awaiting the sensation of a short,
sharp shock,
From a cheap and chippy chopper on
a big black block."

W. S. Gilbert

8.1 Introduction

In chapter 3, we noted that for electric field intensities above about 10^9 V m^{-1}, the non-linear contributions to the induced dipole start to become significant, and that such electric field intensities are associated with the monochromatic radiation produced by giant-pulse lasers. In this chapter, we shall consider four relatively new spectroscopic phenomena which arise from the non-linear interaction of a system with intense monochromatic radiation. All four phenomena involve changes in the wavenumber of laser radiation as a result of its interaction with a system, and so are to be regarded as variants of the Raman effect. This generic relationship is reflected in the names which have been adopted for these new effects; the hyper Raman effect, the inverse Raman effect, the stimulated Raman effect, and coherent anti-Stokes Raman scattering (CARS). Each of these effects involves new principles, and hence novel applications are possible and new information can result. The treatment of these effects, that follows, must necessarily be concise and will, in the main, be qualitative.

8.2 Characteristics of giant-pulse laser systems

It will prove useful, first, to summarize a few of the characteristics of giant-pulse or Q-switched laser systems. Two such systems are commonly used: the ruby system which has an output wavelength of 6943·3 Å (694·33 nm, 14 402 cm^{-1}) and the neodymium-glass system which has an output at 10 600 Å (1060·00 nm, 9434·0 cm^{-1}). By using a frequency doubling crystal, like KDP (potassium dihydrogen phosphate, KH_2PO_4) or CDA (caesium dihydrogen arsenate, CsH_2AsO_4), outputs at half the wavelength (twice the frequency) can be produced. For example, the neodymium glass system will yield an output at 5300·0 Å (530·00 nm, 18 868 cm^{-1}). Throughout this chapter we quote λ_{air} and $\tilde{\nu}_{air}$. (See page 40.)

The output of the Q-switched ruby and neodymium-glass systems consists of a pulse extending over about 10 to 100 nanoseconds. It is this relatively short duration of the pulse which results in outputs of very high power. For example, 1 joule in a 10 nanosecond pulse corresponds to a power of 10^8 watts. It transpires that these pulses of nanosecond duration are made up of overlapping pulses of even shorter duration, of the order of 1 to 10 picoseconds. Special techniques (known as mode-locking) enable these pulses to be disentangled and a train of pulses of picosecond duration to be produced; and other techniques (known as pulse-switching) enable a single picosecond pulse to be isolated. Even with small total energies, the power in such extremely short pulses is very high. The very short duration of such pulses makes them important as probes of short-lived phenomena.

8.3 The hyper Raman effect

8.3.1 *General considerations*

When a system is illuminated with radiation of wavenumber $\tilde{\nu}_0$ from a focused giant-pulse laser giving nanosecond pulses and an irradiance which is just above that for significant non-linear interaction, the scattered radiation is found[1] to include wavenumbers of the type $2\tilde{\nu}_0$ and $2\tilde{\nu}_0 \pm \tilde{\nu}_M$, where $\tilde{\nu}_M$ is a wavenumber associated with a transition between two levels of the scattering molecules. For example, if a giant-pulse ruby laser is used, then as a result of excitation with intense radiation at 14 402 cm^{-1} (6943·3 Å, 694·33 nm), scattered radiation around 28 804 cm^{-1} (3471·7 Å, 347·17 nm) is observed. We shall see subsequently that these wavenumbers arise from the second-order non-linear induced dipole $\boldsymbol{P}^{(2)}$, and so are controlled by the hyperpolarizability tensor $\boldsymbol{\beta}$ (see eq. 3.3). In consequence, the scattering at $2\tilde{\nu}_0$ is called hyper Rayleigh scattering and the scattering at $2\tilde{\nu}_0 - \tilde{\nu}_M$ and $2\tilde{\nu}_0 + \tilde{\nu}_M$ Stokes and anti-Stokes hyper Raman scattering, respectively.

Detailed theoretical treatments of the hyper Raman effect[2,3,4] which parallel those of the Raman effect have been developed: the directional and polarization properties follow from the tensor properties of the hyperpolarizability; the wavenumber dependence and selection rules result from a classical treatment of the induced dipole or a quantum mechanical treatment of the transition moment; and a more complete quantum mechanical treatment establishes the relation of $\boldsymbol{\beta}$ to the energy levels of the system. We shall consider here only a simplified classical treatment to explain the observed wavenumber dependence and establish the general nature of the selection rules, and examine briefly some of the results of a quantum mechanical treatment.

8.3.2 *Classical treatment*

We have seen in chapter 3, eq. (3.3), that $\boldsymbol{P}^{(2)}$, the second-order induced dipole, is given by

$$\boldsymbol{P}^{(2)} = \tfrac{1}{2}\boldsymbol{\beta} : \boldsymbol{EE} \qquad (8.1)$$

Although, in general, a third-rank tensor will have 27 components, if appropriate conditions are satisfied the hyperpolarizability tensor $\boldsymbol{\beta}$ has only ten distinct components: $\beta_{xxx}, \beta_{yyy}, \beta_{zzz}, \beta_{xyy}, \beta_{xzz}, \beta_{yxx}, \beta_{yzz}, \beta_{zxx}, \beta_{zyy}, \beta_{xyz}$. The set of three linear equations implicit in eq. (8.1) is then given in matrix form by:

$$\begin{bmatrix} P_x^{(2)} \\ P_y^{(2)} \\ P_z^{(2)} \end{bmatrix} = \frac{1}{2}\begin{bmatrix} \beta_{xxx} & \beta_{xyy} & \beta_{xzz} & \beta_{yxx} & \beta_{xyz} & \beta_{zxx} \\ \beta_{yxx} & \beta_{yyy} & \beta_{yzz} & \beta_{xyy} & \beta_{zyy} & \beta_{xyz} \\ \beta_{zxx} & \beta_{zyy} & \beta_{zzz} & \beta_{xyz} & \beta_{yzz} & \beta_{xzz} \end{bmatrix}\begin{bmatrix} E_x^2 \\ E_y^2 \\ E_z^2 \\ 2E_xE_y \\ 2E_yE_z \\ 2E_zE_x \end{bmatrix} \tag{8.2}$$

The hyperpolarizability tensor can, in certain circumstances, be non-symmetric (compare the polarizability tensor), but we shall not consider such cases here.

We now proceed to establish the wavenumber dependence of $\boldsymbol{P}^{(2)}$ when radiation of circular frequency ω_0 is incident on a molecular system with a molecular vibration of circular frequency ω_M. To simplify matters, we shall consider a special case, the wavenumber dependence of $\boldsymbol{P}^{(2)}$ when the incident radiation is plane polarized with $E_x \neq 0$ and $E_y = E_z = 0$. We then have simply

$$P_x^{(2)} = \tfrac{1}{2}\beta_{xxx}E_x^2 \tag{8.3}$$

For convenience, in the ensuing mathematics we shall drop the subscripts and write simply

$$P^{(2)} = \tfrac{1}{2}\beta E^2 \tag{8.4}$$

Following the classical treatment for $P^{(1)}$ in chapter 3, we write for radiation of circular frequency ω_0,

$$E = E_0 \cos \omega_0 t \tag{8.5}$$

for a molecular frequency ω_M, in the simple harmonic approximation,

$$Q = Q_0 \cos \omega_M t \tag{8.6}$$

and for the dependence of β on Q to a first approximation,

$$\beta = \beta_0 + \left(\frac{\partial \beta}{\partial Q}\right)_0 Q \tag{8.7}$$

Introducing eqs. (8.5), (8.6), and (8.7) into eq. (8.4), we obtain, after a little rearrangement and the use of the trigonometric identities,

$$2 \cos A \cos B = \cos (A+B) + \cos (A-B) \tag{8.8}$$

$$2 \cos^2 A = 1 + \cos 2A \tag{8.9}$$

the following expression for $P^{(2)}$,

$$P^{(2)} = P^{(2)}(\omega = 0) + P^{(2)}(2\omega_0) + P^{(2)}(2\omega_0 \pm \omega_M) + P^{(2)}(\omega_M) \tag{8.10}$$

where

$$P^{(2)}(\omega = 0) = \tfrac{1}{4}\beta_0 E_0^2 \tag{8.11}$$

$$P^{(2)}(2\omega_0) = \tfrac{1}{4}\beta_0 E_0^2 \cos 2\omega_0 t \tag{8.12}$$

$$P^{(2)}(2\omega_0 \pm \omega_M) = \frac{1}{8}\left(\frac{\partial \beta}{\partial Q}\right)_0 Q_0 E_0^2 \{\cos (2\omega_0 + \omega_M)t + \cos (2\omega_0 - \omega_M)t\} \tag{8.13}$$

and

$$P^{(2)}(\omega_M) = \frac{1}{4}\left(\frac{\partial\beta}{\partial Q}\right)_0 Q_0 E_0^2 \cos\omega_M t \tag{8.14}$$

All these second-order induced dipoles depend, as expected, on the square of the electric field intensity. The non-linear induced dipole $P^{(2)}(\omega=0)$ produces radiation at zero frequency, i.e., a static electric field; $P^{(2)}(\omega_M)$ produces radiation at the molecular frequency ω_M; $P^{(2)}(2\omega_0)$ produces radiation at $2\omega_0$, i.e., hyper Rayleigh scattering, and is also the origin of frequency doubling in crystals; and $P^{(2)}(2\omega_0 \pm \omega_M)$ produces radiation at $2\omega_0 \pm \omega_M$, i.e., Stokes and anti-Stokes hyper Raman scattering.

We have thus accounted for the observed wavenumbers or frequencies of non-linear origin which we have termed hyper Rayleigh and Raman scattering. In addition, we see that for hyper Rayleigh scattering to occur at least one of the components of the equilibrium hyperpolarizability tensor must be non-zero; and for hyper Raman scattering at least one of the components of the derived hyperpolarizability tensor must be non-zero. For insight into the molecular properties determining the $\boldsymbol{\beta}$ tensor, we must turn to time-dependent perturbation theory.

8.3.3 *Quantum mechanical treatment*

When second-order terms are taken into account, time-dependent perturbation theory (chapter 5) relates the x component of the complex amplitude of the second-order transition moment for the transition $f \leftarrow i$, to the complex electric field amplitudes as follows:

$$[\tilde{P}^{(2)}_{xo}]_{fi} = \tfrac{1}{2}\sum[\tilde{\beta}_{xyz}]_{fi}\tilde{E}_{yo}\tilde{E}_{zo} \tag{8.15}$$

where $[\tilde{\beta}_{xyz}]_{fi}$, the xyz component of the transition hyperpolarizability, is given by

$$[\tilde{\beta}_{xyz}]_{fi} = \frac{1}{\hbar^2}\sum\left\{\frac{[\tilde{P}_y]_{fr}[\tilde{P}_x]_{rs}[\tilde{P}_z]_{si}}{(\omega_{si}-\omega_0)(\omega_{rf}+\omega_0)} + \frac{[\tilde{P}_y]_{fr}[\tilde{P}_z]_{rs}[\tilde{P}_x]_{si}}{(\omega_{sf}+2\omega_0)(\omega_{rf}+\omega_0)} + \frac{[\tilde{P}_x]_{fs}[\tilde{P}_y]_{sr}[\tilde{P}_z]_{ri}}{(\omega_{si}-2\omega_0)(\omega_{ri}-\omega_0)}\right\} \tag{8.16}$$

We note that for this tensor component to be non-zero there must be non-zero dipole transition moments involving for example the states, i and r, r and s, and s and f, for at least one pair of states r and s. Also, the frequency denominators can lead to resonance enhancement if either ω_0 or $2\omega_0$ is close to an absorption frequency.

It can be shown that $[\beta_{xyz}]_{fi}$ transforms as

$$\langle f|\beta_{xyz}|i\rangle \tag{8.17}$$

and β_{xyz} as the product xyz. Thus, the general selection rule for hyper Raman activity is that the triple direct product of the species of the wave functions for the states i and f, and xyz must contain the totally symmetric species (compare page 88). For a fundamental vibrational transition associated with the normal coordinate Q_k, we may write

$$[\beta_{xyz}]_{1,0} = \langle\phi_1(Q_k)|\beta_{xyz}|\phi_0(Q_k)\rangle \tag{8.18}$$

and the vibrational selection rule for fundamentals is that Q_k and β_{xyz} (i.e., xyz) must belong to the same symmetry class. The symmetry properties of the components of the $\boldsymbol{\beta}$ tensor (symmetric and non-symmetric) are given in Appendix II.

As with the polarizability tensor, for an essentially non-absorbing system, we may make sweeping simplifications whereby the hyperpolarizability tensor becomes a function of the nuclear coordinates only. Thus, for a vibrational transition, we can expand $[\beta_{xyz}]_{fi}$ in a Taylor series in the normal coordinates Q.

In the approximation of electrical and mechanical harmonicity, we find that for Stokes hyper Raman scattering associated with a normal coordinate Q_k the selection rule $\Delta v_k = +1$ operates and the transition hyperpolarizability is given by

$$\begin{aligned}[\beta_{xyz}]_{v_k^i+1,\, v_k^i} &= \left(\frac{\partial \beta_{xyz}}{\partial Q_k}\right)_0 \langle \phi_{v_k^i+1}(Q_k)|\beta_{xyz}(Q_k)\phi_{v_k^i}(Q_k)\rangle \\ &= \left(\frac{\partial \beta_{xyz}}{\partial Q_k}\right)_0 (v_k^i+1)^{\frac{1}{2}} b_{v_k}\end{aligned} \tag{8.19}$$

where b_{v_k} is the quantum mechanical amplitude factor defined in eq. (4.17). The similarity of eq. (8.19) and eq. (4.19) for Stokes vibrational Raman scattering should be noted.

8.3.4 *Selection rules*

We consider here some qualitative aspects of hyper Rayleigh and vibrational hyper Raman selection rules. Unlike the polarizability $\boldsymbol{\alpha}$, the hyperpolarizability $\boldsymbol{\beta}$ can be zero, for example in molecules with a centre of symmetry, and thus hyper Rayleigh scattering can be absent in some systems. The symmetry factors controlling the components of the derived $\boldsymbol{\beta}$ tensor are not the same as those controlling the components of the derived $\boldsymbol{\alpha}$ tensor, and in consequence the vibrational selection rules for hyper Raman scattering are different from those for linear Raman scattering. In particular, some vibrations which are both infrared and Raman inactive, and so had been termed spectroscopically inaccessible, are hyper Raman active.

To illustrate this, Table 8.1 summarizes the infrared, Raman, and hyper Raman activity of the fundamental vibrations of cyclopropane, C_3H_6, sulphur hexafluoride, SF_6, ethane, C_2H_6, and benzene, C_6H_6. It can be seen that, for these molecules, the following vibrations which are infrared and Raman inactive are hyper Raman active: $1a_2'$ for C_3H_6; $1f_{2u}$ for SF_6; $1a_{1u}$ for C_2H_6; and $2b_{1u}+2b_{2u}+2e_{2u}$ for C_6H_6. It can also be seen that for these molecules, whereas some Raman active bands are not hyper Raman active, *all infrared active bands are also hyper Raman active.* This latter statement is a general theorem which can be established by symmetry arguments. Furthermore, those hyper Raman active vibrations which are also infrared active can be distinguished in a hyper Raman spectrum by the fact that they are *always polarized.*

For activity of rotational transitions in the hyper Raman effect, the $\boldsymbol{\beta}$ tensor must be anisotropic. It should be noted that for some spherically symmetrical systems like CH_4, although the $\boldsymbol{\alpha}$ tensor is isotropic, the $\boldsymbol{\beta}$ tensor is not. Thus, pure rotational hyper Raman spectra can be observed in principle from such systems. This is of potential importance since pure rotational spectra cannot be observed in the infrared or Raman effect for such systems. The J selection rules for pure rotational hyper Raman scattering are complicated,[5] and in the most general case we have

Table 8.1 Infrared, Raman, and hyper Raman activity of fundamental vibrations in (a) cyclopropane (C_3H_6), (b) sulphur hexafluoride (SF_6), (c) ethane (C_2H_6), and (d) benzene (C_6H_6)

Symmetry species	x, y, z	Components of α	Components of β	Number of distinct frequencies
(a) Cyclopropane, C_3H_6 (point group D_{3h})				
A'_1		$\alpha_{xx}+\alpha_{yy}, \alpha_{zz}$	$\beta_{xxx}-3\beta_{xyy}$	3
A'_2			$\beta_{yyy}-3\beta_{xxy}$	1
E'	(x, y)	$(\alpha_{xx}-\alpha_{yy}, \alpha_{xy})$	$(\beta_{xxx}+\beta_{xyy},$ $\beta_{yyy}+\beta_{xxy}),$ $(\beta_{zzx}, \beta_{yzz})$	4
A''_1				1
A''_2	z		$\beta_{yyz}+\beta_{zxx}, \beta_{zzz}$	2
E''		$(\alpha_{yz}, \alpha_{zx})$	$(\beta_{yyz}-\beta_{zxx}, \beta_{xyz})$	3
(b) Sulphur hexafluoride, SF_6 (point group O_h)				
A_{1g}		$\alpha_{xx}+\alpha_{yy}+\alpha_{zz}$		1
A_{2g}				0
E_g		$(\alpha_{xx}+\alpha_{yy}-2\alpha_{zz},$ $\alpha_{xx}-\alpha_{yy})$		1
F_{1g}				0
F_{2g}		$(\alpha_{xy}, \alpha_{xz}, \alpha_{yz})$		1
A_{1u}				0
A_{2u}			β_{xyz}	0
E_u				0
F_{1u}	(x, y, z)		$(\beta_{xxx}, \beta_{yyy}, \beta_{zzz}),$ $(\beta_{xyy}+\beta_{zzx},$ $\beta_{yzz}+\beta_{xxy},$ $\beta_{zxx}+\beta_{yyz})$	2
F_{2u}			$(\beta_{xyy}-\beta_{zzx},$ $\beta_{yzz}-\beta_{xxy},$ $\beta_{zxx}-\beta_{yyz})$	1
(c) Ethane, C_2H_6 (point group D_{3d})				
A_{1g}		$\alpha_{xx}+\alpha_{yy}, \alpha_{zz}$		3
A_{2g}				
E_g		$(\alpha_{xx}-\alpha_{yy}, \alpha_{xy})\,(\alpha_{yz}, \alpha_{zx})$		3
A_{1u}			$\beta_{xxx}-3\beta_{xyy}$	1
A_{2u}	z		$\beta_{yyy}-3\beta_{xxy}, \beta_{yyz}+\beta_{zxx}, \beta_{zzz}$	2
E_u	(x, y)		$(\beta_{xxx}+\beta_{xyy}, \beta_{yyy}+\beta_{xxy}), (\beta_{yyz}-\beta_{zxx},$ $\beta_{xyz})\,(\beta_{zzx}, \beta_{yzz})$	3
(d) Benzene, C_6H_6(point group D_{6h})				
A_{1g}		$\alpha_{xx}+\alpha_{yy}, \alpha_{zz}$		2
A_{2g}				1
B_{1g}				0
B_{2g}				2
E_{1g}		$(\alpha_{yz}, \alpha_{zx})$		1
E_{2g}		$(\alpha_{xx}-\alpha_{yy}, \alpha_{xy})$		4
A_{1u}				0
A_{2u}	z		$\beta_{yyz}+\beta_{zxx}, \beta_{zzz}$	1
B_{1u}			$\beta_{xxx}-3\beta_{xyy}$	2
B_{2u}			$\beta_{yyy}-3\beta_{xxy}$	2
E_{1u}	(x, y)		$(\beta_{xxx}+\beta_{xyy},$ $\beta_{yyy}+\beta_{xxy}),$ $(\beta_{zzx}, \beta_{yzz})$	3
E_{2u}			$(\beta_{yyz}-\beta_{zxx}, \beta_{xyz})$	2

$\Delta J = 0, \pm 1, \pm 2$, and ± 3, leading to *NOPQRS* and *T* branches! In specific cases, there will, of course, be further restrictions, e.g., for a diatomic molecule only $\Delta J = \pm 1, \pm 3$ are permitted.

8.3.5 *Experimental techniques*

As might be expected, hyper Rayleigh and Raman scattering are many orders of magnitude less intense than their linear counterparts. For example, a typical nanosecond pulse from a giant-pulse laser would deliver about 3×10^{17} photons to the sample at frequency ω_0, but the number of photons of frequency $2\omega_0 - \omega_M$ which the detector would receive ranges from a few photons per laser pulse down to one photon every hundred laser pulses, depending on the sample. There are two methods[6] of tackling the problem of observing scattered radiation of such low intensities: single-channel detection or multichannel detection. In single-channel detection, the number of photons in a small wavenumber range $\Delta\tilde{\nu}$, averaged over a large number of laser pulses, is determined using photon counting equipment, and the procedure repeated over successive $\Delta\tilde{\nu}$ until the desired wavenumber range is covered. This is relatively inexpensive but time-consuming. In multichannel detection, an image intensifier is used to intensify the complete spectrum to a level at which it can be scanned by a television camera. In this way, all the spectral information can be recorded for each laser pulse and the observation time much reduced. Although this is a technically sophisticated and costly procedure, it is often the only practicable technique. For many samples, it is necessary to keep the electric field intensity low enough to avoid dielectric breakdown. The consequent low level of signal usually means it is necessary to average over at least 6000 laser shots. The repetition rate of a ruby laser is restricted to about one pulse per second, and thus 6000 laser shots require 100 minutes. Clearly, with single-channel detection, the total time involved would be prohibitively long. A 200-channel spectrum, for example, would involve over 300 hours! A block diagram of a versatile single-multichannel hyper Raman spectrometer is shown in Fig. 8.1.

8.3.6 *Typical hyper Raman spectra*

Some typical hyper Raman spectra are presented in Figs. 8.2 to 8.7. Figure 8.2 shows the vibrational spectrum of water.[6] This spectrum contains three broad bands centred at around 600 cm^{-1} (librational mode), 1600 cm^{-1} (deformation mode), and 3400 cm^{-1} (stretching mode), respectively. Figure 8.3 shows the vibrational hyper Raman spectrum of methanol (liquid) and Fig. 8.3(b) the low resolution vibrational infrared spectrum.[7] Comparison of the two spectra will show that all the infrared wavenumbers are present in the hyper Raman spectrum. Figure 8.4(a) shows the hyper Raman spectrum of carbon tetrachloride (liquid);[6] the normal Raman spectrum is shown in Fig. 8.5 for comparison. The differences between the two spectra should be noted. In the hyper Raman spectrum, the e class band at 217 cm^{-1} is inactive, the f_2 band at 314 cm^{-1} and the a_1 band at 459 cm^{-1} are very weak, and the f_2 band centred at 776 cm^{-1} is very strong. Figure 8.6 shows the hyper Raman spectrum of methane gas[8] at high pressure. The contour of the low wavenumber band centred around $2\tilde{\nu}_0$ is consistent with its attribution to unresolved pure rotational hyper Raman scattering; the strong band at 3050 cm^{-1} corresponds to a C—H stretching mode. Figure 8.7 shows the vibrational hyper Raman spectrum of ethane

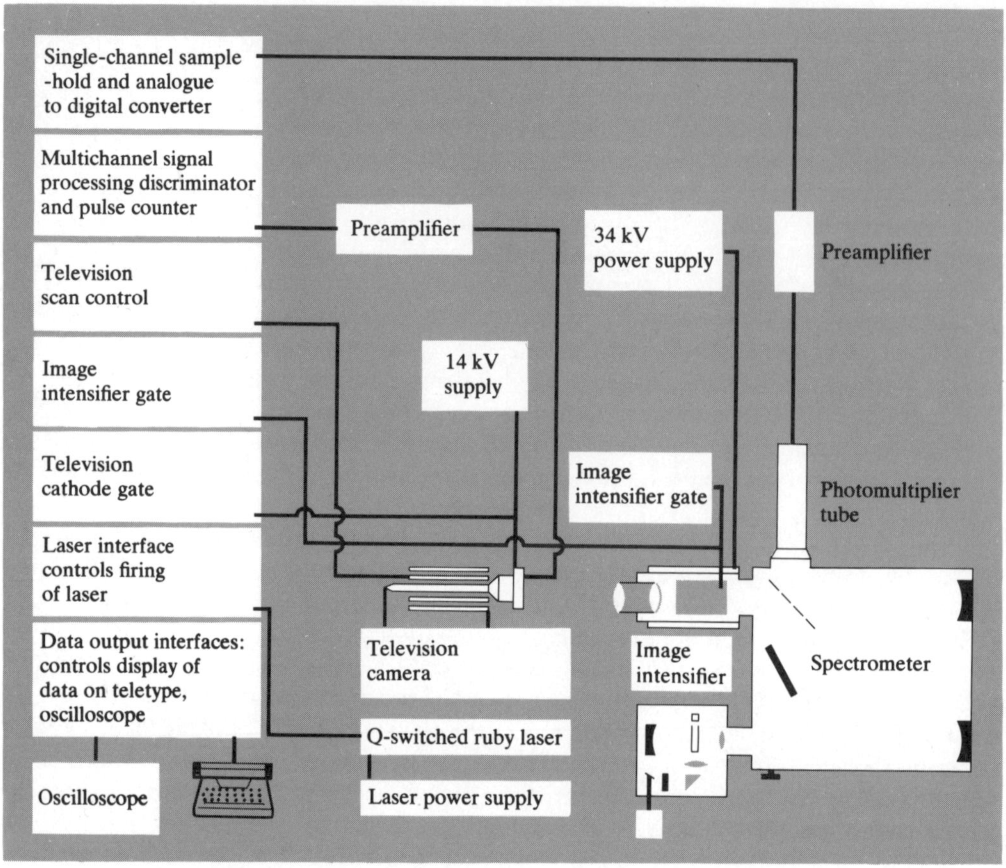

Fig. 8.1 Block diagram of a versatile single-multichannel hyper Raman spectrometer

gas[8] at high pressure. The hyper Rayleigh line is absent since the molecule has a centre of symmetry. Of special interest in this spectrum is the band at 300 cm^{-1} which is assigned to the torsional mode of a_{1u} symmetry which is infrared and Raman inactive. Figure 8.8 shows the low wavenumber region of the hyper Raman spectrum[9] of a single crystal of NH_4Cl. The band at 360 cm^{-1} on the Stokes side is assigned to a librational motion of the ammonium ion. In a symmetrical environment, this mode is not infrared active since the ammonium ion has no dipole moment, nor is it Raman active since the polarizability is isotropic.

These examples illustrate how hyper Raman spectroscopy can provide novel spectroscopic information.

8.4 The stimulated Raman effect

8.4.1 *General characteristics*

When monochromatic radiation from a giant-pulse laser of sufficiently large irradiance is incident upon a scattering system, hyper Raman scattering is superseded by a new phenomenon—*stimulated Raman scattering*. Since some secondary

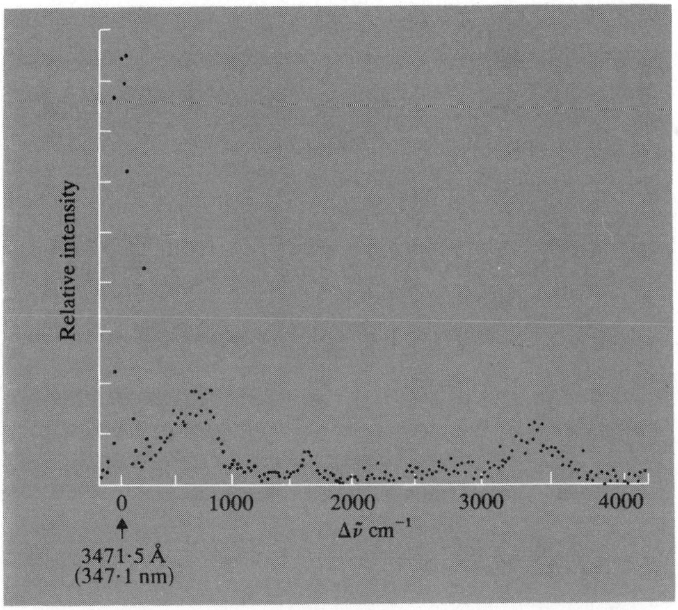

Fig. 8.2 Hyper Rayleigh and Stokes hyper Raman spectra ($\Delta\tilde{\nu}$, 0–4000 cm^{-1}) of water obtained with single-channel operation (200 channels; 60 laser shots per channel; slits 70 cm^{-1} (1 mm); laser energy 100 mJ per pulse; total recording time 3h 20 min)

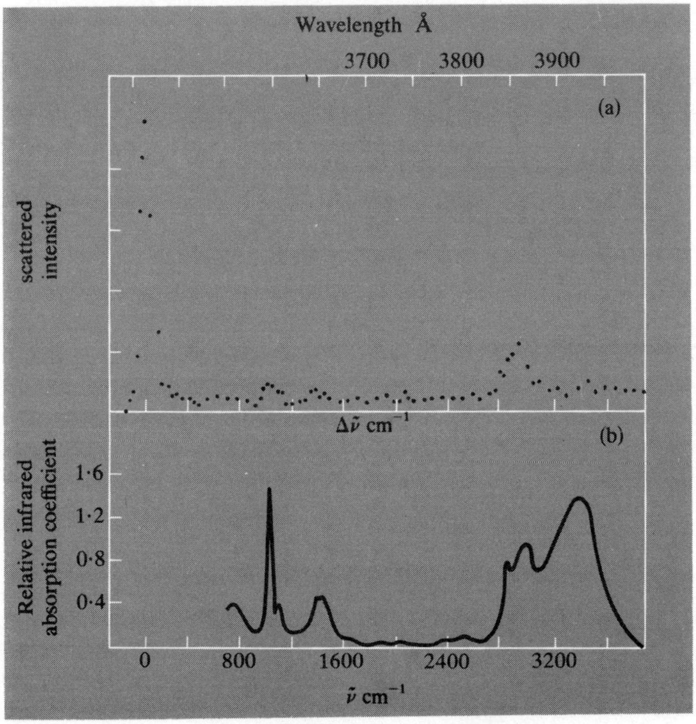

Fig. 8.3 (a) The Stokes hyper Raman ($\Delta\tilde{\nu}$, 0–3700 cm^{-1}) and (b) the infrared (650–3700 cm^{-1}) spectra of methanol

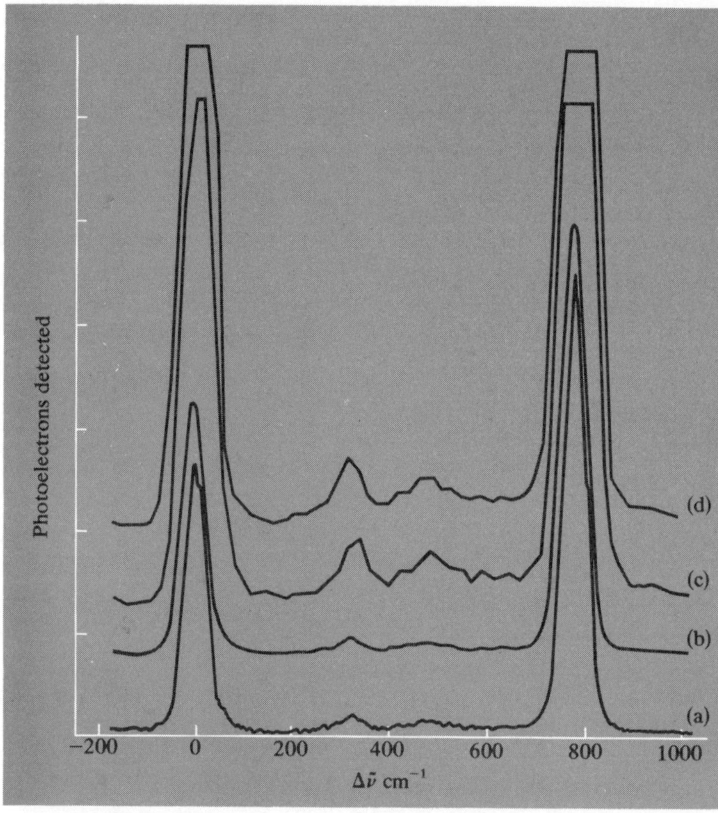

Fig. 8.4 Hyper Rayleigh and Stokes hyper Raman spectra ($\Delta\tilde{\nu}$, 0–1000 cm^{-1}) of carbon tetrachloride (liquid) with multichannel operation: (a) virgin data in 256 channels; (b) four-point running average of virgin data; (c) compression of virgin data to 64 channels; (d) compression of four-point running average of virgin data to 64 channels (8153 laser shots; slits 70 cm^{-1} (1 mm); laser energy 50 mJ per pulse; total recording time 2h 16 min)

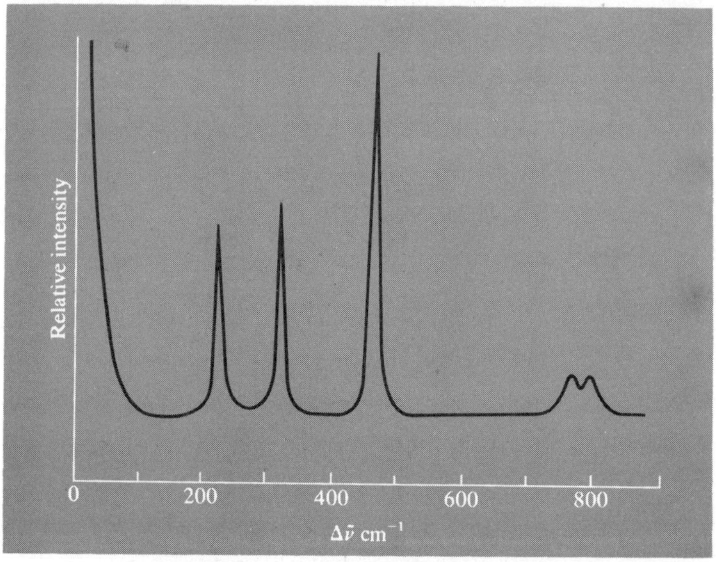

Fig. 8.5 The normal Raman spectrum ($\Delta\tilde{\nu}$, 0–900 cm^{-1}) of carbon tetrachloride (liquid)

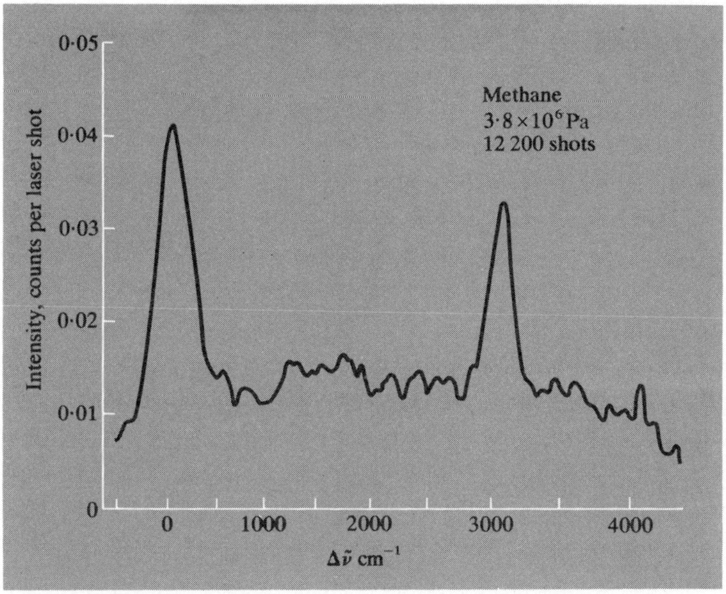

Fig. 8.6 Hyper Raman spectrum of methane, $3{\cdot}8 \times 10^6$ Pa, 12 200 laser shots

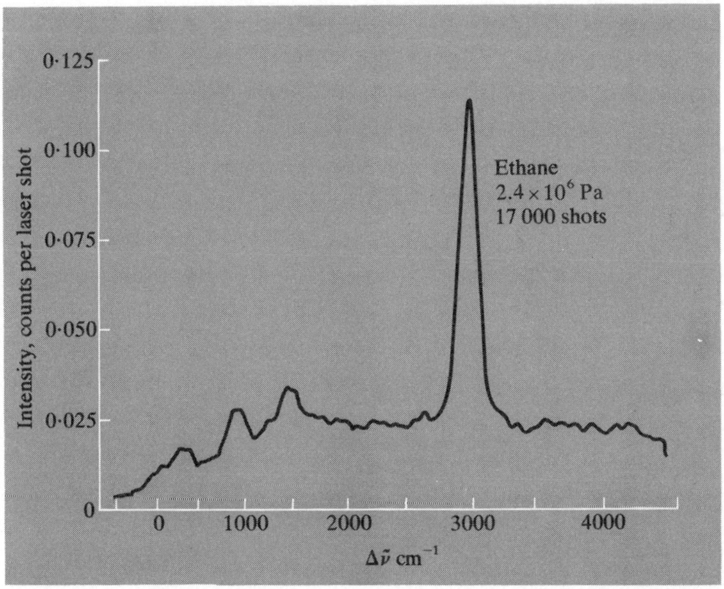

Fig. 8.7 Hyper Raman spectrum of ethane, $2{\cdot}4 \times 10^6$ Pa, 17 000 laser shots

characteristics of the stimulated Raman effect depend on the direction and manner of both illumination and observation, the account of the properties of the stimulated Raman effect given below will be confined to one particular configuration of the laser–sample–detection system. This arrangment, which is a very typical one, is shown in Fig. 8.9(a). The giant-pulse laser radiation is focused into the sample and the scattering observed along the laser beam direction (the forward direction) and at small angles to this direction.

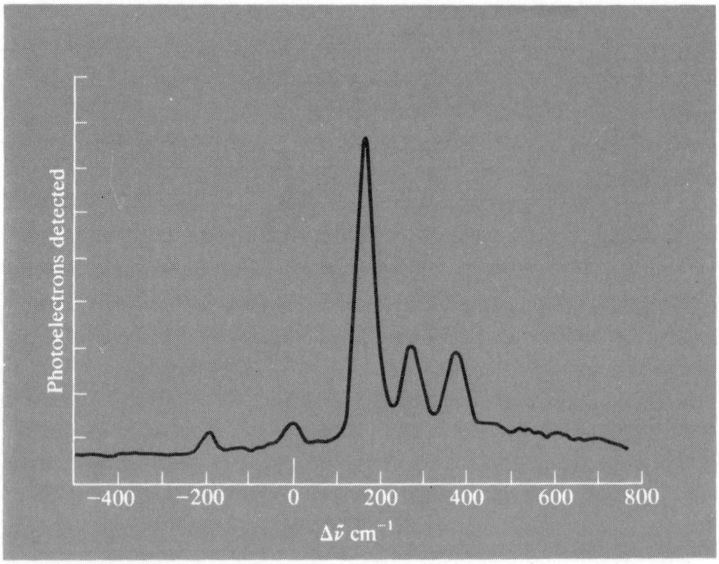

Fig. 8.8 The hyper Rayleigh and low frequency hyper Raman spectrum of a solution-grown single crystal of ammonium chloride, NH_4Cl, at room temperature. The very weak hyper Rayleigh band at $\Delta\tilde{\nu} = 0$ cm^{-1} indicates the essential absence, statistically, of a centre of symmetry in the crystal. The three low frequency hyper Raman bands occur at about $\Delta\tilde{\nu} = 170$ cm^{-1}, $\Delta\tilde{\nu} = 280$ cm^{-1}, and $\Delta\tilde{\nu} = 360$ cm^{-1}. The band at $\Delta\tilde{\nu} = 360$ cm^{-1} is associated with NH_4^+ ion librations and is infrared and Raman *inactive*

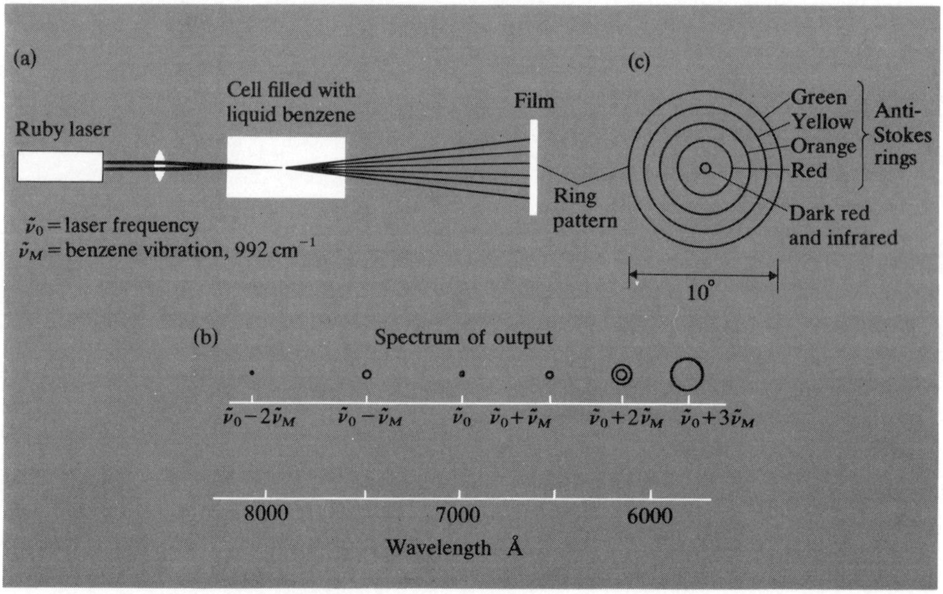

Fig. 8.9 (a) Experimental arrangement for the stimulated Raman effect; (b) diagrammatic representation of the stimulated Raman spectrum of benzene; (c) anti-Stokes rings in the stimulated Raman spectrum of benzene

If the forward scattered radiation is dispersed and photographically recorded, it is found to consist of the incident wavenumber $\tilde{\nu}_0$ and Stokes and anti-Stokes wavenumbers of general formula $\tilde{\nu}_0 \pm n\tilde{\nu}_M$, where $\tilde{\nu}_M$ is usually associated with just *one* Raman active vibration of the scattering molecules and the values of n are restricted to the integers 1, 2, 3, etc. For example, if liquid benzene is illuminated with the focused output of a giant-pulse ruby laser ($\lambda_0 = 6943{\cdot}3$ Å, 694·33 nm, $\tilde{\nu}_0 = 14\,402\,\mathrm{cm}^{-1}$), the Stokes and anti-Stokes shifts are all *exact* multiples of the wavenumber 992 cm^{-1} which is the strongest vibrational band in the normal Raman spectrum of benzene (Fig. 8.9b). The wavenumbers associated with $n > 1$ are not overtones, since anharmonicity would result in wavenumber shifts which did not stand in the exact ratio of 1:2:3, etc.

If the forward scattered radiation is allowed to fall on a colour-sensitive film at right angles to the laser beam, a striking pattern of concentric coloured rings is obtained (see Fig. 8.9(c) and Plate 2 (facing page 232)). This reveals that stimulated Raman scattering has a special angular dependence. The central red spot corresponds to the ruby laser wavelength and also to the Stokes bands which lie at higher wavelengths (lower wavenumbers) and are emitted essentially along the laser beam direction. The coloured rings correspond to successive anti-Stokes bands at lower wavelengths (higher wavenumbers), each of which is emitted only along directions which make a specific small angle with the laser beam direction. The first ring, which is red, comes from the first anti-Stokes band at 15 394 cm^{-1} (6496 Å, 649·6 nm); the second ring, orange, corresponds to the second anti-Stokes band at 16 386 cm^{-1} (6103 Å, 610·3 nm), and so on.

Stimulated Raman scattering differs from normal Raman scattering not only in its wavenumber pattern and angular dependence but also in its intensity. A very substantial fraction of the exciting radiation at $\tilde{\nu}_0$ is converted into radiation at $\tilde{\nu}_0 \pm n\tilde{\nu}_M$. For example, in benzene, about 50 per cent of the incident radiation at $\tilde{\nu}_0$ may be converted to Stokes radiation at $\tilde{\nu}_0 - \tilde{\nu}_M$. For successive Stokes and anti-Stokes lines of general formula $\tilde{\nu}_0 \pm n\tilde{\nu}_M$, the intensity decreases as n increases, but, up to $n = 3$ at least, it is still larger than that found in normal Raman scattering associated with fundamental vibrations. The width of lines in stimulated Raman scattering is also generally much less than in normal Raman scattering, but the inherent narrowness may be obscured by other effects.

The dramatically high intensities of stimulated Raman scattering arise because, at the high irradiances associated with giant-pulse lasers, the power transferred to the first Stokes line at $\tilde{\nu}_0 - \tilde{\nu}_M$ is related *exponentially* to the power in the laser at $\tilde{\nu}_0$. The selective excitation of one molecular vibration at $\tilde{\nu}_M$ arises because there is also an *exponential* dependence on the *intensity per unit bandwidth* associated with the wavenumber $\tilde{\nu}_M$ when observed in the *normal* Raman effect. Thus, the transfer of power in the stimulated Raman spectrum from $\tilde{\nu}_0$ to the Stokes line which has the largest intensity per unit bandwidth in the normal Raman effect generally outstrips that for other Raman lines. Only if two or more vibrational modes have closely similar intensities per unit bandwidth in the normal Raman effect will the stimulated Raman spectrum contain lines associated with more than one value of $\tilde{\nu}_M$.

The first Stokes line in the stimulated Raman effect rapidly becomes intense enough to act as a powerful source of wavenumber $\tilde{\nu}_0 - \tilde{\nu}_M$; thus, a Stokes line at

$$(\tilde{\nu}_0 - \tilde{\nu}_M) - \tilde{\nu}_M = \tilde{\nu}_0 - 2\tilde{\nu}_M$$

is generated, and as this gains in intensity it acts as another source, and so on. Sophisticated experiments show that there is a very small time lag between the generation of successive Stokes lines, as might be expected for this mechanism.

The generation of the anti-Stokes wavenumbers does not arise as a result of downward transitions from a populated upper state as in the normal Raman effect. Formally, the anti-Stokes photons may be regarded as being created at the expense of the two laser photons, and conservation of energy according to

$$2\tilde{\nu}_0 = (\tilde{\nu}_0 + \tilde{\nu}_M) + (\tilde{\nu}_0 - \tilde{\nu}_M)$$

dictates that a Stokes photon is also generated. Conservation of momentum leads to the anti-Stokes photons being restricted to a particular angle with the laser beam direction.

The high conversion efficiency of stimulated Raman scattering means that it can be used to generate intense, coherent, laser-like sources over a wide range of wavenumbers, simply by judicious choice of scattering material. A giant-pulse laser of wavenumber $\tilde{\nu}_0$ is thus 'chemically tunable' over the range $\tilde{\nu}_0 \pm 3000\ \text{cm}^{-1}$ through the stimulated Raman effect. With a few laser wavenumbers and the resources of a modest chemical store, wavenumbers ranging from the ultraviolet through the visible to the infrared can be produced. One interesting application of the availability of specific wavenumbers is in resonance absorption studies.

8.4.2 *Application to the study of vibrational lifetimes*

The stimulated Raman effect also provides the basis of a very important method of studying the lifetimes of vibrational states. The principle of the method is as follows. Since in the stimulated Raman effect a large fraction of the incident radiation at wavenumber $\tilde{\nu}_0$ is converted to the first Stokes wavenumber, $\tilde{\nu}_0 - \tilde{\nu}_M$, a substantial population of the first vibrational level associated with $\tilde{\nu}_M$ results. In the jargon of the subject, the giant-pulse laser radiation is said to pump the scattering system up to the first vibrational level through the mechanism of the stimulated Raman effect. Provided the pumping can be done rapidly compared with the relaxation of the population of the vibrational level in question, the decrease in the population of this level can then be monitored by measuring, at successive intervals of time, the intensity of the *normal anti-Stokes* Raman scattering at $\tilde{\nu}_0' + \tilde{\nu}_M$, where $\tilde{\nu}_0'$ is the wavenumber of another laser whose power is below the threshold for stimulated Raman scattering. In the jargon of the subject, this laser wavenumber is said to act as a population probe through the mechanism of the normal anti-Stokes Raman effect. Generally, the pump and probe wavenumbers $\tilde{\nu}_0$ and $\tilde{\nu}_0'$ are chosen to be different, but this is not always necessary.

The first experiments of this kind were made on hydrogen gas[10] at about 25 atmospheres pressure and 300 K, where the lifetime for the vibrational state ($\tilde{\nu}_M = 4156\ \text{cm}^{-1}$) is of the order of 10^{-5} s. A giant-pulse ruby laser (output wavenumber, $14\,402\ \text{cm}^{-1}$) was used as a pump. The pulse duration was about 20×10^{-9} s, which is sufficiently short compared with the vibrational relaxation time. Another ruby laser (output also $14\,402\ \text{cm}^{-1}$), but of much lower power and longer pulse duration, was used as the probe, and the intensity of the anti-Stokes Raman scattering which occurs at $18\,558\ \text{cm}^{-1}$ ($14\,402\ \text{cm}^{-1} + 4156\ \text{cm}^{-1}$) was measured as a function of time. There would, of course, be no detectable normal anti-Stokes Raman scattering if the population of the first vibrational level had not been artificially increased through the

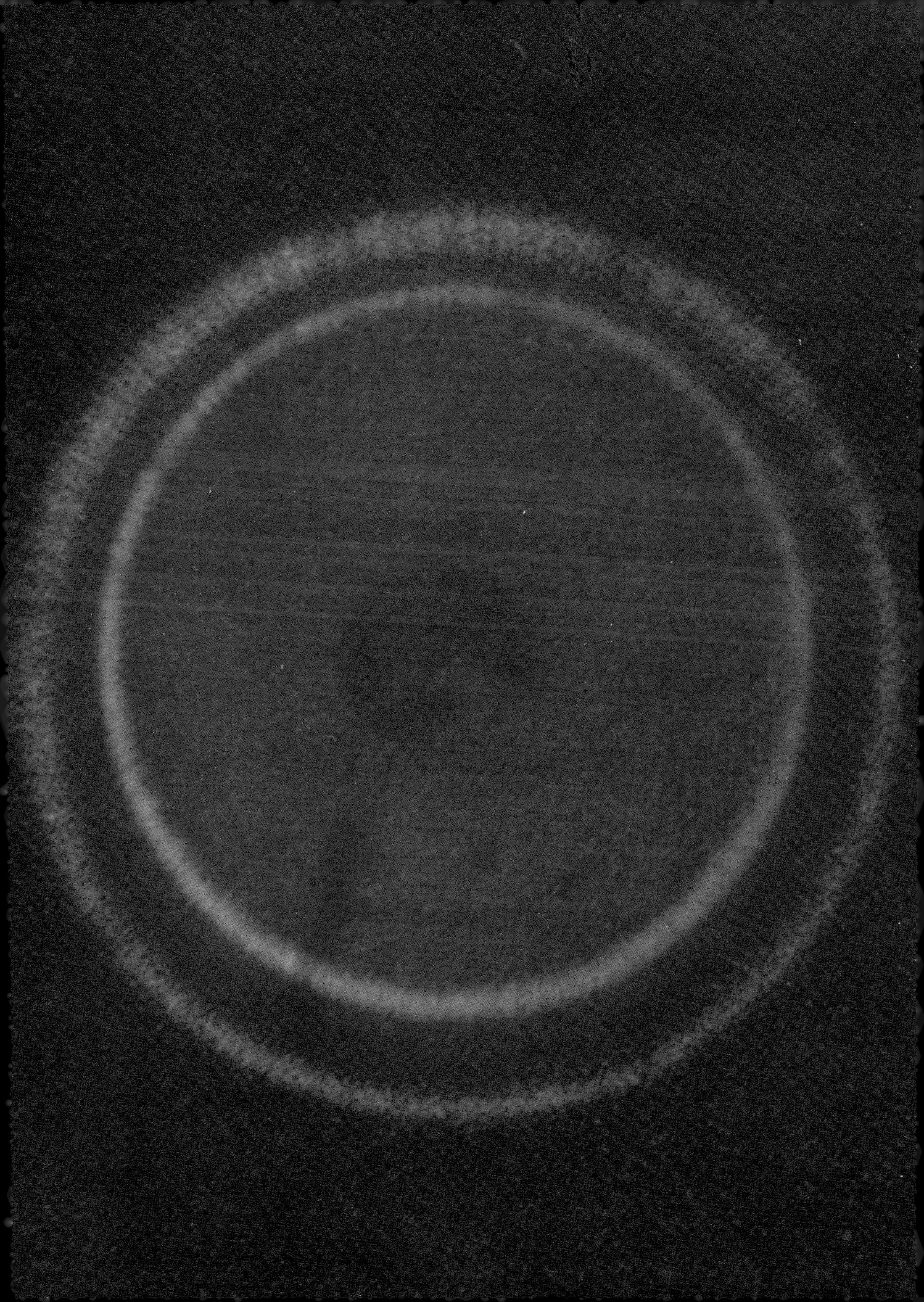

pumping effect of the stimulated Raman effect; for a vibrational wavenumber $\tilde{\nu}_M = 4156\ cm^{-1}$, the equilibrium population of the first vibrational level at 300 K is vanishingly small. In these experiments, pump and probe radiations of the same wavenumber were used. This was possible because any stimulated anti-Stokes Raman scattering produced by the pump radiation, which would also occur at $18\,558\ cm^{-1}$, was only of nanosecond duration and could be separated from the normal anti-Stokes Raman scattering by gating. Measurements of this kind yielded a pressure relaxation-time product at 300 K of $(1060 \pm 100) \times 10^{-6}$ atm s.

In liquids, vibrational relaxation times are much shorter than in gases, and are of the order of tens of picoseconds. Their study by this method therefore requires pump and probe pulses of picosecond duration. A typical experimental arrangement[11] is shown diagrammatically in Fig. 8.10. An optical switching device is used to isolate a

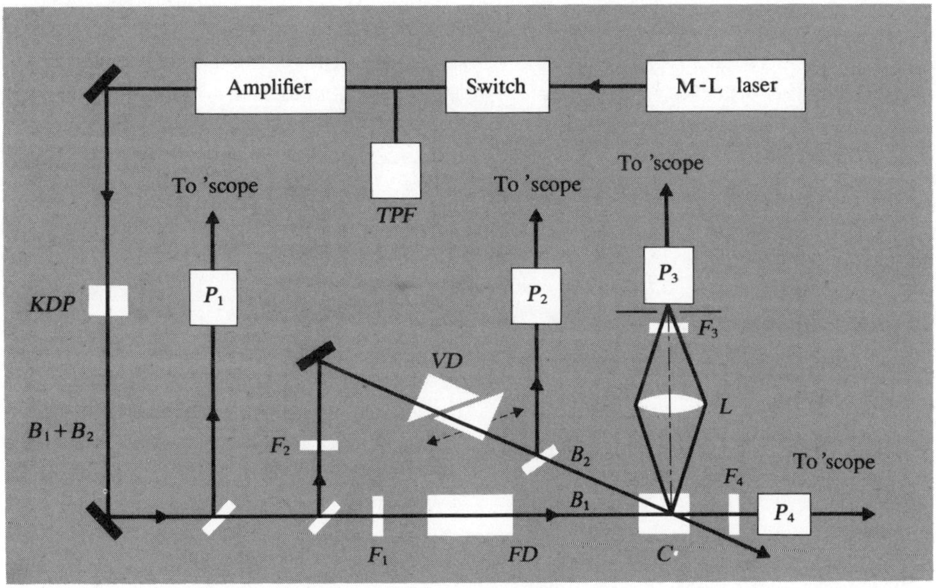

Fig. 8.10 Schematic diagram of apparatus for measurement of vibrational relaxation times (KDP=frequency doubling crystal)

single pulse of about 5 picosecond duration and wavenumber $9434{\cdot}0\ cm^{-1}$ (10 600 Å, 1060·00 nm) from a train of such picosecond pulses produced by a mode-locked neodymium-glass laser. The single pulse is then amplified in a laser amplifier system to give an energy of about 10 mJ. The amplified pulse passes through a frequency doubling crystal when about five per cent is converted to radiation at $18\,868\ cm^{-1}$ (5300 Å, 530·00 nm). The 10 mJ picosecond pulse at $9434{\cdot}0\ cm^{-1}$ constitutes the pump radiation and the 0·5 mJ picosecond pulse at $18\,868\ cm^{-1}$ the probe radiation. Both eventually arrive at the sample cell *C*, but they follow different routes, the lengths of which may be varied so that the probe pulse arrives in the cell some picoseconds after the pump pulse. Since light travels 0·3 mm in 10^{-12} s, path differences of the order of mm are involved. The filter F_1 passes only the pump radiation, and the glass rod *FD* provides a fixed optical delay. The detector P_1 monitors the pump radiation. The filter F_2 passes only the probe radiation which is monitored by the detector P_2, and the prism system *VD* acts as a variable optical delay permitting the probe radiation to arrive in the cell a predetermined number of picoseconds after the pump pulse. The normal anti-Stokes scattering excited by the probe radiation is collected by the lens L, and after passage through an appropriate filter F_3 to remove other radiation, it is monitored by the detector P_3. The stimulated Stokes Raman scattering produced by the pump pulse is monitored by the detector P_4.

In a typical experiment,[11] the vibrational relaxation of the symmetric C—H stretching mode ($\tilde{\nu}_M = 2939\ \mathrm{cm}^{-1}$) in 1,1,1-trichloroethane, CCl_3CH_3, was studied. The 10 mJ pump pulse is estimated to cause about 5×10^{15} molecules to make the vibrational transition $1 \leftarrow 0$ for the symmetric C—H stretching mode. This represents an excitation of about 10^{-3} of the molecules present in the illuminated volume of about $10^{-3}\ \mathrm{cm}^3$ in the liquid sample, for which the number density is about $5 \times 10^{21}\ \mathrm{cm}^{-3}$. This number of excited molecules is 10^3 greater than for thermal equilibrium. With a probe pulse of 0.5 mJ, the 5×10^{15} excited molecules produce a maximum anti-Stokes Raman signal of about 10^4 photons scattered into a solid angle of about 0·2 steradian. This is quite a large signal and is readily detected; this anti-Stokes Raman signal occurs at an absolute wavenumber of $21\,807\ \mathrm{cm}^{-1}$ ($18\,868 + 2939 = 21\,807$) or 4585·7 Å. The stimulated Stokes Raman signal excited by the pump radiation occurs at $6495{\cdot}0\ \mathrm{cm}^{-1}$ ($9434{\cdot}0 - 2939 = 6495{\cdot}0$) or 15 396 Å. The ratio of this signal to the pump radiation signal serves to measure the number of molecules excited into the first vibrational state. The results obtained for CH_3CCl_3 are shown graphically in Fig. 8.11. The broken curve shows the rise and fall of the pump pulse signal and the black line curve shows how the anti-Stokes signal varies as the delay time of the probe pulse is increased. This latter curve indicates that the anti-Stokes signal first increases and reaches a maximum shortly after the maximum in the pump pulse, showing that time is required for the molecular excitation. After the maximum is reached, the signal falls off more slowly than the pump pulse in a manner which is dependent on the vibrational relaxation time. An analysis of the

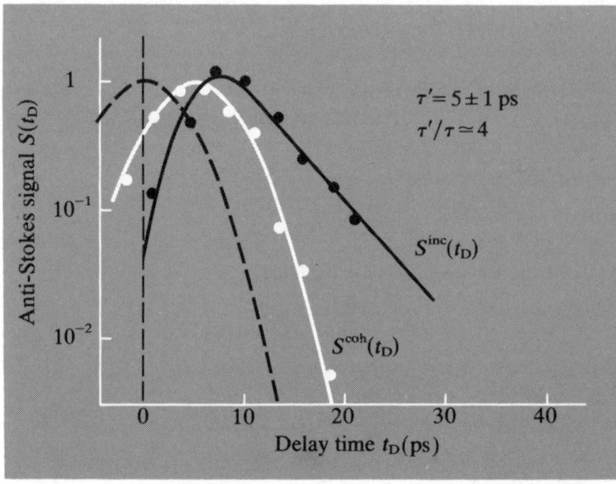

Fig. 8.11 Measured incoherent scattering signal $S^{inc}(t_D)/S^{inc}_{max}$ (black dots) and coherent scattering $S^{coh}(t_D)/S^{coh}_{max}$ (white dots) versus delay time t_D between pump pulse and probe pulse for 1,1,1-trichlorethane. The black and white curves are calculated. The broken curve indicates the pump pulse

data shows that the curve can be fitted with a vibrational relaxation time of $(5 \pm 1) \times 10^{-12}$ s. The white curve in Fig. 8.11 shows how the phase-matched forward anti-Stokes Raman signal varies with time. This falls off very rapidly compared with the normal anti-Stokes Raman radiation. This shows that the phase relationships upon which this forward Raman scattering depends relax much more rapidly than the vibrational energy. The dephasing time is estimated to be about 1 picosecond, so that the ratio of the vibrational relaxation time to the dephasing time is about 4. This emphasizes that the mechanisms of the two relaxations are different. Figure 8.12 shows results obtained for ethanol, C_2H_5OH. In this molecule, the vibrational relaxation time is much longer, of the order of 20 ± 5 picoseconds.

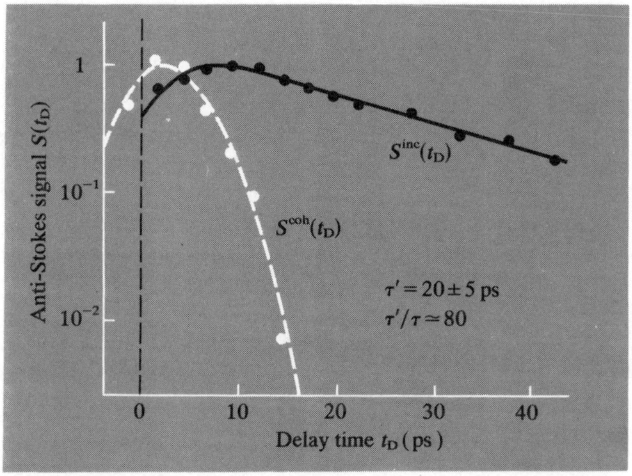

Fig. 8.12 Measured incoherent scattering signal $S^{inc}(t_D)/S^{inc}_{max}$ (black dots) and coherent scattering $S^{coh}(t_D)/S^{coh}_{max}$ (white dots) versus delay time t_D for ethanol. The solid and dashed curves are calculated

If the filter F_3 is replaced by a spectrometer, the frequency content of the normal anti-Stokes Raman scattering may be analysed under reasonable resolution at specific intervals of time after the initial excitation. Very interesting results[12] relating to the C—H stretching modes in ethanol have been obtained in this way. Although only *one* of the C—H modes at 2939 cm^{-1} is populated by the stimulated Raman effect, a few picoseconds after this event, the frequency profile of the anti-Stokes normal Raman scattering shows that more than one C—H stretching mode in this region has become populated, presumably as a result of rapid energy transfer between the various C—H modes around 2900 cm^{-1} (see Fig. 8.13(a)). The decay from these levels is much slower and, in consequence, a quasi-equilibrium for the excess population develops. In addition, it has been established that the process $v = 2 \leftarrow v = 1$ for the C—H bending mode, whose fundamental frequency is around 1450 cm^{-1}, plays a significant role in the rapid relaxation of the upper levels. This was deduced from a careful comparison of the frequency profiles in the 1450 cm^{-1} region of the spontaneous Stokes Raman scattering for the C—H bending mode and the spontaneous anti-Stokes Raman scattering after the population changes induced by the stimulated Raman effect (see Fig. 8.13(b)). A quantitative analysis of these results leads to the relaxation scheme shown in Fig. 8.13(c).

8.5 The inverse Raman effect

If a system is illuminated simultaneously with a giant-pulse laser beam of wavenumber $\tilde{\nu}_0$ and a continuum covering the wavenumber range $\tilde{\nu}_0$ to $\tilde{\nu}_0 + 3500\ \mathrm{cm}^{-1}$, absorptions are observed in the continuum at wavenumbers $\tilde{\nu}_0 + \tilde{\nu}_M$ (where $\tilde{\nu}_M$ is associated with a Raman active vibration of the system) and, in addition, there is emission at the laser wavenumber $\tilde{\nu}_0$. Simple considerations of conservation of energy, as illustrated in Fig. 8.14, show that this is an inverse Raman effect. Under the influence of the continuum and the laser, a quantum of energy $hc(\tilde{\nu}_0 + \tilde{\nu}_M)$ is absorbed from the continuum; the system undergoes an upward transition between two levels of energy E_1 and E_2 with an uptake of energy $E_2 - E_1 = hc\tilde{\nu}_M$, and the balance of energy $hc(\tilde{\nu}_0 + \tilde{\nu}_M) - hc\tilde{\nu}_M$ is emitted as a quantum of energy $hc\tilde{\nu}_0$. Absorptions at $\tilde{\nu}_0 - \tilde{\nu}_M$ with an accompanying emission at $\tilde{\nu}_0$ can also be observed if the continuum is extended to include the low wavenumber side of $\tilde{\nu}_0$. In this case, the system undergoes a downward transition between two levels, resulting in the release of energy $E_2 - E_1 = hc\tilde{\nu}_M$; combination of this with energy absorbed from the continuum at $\tilde{\nu}_0 - \tilde{\nu}_M$ results in an emission at $\tilde{\nu}_0$, since $hc(\tilde{\nu}_0 - \tilde{\nu}_M) + hc\tilde{\nu}_M = hc\tilde{\nu}_0$. This complementary effect will require appropriate population of the upper state.

In practice, conditions necessary for the production of the inverse Raman effect are not easy to achieve. The continuum must coincide with the laser pulse as it passes

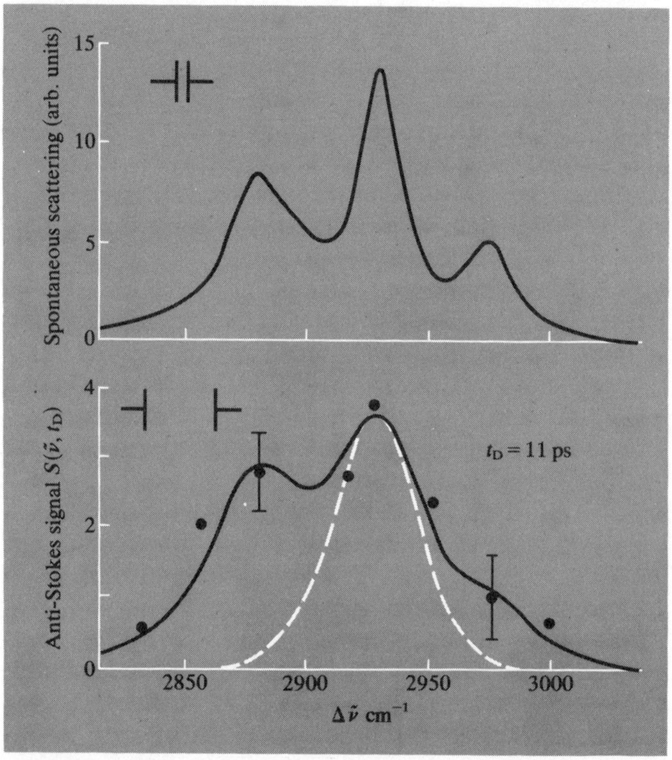

Fig. 8.13(a) Comparison of spontaneous Stokes spectrum of the C—H stretching vibrations and anti-Stokes probe scattering of C—H stretching vibrations for ethanol at $t_D = 11$ ps; broken line indicates spectral profile of the pumped mode at 2928 cm^{-1}; lower solid curve is calculated for quasi-equilibrium between levels around 2900 cm^{-1}

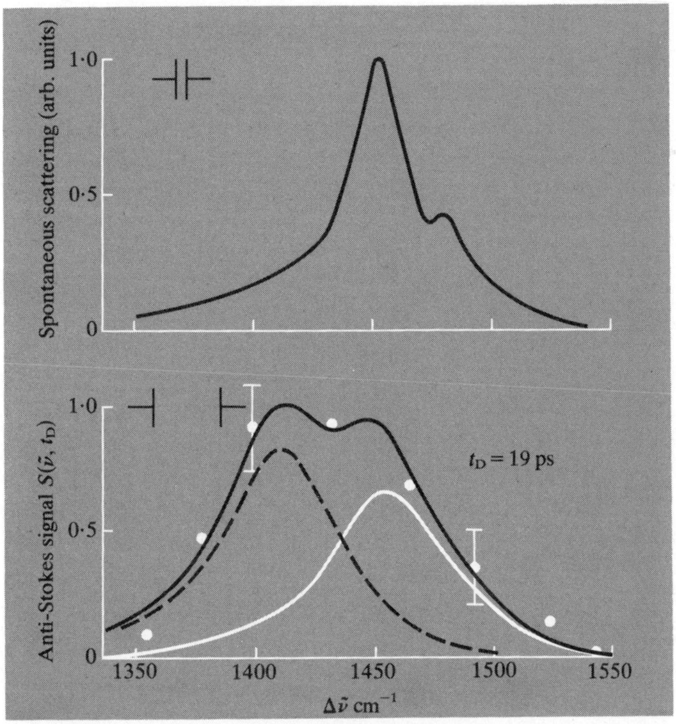

Fig. 8.13(b) Comparison of spontaneous Stokes spectrum of the C—H bending modes $\delta_H(\nu=1)$ and anti-Stokes probe scattering around 1400 cm^{-1} for ethanol; and calculated curves for $\delta_H(v=1)$ scattering (white line) and $\delta_H(v=2)$ scattering (broken line)

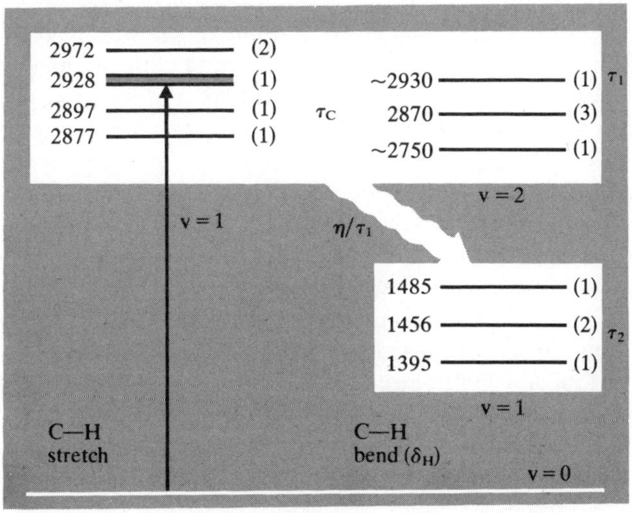

Fig. 8.13(c) Schematic diagram of relevant vibrational states, indicating relaxation routes for ethanol

through the system. It must be of adequate intensity for detection in the duration of the experiment, which is the duration of the giant pulse (typically 10^{-8} to 10^{-7} s). If photographic detection is used, the continuum must have virtually the same lifetime

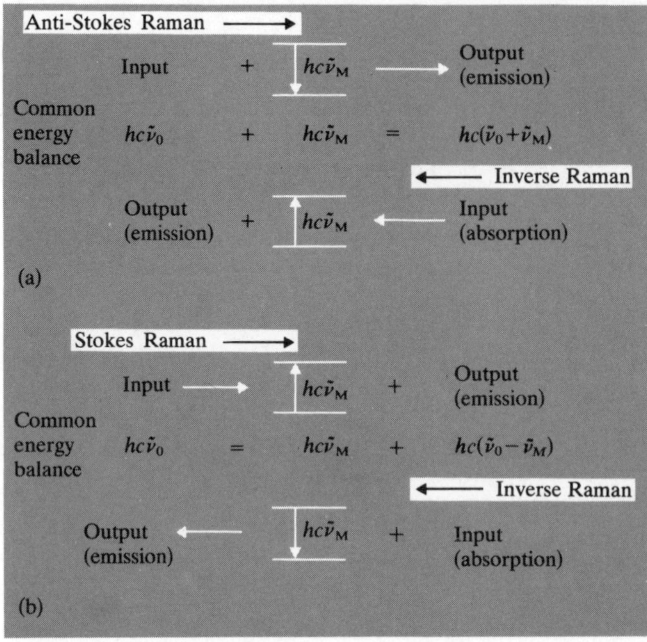

Fig. 8.14 Comparison of energy balance in (a) anti-Stokes Raman and inverse Raman effects, and (b) Stokes Raman and inverse Raman effects

as the giant pulse itself. It should also cover as much as possible of the wavenumber range $\tilde{\nu}_0$ to $\tilde{\nu}_0 + 3500 \text{ cm}^{-1}$.

The first observation of the inverse Raman effect was made by Jones and Stoicheff.[13] Their procedure is represented diagrammatically in Fig. 8.15. A cell containing benzene was illuminated with the radiation from a giant-pulse ruby laser which had first passed through a cell containing toluene. Under the particular conditions of this experiment, the stimulated first anti-Stokes Raman line of toluene generated by the laser in the toluene cell at $\tilde{\nu}_0 + 1003 \text{ cm}^{-1}$ was somewhat broadened. Thus the radiation incident on the benzene cell was not only the laser radiation beam at $\tilde{\nu}_0$, but also a continuum of limited extent ranging from $\tilde{\nu}_0 + 1003 + \Delta \text{ cm}^{-1}$ to $\tilde{\nu}_0 + 1003 - \Delta \text{ cm}^{-1}$, where Δ exceeds 20 cm^{-1}. A sharp absorption at $\tilde{\nu}_0 + 992 \text{ cm}^{-1}$ was observed in this limited continuum when the radiation emerging from the benzene cell was dispersed and photographically recorded.

While use of the giant-pulse laser itself to generate the continuum ensures the spatial and temporal coincidence of the laser and continuum in the sample, the particular method described above is of very limited applicability, since it gives such a restricted continuum. Some effort has been devoted to other methods of continuum generation by the laser. One interesting technique has been that developed by McLaren and Stoicheff.[14] Their experimental arrangement is shown in Fig. 8.16.

The radiation from a giant-pulse ruby laser is passed through a frequency doubling crystal; about three per cent of the laser radiation ($\lambda = 6943{\cdot}3$ Å, $694{\cdot}33$ nm; $\tilde{\nu}_0 = 14\,402 \text{ cm}^{-1}$) is converted into the harmonic at $2\nu_0$. The two wavenumbers are separated into two beams by a prism and the harmonic component ($2\tilde{\nu}_0 =$

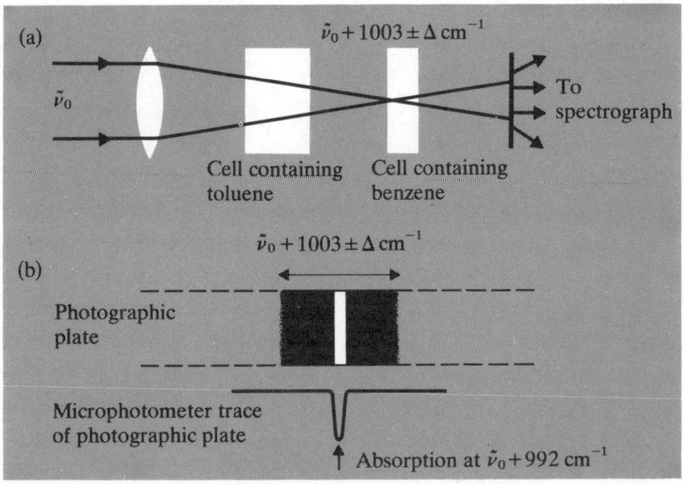

Fig. 8.15(a) The initial experimental arrangement which Jones and Stoicheff used to observe the inverse Raman effect; and (b) a diagrammatic representation of part of the inverse Raman spectrum of benzene

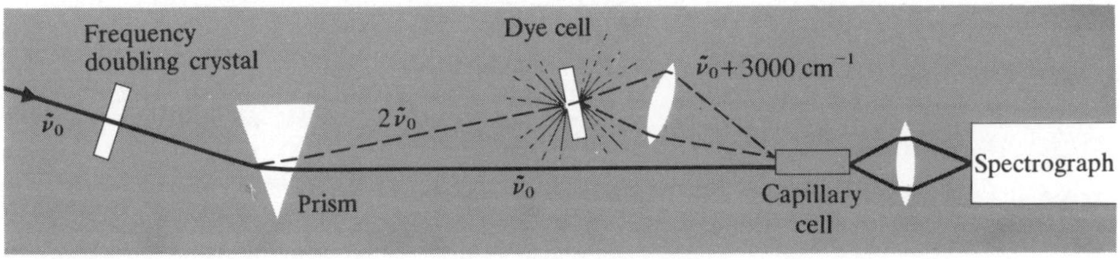

Fig. 8.16 The experimental arrangement with which McLaren and Stoicheff observed the inverse Raman effect

28 804 cm^{-1}; $\lambda = 3471{\cdot}7$ Å, 347·17 nm) is allowed to fall on a 1 mm thick cell containing a solution of a dyestuff with appropriate fluorescence characteristics. For example, a $2{\cdot}0 \times 10^{-3}$ M solution of rhodamine 6G in methanol absorbs strongly at 3471 Å and has intense fluorescence over the wavelength range 5600 to 6800 Å (560·0 to 680·0 nm), corresponding to the high frequency side of the laser frequency. The laser beam and the fluorescence continuum are recombined in a glass capillary cell containing the liquid under investigation. Spatial coincidence of the two beams is achieved by trapping through total internal reflection. The fluorescent lifetime of rhodamine 6G is only $5{\cdot}5 \times 10^{-9}$ s, so that good temporal coincidence of the fluorescence continuum with the laser pulse is achieved. The radiation from within the capillary cell is optically coupled to an appropriate spectrograph and photographically recorded.

Another technique[15] for continuum generation in the nanosecond region involves focusing the giant-pulse laser radiation into a cell containing krypton at high pressure, and this has been used successfully by Dumartin, Oksengorn, and Vodar (Fig. 8.17).

Alfano and Shapiro[16] have obtained inverse Raman spectra using laser pulses of picosecond duration. When a picosecond pulse of high intensity passes through a

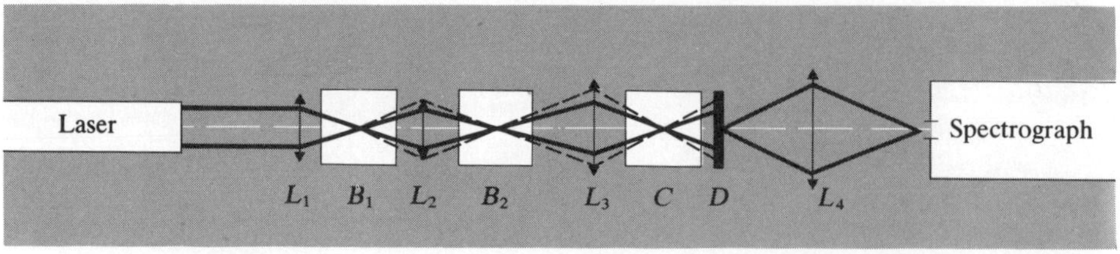

Fig. 8.17 Experimental arrangement for continuum generation used by Dumartin, Oksengorn, and Vodar. ***L*****, lens;** ***B*****, pressure cell;** ***C*****, sample cell;** ***D*****, diffusing screen.**

solid or liquid, complex non-linear processes lead to the generation of a continuum over the entire visible region. Picosecond pulses of radiation at 10 600 Å (1060·0 nm, 9434·0 cm^{-1}) produced by a mode-locked neodymium-glass laser were frequency doubled to give radiation at 5300 Å (530·0 nm, 18 868 cm^{-1}). These picosecond pulses, each of which had a power of about 0·5 GW, were focused into a 5 cm sample of a special borosilicate glass when about 10^{-2} to 10^{-3} of the laser radiation was converted into a continuum extending over about 10 000 cm^{-1}. The laser radiation at 5300 Å (530·0 nm, 18 868 cm^{-1}) (attenuated as necessary), together with the continuum, then passed into the sample, and the emerging radiation was collected by a lens and imaged into a spectrograph. The resulting inverse Raman spectra were then recorded photographically.

Inverse Raman spectroscopy offers a method of recording Raman spectra in periods as short as 10^{-8} to 10^{-11} seconds. Also, it does not suffer from the restrictions of the stimulated Raman effect in which only one or two vibrations are excited. It therefore has promise as a technique for studying short-lived species. However, the experimental procedures are not easy and are only just being established.

8.6 Coherent anti-Stokes Raman scattering (CARS)

If coherent radiation of wavenumber $\tilde{\nu}_1$ is 'mixed' in a molecular medium with coherent radiation of wavenumber $\tilde{\nu}_2$ and the irradiances of the two radiations are sufficiently great, then *inter alia* coherent radiation of wavenumber $\tilde{\nu}_3$ can result where

$$\tilde{\nu}_3 = \tilde{\nu}_1 + (\tilde{\nu}_1 - \tilde{\nu}_2) \tag{8.20}$$

Here, 'mixing' implies spatial and temporal coincidence of the two beams. If $\tilde{\nu}_1$ is kept fixed and $\tilde{\nu}_2$ is varied so that the condition

$$\tilde{\nu}_1 - \tilde{\nu}_2 = \tilde{\nu}_M \tag{8.21}$$

is achieved, where $\tilde{\nu}_M$ is some molecular wavenumber of the medium, then

$$\tilde{\nu}_3 = \tilde{\nu}_1 + \tilde{\nu}_M \tag{8.22}$$

Thus $\tilde{\nu}_3$ is coincident in wavenumber with anti-Stokes Raman scattering associated with the molecular wavenumber $\tilde{\nu}_M$ when excited by radiation of wavenumber $\tilde{\nu}_1$. Radiation produced in this way is termed coherent anti-Stokes scattering (CARS) to emphasize its different origin and properties.

The possibility of non-linear mixing of this kind was first discovered by Terhune[17] in 1963. However, it is only very recently, with the availability of sufficiently powerful continuously tunable lasers, that the practical exploitation of this interesting phenomenon has become possible.[18] We shall conclude this chapter on non-linear effects with a short qualitative account of the properties of coherent anti-Stokes Raman scattering.

With CARS, the conversion efficiency to $\tilde{\nu}_3$ is several orders of magnitude greater than conversion efficiencies in the normal Raman effect. The radiation at $\tilde{\nu}_3$ resulting from CARS forms a coherent, highly collimated beam, whereas normal Raman scattering is incoherent and extends over a solid angle of 4π. It follows that collection efficiencies of radiation from CARS can be very high, since at least 50 per cent of the radiation can fall on the detector. Further, fluorescence and thermal radiation from hot samples, which can be very troublesome in normal Raman scattering, can be simply and effectively discriminated against by interposing between the sample and the detector, a screen with a hole of a few mm diameter which just passes the CARS beam. Since fluorescence and thermal radiation extend over a large solid angle whereas radiation from CARS is highly collimated, the spatial filter effects almost complete discrimination.

In principle, no dispersing device is necessary for CARS experiments since, provided the wavenumbers $\tilde{\nu}_1$ and $\tilde{\nu}_2$ are known with sufficient accuracy, each $\tilde{\nu}_M$ can be found from eq. (8.21) using the values of $\tilde{\nu}_2$ which produce, in turn, peak signals at the detector. Interference filters are normally adequate to reject $\tilde{\nu}_1$ and $\tilde{\nu}_2$. A typical experimental arrangement is shown in Fig. 8.18.

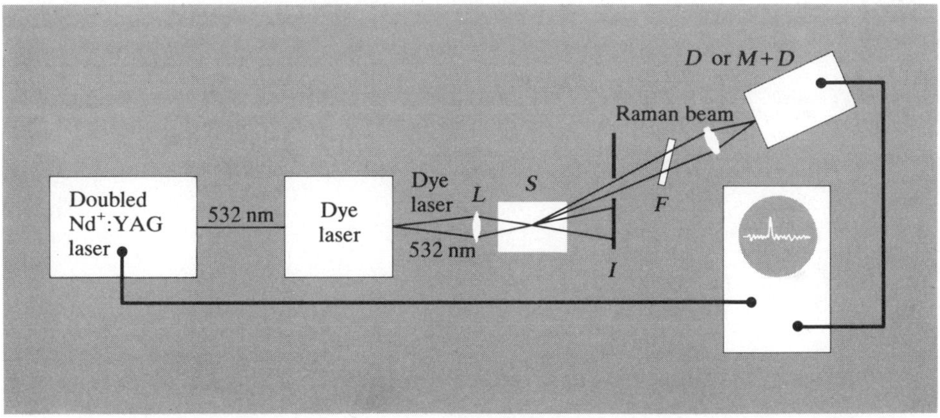

Fig. 8.18 Typical arrangements for CARS using a frequency doubled Nd: YAG pumped dye laser. *L* is a short focal lens (3–4 cm); *S* is the sample; *I* is an iris for spatially filtering the two exciting beams; *F* is a wideband interference filter; *D* is the detector (usually a PIN diode); and *M* is a monochromator (not usually necessary)

The dependence of CARS on various parameters is quite different from that of normal Raman scattering. Thus the power associated with CARS depends on the *square* of the normal Raman scattering cross section; on $\mathcal{I}^2_{\tilde{\nu}_1}\mathcal{I}_{\tilde{\nu}_2}$ where $\mathcal{I}_{\tilde{\nu}_1}$ and $\mathcal{I}_{\tilde{\nu}_2}$ are the irradiances of the incident radiation at $\tilde{\nu}_1$ and $\tilde{\nu}_2$; and on the *square* of the number of scattering molecules.

CARS is an exciting new development which promises to have widespread applications. For example it has already been used as an effective spatial probe of concentrations of gases in flames. Figure 8.19 shows the distribution of hydrogen in a horizontal gas flame obtained using the CARS technique[19]. CARS has also been reported for pure rotation and rotation–vibration Raman transitions[20,21] and for biological chromophores.[22]

There seems no doubt that this latest addition to the family of light scattering phenomena generically related to the Raman effect will become an important structural and analytical tool.

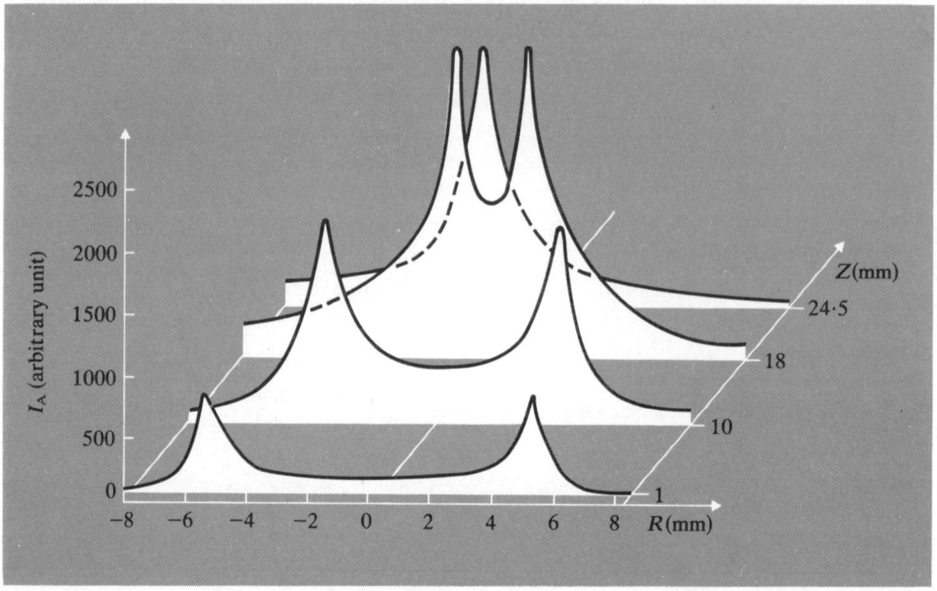

Fig. 8.19 Distribution of hydrogen in a horizontal gas flame obtained using the CARS technique

References

1. R. W. Terhune, P. D. Maker, and C. M. Savage, *Phys. Rev. Letters*, **14**, 681, 1965.
2. D. A. Long and L. Stanton, *Proc. Roy. Soc.*, **A318**, 441, 1970.
3. S. J. Cyvin, J. E. Rauch, and J. C. Decius, *J. Chem. Phys.*, **43**, 4083, 1965.
4. J. H. Christie and D. J. Lockwood, *J. Chem. Phys.*, **54**, 1141, 1971.
5. L. Stanton, *J. Raman Spectroscopy*, **1**, 53, 1973.
6. M. J. French and D. A. Long, *J. Raman Spectroscopy*, **3**, 391, 1975.
7. P. D. Maker and C. M. Savage, unpublished work.
8. J. F. Verdieck, S. H. Peterson, C. M. Savage, and P. D. Maker, *Chem. Phys. Letters*, **7**, 219, 1970.
9. T. J. Dines, M. J. French, R. J. B. Hall and D. A. Long, page 707 in E. D. Schmid (Ed.), *Proceedings of the Fifth International Conference on Raman Spectroscopy*, Schulz Verlag, Freiburg, 1976.
10. J. Ducuing and F. de Martini, *Phys. Rev. Letters*, **17**, 117, 1966.
11. A. Laubereau, D. von der Linde, and W. Kaiser, *Phys. Rev. Letters*, **28**, 1162, 1972.
12. A. Laubereau, G. Kehl, W. Kaiser, *Optics Comm.*, **11**, 74, 1974.
13. W. J. Jones and B. P. Stoicheff, *Phys. Rev. Letters*, **13**, 657, 1964.
14. R. A. McLaren and B. P. Stoicheff, *Appl. Phys. Letters*, **16**, 140, 1970.
15. S. Dumartin, B. Oksengorn, and B. Vodar, *C. R. Acad. Sci. Paris*, **261**, 3767, 1965.
16. R. R. Alfano and S. L. Shapiro, *Chem. Phys. Letters*, **8**, 631, 1971.

17. P. D. Maker and R. W. Terhune, *Phys. Rev.*, **137A**, 801, 1965.

18. R. F. Begley, A. B. Harvey, R. L. Byer, and B. S. Hudson, *J. Chem. Phys.*, **61**, 2466, 1974; J. J. Barrett and R. F. Begley, *Appl. Phys. Letters*, **27**, 129, 1975; I. Itzkan and D. A. Leonard, *Appl. Phys. Letters*, **26**, 106, 1975; F. Moya, S. A. J. Druet, and J. P. E. Taran, *Optics Comm.*, **13**, 169, 1975.

19. P. R. Regnier and J. P. E. Taran, *A.I.A.A. Journal*, **12**, 826, 1974.

20. J. W. Nibler, J. R. McDonald and A. B. Harvey, page 717 in E. D. Schmid (Ed.), *Proceedings of the Fifth International Conference on Raman Spectroscopy*, Schulz Verlag, Freiburg, 1976.

21. J. J. Barrett, page 732 in E. D. Schmid (Ed.), *Proceedings of the Fifth International Conference on Raman Spectroscopy*, Schulz Verlag, Freiburg, 1976.

22. J. R. Nestor, T. G. Spiro and G. L. Klauminzer, page 738 in E. D. Schmid (Ed.), *Proceedings of the Fifth International Conference on Raman Spectroscopy*, Schulz Verlag, Freiburg, 1976.

Appendix I

Guide to the literature of Raman spectroscopy

The following list of books, review articles, and key papers is intended to provide the reader with a guide to the now extensive literature of Raman spectroscopy. General items are listed first; the rest of the material is arranged according to the subject matter of each chapter or section.

General

K. W. F. Kohlrausch, *Der Smekal-Raman-Effekt*, Springer-Verlag, Berlin, 1931, and *Der Smekal-Raman-Effekt, Ergänzungstand, 1931–7,* Springer-Verlag, Berlin, 1938.
(Now available as a limited edition reprint from Heyden, London.) This is the first book on Raman spectroscopy and is still a very valuable source book.

G. Placzek, Rayleigh-Streuung und Raman-Effekt, in E. Marx (Ed.), *Handbuch der Radiologie*, vol. VI, 2, pp. 205–374, Akademische Verlag, Leipzig, 1934.
(An English translation is available: UCRL-Trans-526(L), United States Atomic Energy Commission, Division of Technical Information, 1962.) The most important single article on the theory of Rayleigh and Raman scattering. A masterly treatment.

G. Placzek, Ramaneffekt und Molekülbau, in P. Debye (Ed.), *Molekülstruktur, Leipziger Vorträge 1931,* Verlag S. Hirzel, Leipzig, 1931.
An earlier article by Placzek.

J. H. Hibben, *The Raman Effect and its Chemical Applications*, Reinhold, New York, 1939.
A detailed survey of early work which is still valuable.

S. Bhagavantam, *Scattering of Light and the Raman Effect*, Chemical Publishing Co., New York, 1942.
This is still a useful introductory text. Classical work on Rayleigh scattering is well covered.

G. Herzberg, *Infrared and Raman Spectra*, Van Nostrand, New York, 1945.
This volume deals with polyatomic molecules. It is a work of great thoroughness and authority.

G. Herzberg, *Spectra of Diatomic Molecules*, 2nd edn., Van Nostrand, New York, 1950.
This volume deals with infrared, Raman and electronic spectra of diatomic molecules. It is no less thorough and authoritative than its companion volume.

J. Brandmüller and H. Moser, *Einführung in die Ramanspektroskopie*, Steinkopff Verlag, Darmstadt, 1962.
A comprehensive account of theory, experimental methods, and results, but compiled before laser sources were available.

H. A. Szymanski (Ed.), *Raman Spectroscopy, Theory and Practice*, Plenum Press, New York, vol. 1, 1967, vol. 2, 1970.
These two volumes contain a number of reviews, which are still of considerable value.

C. J. Schuler, Laser-induced spontaneous and stimulated Raman scattering, *Progress in Nuclear Energy*, series IX, vol. 8, pt. 2, Pergamon, Oxford, 1968.
The account of stimulated Raman scattering is particularly useful.

T. R. Gilson and P. J. Hendra, *Laser Raman Spectroscopy*, Wiley, London, 1970.
This book is particularly valuable for its detailed account of Raman spectroscopy of single crystals.

J. Loader, *Basic Laser Raman Spectroscopy*, Heyden, London, 1970.
Deals mainly with experimental techniques.

D. A. Long, A. J. Downs, and L. A. K. Staveley (Eds.), *Essays in Structural Chemistry*, Macmillan, London, 1971.
Most of the chapters in this *Festschrift* for L. A. Woodward deal with aspects of Raman spectroscopy.

M. C. Tobin, *Laser Raman Spectroscopy*, vol. 35 in P. J. Elving and I. M. Kolthoff (Eds.), *Chemical Analysis*, a series of Monographs on Analytical Chemistry and its Applications, Wiley–Interscience, New York, 1971.
Emphasizes experimental techniques and also includes extensive tables of group wavenumbers.

A. Anderson (Ed.), *The Raman Effect, Principles*, vol. 1, and *Applications*, vol. 2, Marcel Dekker, New York, vol. 1, 1971, vol. 2, 1973.
Contains a number of authoritative review articles with detailed bibliographies.

M. M. Sushchinskii, *Raman Spectra of Molecules and Crystals*, Israel Programme for Scientific Translations, New York, 1972.
Includes detailed treatments of the theory of both Raman and stimulated Raman scattering.

J. A. Koningstein, *Introduction to the Theory of the Raman Effect*, Reidel, Dordrecht, Holland, 1972.
Includes aspects of the electronic Raman effect and the stimulated, hyper, and inverse Raman effects.

J. P. Mathieu (Ed.), *Advances in Raman Spectroscopy*, vol. 1, Heyden, London, 1973.
This collection of plenary lectures and papers presented at the Third International Conference on Raman Spectroscopy held at Reims in 1972 provides a very good survey of the state of the subject at that time.

N. B. Colthup, L. H. Daly, and S. E. Wiberley, *Introduction to Infrared and Raman Spectroscopy*, 2nd edn., Academic Press, New York, 1974.
A useful survey of theory, experimental techniques, and results.

E. D. Schmid (Ed.), *Proceedings of the Fifth International Conference on Raman Spectroscopy*, Schulz Verlag, Freiburg, 1976.
These proceedings provide a good account of current work in the subject.

Chapter 1

R. S. Krishnan, Historical Introduction, in A. Anderson (Ed.), *The Raman Effect*, vol. 1, Marcel Dekker, New York, 1971.

The books by Kohlrausch, Hibben, and Brandmüller and Moser also deal with aspects of the early development of the Raman effect.

Chapter 2

D. R. Corson and P. Lorrain, *Introduction to Electromagnetic Fields and Waves*, Freeman, San Francisco, 1962.
A most useful standard text on electromagnetic theory.

J. P. Mathieu, *Optics*, Pergamon Press, Oxford, 1975.
A wide-ranging book covering classical optics, electromagnetic theory, and aspects of quantum electrodynamics and spectroscopy.

F. S. Crawford, *Waves*, Berkeley Physics Course, vol. 3, McGraw-Hill, New York, 1965.
An attractive book dealing with many aspects of wave phenomena including electromagnetic waves.

M. Born and E. Wolf, *Principles of Optics*, Fifth Edition, Pergamon Press, Oxford, 1975.
A comprehensive and authoritative treatise.

A. L. Schawlow (Ed.), *Lasers and Light, Readings from Scientific American*, Freeman, San Francisco, 1969.
A collection of articles from *Scientific American*, written with characteristic clarity and beautifully illustrated.

R. Loudon, *The Quantum Theory of Light*, Oxford University Press, Oxford, 1973.
A valuable text which covers many aspects of light scattering.

E. Hecht and A. Zajac, *Optics*, Addison-Wesley, Reading, Massachusetts, 1974.

Chapters 3, 4, and 5

General

The theory of Rayleigh and Raman scattering is treated to varying extents in most of the books listed at the beginning of this appendix. Particularly valuable are the articles by Placzek and the books by Herzberg, Brandmüller and Moser, and Sushchinskii. The theory is also dealt with in

D. Marcuse, *Engineering Quantum Electrodynamics*, Harcourt, Brace and World, New York, 1970.

In the volumes edited by Szymanski and by Anderson the following articles deal particularly with theory:

L. A. Woodward, General Introduction (Szymanski, vol. 1).
R. E. Hester, Raman Intensities and the Nature of the Chemical Bond (Szymanski, vol. 1).
J. Tang and A. C. Albrecht, Developments in the Theories of Vibrational Raman Intensities (Szymanski, vol. 2).
G. P. Barnett and A. C. Albrecht, Comments on the Derivation of the Dispersion Equation for Molecules (Szymanski, vol. 2).
G. W. Chantry, Polarizability Theory of the Raman Effect (Anderson, vol. 1).
R. A. Cowley, The Theory of Raman Scattering from Crystals (Anderson, vol. 1).
J. A. Koningstein and O. S. Mortensen, Electronic Raman Transitions (Anderson, vol. 2).
A. Weber, High Resolution Raman Studies of Gases (Anderson, vol. 2).

The theory of rotational and vibration–rotation Raman scattering is also dealt with by

B. P. Stoicheff, High Resolution Raman Spectroscopy, in H. W. Thompson (Ed.), *Advances in Spectroscopy*, vol. 1, Interscience Publishers, New York, 1959.

The theory of vibrational Rayleigh and Raman scattering from gases and liquids is also treated by

L. D. Barron, Rayleigh and Raman Scattering of Polarized Light, in D. A. Long, R. F. Barrow, and J. Sheridan (Eds.), *Chemical Society Specialist Periodical Reports*, No. 29, *Molecular Spectroscopy*, vol. 4, Chemical Society, London, 1976.

The symmetry of molecular vibrations, selection rules, and normal coordinate calculations are fully treated in

E. B. Wilson, J. C. Decius and P. C. Cross, *Molecular Vibrations*, McGraw-Hill, London, 1955.
L. A. Woodward, *Introduction to the Theory of Molecular Vibrations and Vibrational Spectroscopy*, Oxford University Press, Oxford, 1972.

Chapter 6

G. R. Harrison, R. C. Lord, and J. R. Loofbourow, *Practical Spectroscopy*, Prentice-Hall, New Jersey, 1948.
R. A. Sawyer, *Experimental Spectroscopy*, Dover, New York, 1963.

J. R. Ferraro, Advances in Raman Instrumentation and Sampling Techniques, in H. A. Szymanski (Ed.), *Raman Spectroscopy, Theory and Practice*, vol. 1, Plenum Press, New York, 1967.
J. A. Koningstein, Laser Raman Spectroscopy, in H. A. Szymanski (Ed.), *Raman Spectroscopy, Theory and Practice*, vol. 1, Plenum Press, New York, 1967.
H. W. Schrötter, Raman Spectroscopy with Laser Excitation, in H. A. Szymanski (Ed.), *Raman Spectroscopy, Theory and Practice*, vol. 2, Plenum Press, New York, 1970.
E. Steger, Raman Spectroscopy with Poor Scatterers, in H. A. Szymanski (Ed.), *Raman Spectroscopy, Theory and Practice*, vol. 2, Plenum Press, New York, 1970.
H. J. Sloane, Technique of Raman Spectroscopy: Comparison with Infrared, *Applied Spectroscopy*, **25**, 430, 1971.
C. E. Hathaway, Raman Instrumentation and Techniques, in A. Anderson (Ed.), *The Raman Effect*, vol. 1, Marcel Dekker, New York, 1971.
A detailed article with an extensive bibliography.
J. R. Ferraro and L. J. Basile, Spectroscopy at High Pressures, *Applied Spectroscopy*, **28**, 256, 1974.
J. R. Downey and G. J. Janz, Digital Methods in Raman Spectroscopy, in R. J. H. Clark and R. E. Hester (Eds.), *Advances in Infrared and Raman Spectroscopy*, vol. 1, Heyden, London, 1976.
A critical survey of the advantages and problems of coupling a computer to a Raman spectrometer.
W. J. Jones, Lasers, *Quarterly Reviews*, **23**, 73, 1969.

Chapter 7

7.1 *Introduction*

The development of Raman spectroscopy as a structural and analytical tool can be followed in the reviews which have appeared from time to time in the *Annual Reports* of the Chemical Society of London. For example:

L. A. Woodward, *Annual Reports (London)*, **31**, 21, 1934.
L. A. Woodward, *Annual Reports (London)*, **56**, 67, 1959.
D. A. Long, *Annual Reports (London)*, **60**, 120, 1963.
D. B. Powell, *Annual Reports (London)*, **63**, 112, 1966.
D. A. Long, *Annual Reports (London)*, **65A**, 83, 1968.

A critical account of the uses and limitations of vibrational selection rules has been given by

L. A. Woodward, in H. A. Szymanski (Ed.), *Raman Spectroscopy, Theory and Practice*, vol. 2, Plenum Press, New York, 1970.

7.2 *Rotation and vibration–rotation Raman spectra*

B. P. Stoicheff, High Resolution Raman Spectroscopy, in H. W. Thompson (Ed.), *Advances in Spectroscopy*, vol. 1, Interscience Publishers, New York, 1959.
A comprehensive account of rotation and vibration–rotation Raman spectra of gases investigated using mercury arc excitation. Theory, experimental techniques, and results are covered up to 1959.
H. G. M. Edwards, High Resolution Raman Spectroscopy of Gases, in D. A. Long, L. A. K. Staveley, and A. J. Downs (Eds.), *Essays in Structural Chemistry*, Macmillan, London, 1971.
Covers rotation and vibration–rotation Raman spectra for the period from 1959 to 1970 and includes an account of the advantages of laser excitation.
C. G. Gray and H. L. Welsh, Intermolecular Force Effects in the Raman Spectra of Gases, in D. A. Long, L. A. K. Staveley, and A. J. Downs (Eds.), *Essays in Structural Chemistry*, Macmillan, London, 1971.
A review of the effects of pressure and intermolecular forces on band shapes in the rotation and vibration–rotation Raman spectra of gases.
W. J. Jones, On the use of a Fabry–Perot interferometer for the study of Raman spectra of gases under high resolution, *Contemporary Physics*, **13**, 419, 1972.

A. Weber, Recent Developments in High-Resolution Raman Spectroscopy of Gases, *Developments in Applied Spectroscopy*, **10**, 137, 1972.
This review of rotation and vibration–rotation Raman spectroscopy includes work not published elsewhere.

A. Weber, High-Resolution Raman Spectra of Gases, in A. Anderson (Ed.), *The Raman Effect*, vol. 2, Marcel Dekker, New York, 1973.
A very detailed account of theory, experimental techniques, and results. A valuable section considers critically the analysis of rotational constants to yield information on molecular structures.

H. G. M. Edwards and D. A. Long, Rotation and Vibration–Rotation Raman and Infrared Spectra of Gases, in D. A. Long, R. F. Barrow, and D. J. Millen (Eds.), *Chemical Society Specialist Periodical Reports*, No. 29, *Molecular Spectroscopy*, vol. 1, Chemical Society, London, 1973.
Includes a survey of rotation and vibration–rotation Raman spectra for the period from 1970 to 1973.

H. G. M. Edwards, Pure Rotation and Vibration–Rotation Raman and Infrared Spectra of Gases, in D. A. Long, R. F. Barrow, and D. J. Millen (Eds.), *Chemical Society Specialist Periodical Reports*, No. 29, *Molecular Spectroscopy*, vol. 3, Chemical Society, London, 1975.
Includes a survey of rotation and vibration–rotation Raman spectra for the period from 1973 to 1975.

7.3 *Organic chemistry*

B. Schrader, Chemical Applications of Raman Spectroscopy, *Angewandte chemie* (*Int. Ed.*), **12**, 884, 1973.
This review contains many examples of applications of Raman spectroscopy to organic chemistry.

B. Schrader and W. Meier (Eds.), *Raman and Infrared Atlas of Organic Compounds*, vols. 1 and 2, Verlag Chemie, Weinheim, vol. 1, 1974, vol. 2, 1975.

F. R. Dollish, W. G. Fateley, and F. F. Bentley, *Characteristic Raman Frequencies of Organic Compounds*, Wiley–Interscience, New York, 1974.

7.4 *Inorganic chemistry*

L. A. Woodward, Application of Raman Spectroscopy to Inorganic Chemistry, *Quarterly Reviews*, **10**, 185, 1956.
Although written 20 years ago, this is a classic review which still has much to say.

D. W. James and M. J. Nolan, Vibrational Spectra of Transition Metal Complexes and the Nature of the Metal–Ligand Bond, in F. A. Cotton (Ed.), *Progress in Inorganic Chemistry*, vol. 9, Wiley, New York, 1968.

I. R. Beattie, Single Crystal and High Temperature Gas Phase Raman Spectroscopy, in D. A. Long, L. A. K. Staveley, and A. J. Downs (Eds.), *Essays in Structural Chemistry*, Macmillan, London, 1971.

W. H. Fletcher, Vibrational Assignments in Small Molecules, in D. A. Long, L. A. K. Staveley, and A. J. Downs (Eds.), *Essays in Structural Chemistry*, Macmillan, London, 1971.

J. R. Hall, Infrared and Raman Spectra of Organometallic and Related Compounds, in D. A. Long, L. A. K. Staveley, and A. J. Downs (Eds.), *Essays in Structural Chemistry*, Macmillan, London, 1971.

M. J. Ware, Vibrational Studies of Metal–Metal Bonding, in D. A. Long, L. A. K. Staveley, and A. J. Downs (Eds.), *Essays in Structural Chemistry*, Macmillan, London, 1971.

G. A. Ozin, Single Crystal and Gas Phase Raman Spectroscopy in Inorganic Chemistry, in S. J. Lippard (Ed.), *Progress in Inorganic Chemistry*, **14**, 173, 1971.

D. M. Adams, Inorganic Vibrational Spectroscopy, *Annual Reports* (*London*), **68A**, 47, 1971.

S. D. Ross, *Inorganic Infrared and Raman Spectra*, McGraw-Hill, London, 1972.
A very useful compilation.

R. S. Tobias, Applications (of Raman spectroscopy) to Inorganic Chemistry, in A. Anderson, *The Raman Effect*, vol. 2, Marcel Dekker, New York, 1973.
A wide-ranging review with an extensive bibliography.

N. N. Greenwood (Ed.), *Chemical Society Specialist Periodical Reports* No. 2, *Spectroscopic Properties of Inorganic and Organometallic Compounds*, vols. 1–10, Chemical Society, London, 1967–76.
This valuable series provides detailed annual coverage of the literature relating to infrared and Raman spectroscopy of inorganic and organometallic compounds.

N. N. Greenwood, E. J. F. Ross, and B. P. Straughan, *Index of Vibrational Spectra of Inorganic and Organometallic Compounds, 1935–1960*, vol. 1, Butterworths, London, 1972.

N. N. Greenwood, E. J. F. Ross, and B. P. Straughan, *Index of Vibrational Spectra of Inorganic and Organometallic Compounds, 1961–1963*, vol. 2, Butterworths, London, 1975.

N. N. Greenwood, E. J. F. Ross, and B. P. Straughan, *Index of Vibrational Spectra of Inorganic and Organometallic Compounds, 1964–1966*, vol. 3, Butterworths, London, 1977.

R. J. Gillespie and J. Passmore, Homopolyatomic Cations of the Elements, in H. J. Emeléus and A. G. Sharpe (Eds.), *Advances in Inorganic Chemistry and Radiochemistry*, vol. 17, Academic Press, New York, 1975.
Includes vibrational spectroscopic studies of these species.

I. R. Beattie, Vibrational Infrared and Raman Spectroscopy in Inorganic Chemistry, *Chemical Society Reviews*, **4**, 167, 1975.
A very useful survey with broad coverage.

R. J. H. Clark, Resonance Raman Spectra of Inorganic Molecules and Ions, in R. J. H. Clark and R. E. Hester (Eds.), *Advances in Infrared and Raman Spectroscopy*, vol. 1, Heyden, London, 1976.

7.5 *Matrix isolation Raman spectroscopy*

H. E. Hallam, Infrared and Raman Spectra of Trapped Species, *Annual Reports* (*London*), **67**, 117, 1970.

G. A. Ozin, Matrix Isolation Raman Spectroscopy, in H. E. Hallam (Ed.), *Vibrational Spectroscopy of Trapped Species*, Wiley and Sons, London, 1973.

A. J. Downs and S. C. Peake, Matrix Isolation, in D. A. Long, R. F. Barrow, and D. J. Millen (Eds.), *Chemical Society Specialist Reports*, No. 29, *Molecular Spectroscopy*, vol. 1, Chemical Society, London, 1973.
A comprehensive account of matrix isolation spectroscopy up to July 1972. A detailed tabulation of results is included.

B. M. Chadwick, Matrix Isolation, in D. A. Long, R. F. Barrow, and D. J. Millen (Eds.), *Chemical Society Specialist Periodical Reports*, No. 29, *Molecular Spectroscopy*, vol. 3, Chemical Society, London, 1975.
A comprehensive account of matrix isolation spectroscopy from 1972 to 1974. A detailed tabulation of results is included.

G. A. Ozin and A. van der Voet, Cryogenic Inorganic Chemistry, in S. J. Lippard (Ed.), *Progress in Inorganic Chemistry*, **19**, 105, 1975.
A review of metal–gas reactions as studied by matrix isolation, infrared, and Raman spectroscopic techniques.

7.6 and 7.7 *Ionic equilibria and ionic interactions*

D. E. Irish, Raman Spectroscopy of Complex Ions in Solution, in H. A. Szymanski (Ed.), *Raman Spectroscopy, Theory and Practice*, vol. 1, Plenum Press, New York, 1967.

G. J. Janz and S. C. Wait, Ionic Melts, in H. A. Szymanski (Ed.), *Raman Spectroscopy, Theory and Practice*, vol. 1, Plenum Press, New York, 1967.

R. E. Hester, Raman and Infrared Spectra of Concentrated Electrolytic Solutions and Fused Salts, *Annual Reports* (*London*), **66A**, 79, 1969.

D. E. Irish, Vibrational Spectral Studies of Electrolytic Solutions and Fused Salts, in S. Petrucci (Ed.), *Ionic Interactions; From Dilute Solutions to Fused Salts*, vol. 2, Academic Press, New York, 1971.

7.8 *Resonance Raman spectroscopy*

J. Behringer, Observed Resonance Raman Spectra, in H. A. Szymanski (Ed.), *Raman Spectroscopy*, vol. 1, Plenum Press, New York, 1967.
Covers earlier work in this area.

J. Behringer, Theories of Resonance Raman Scattering, in D. A. Long, R. F. Barrow, and D. J. Millen (Eds.), *Chemical Society Specialist Periodical Reports*, No. 29, *Molecular Spectroscopy*, vol. 2, Chemical Society, London, 1974.
A thorough account of the theory.

J. Behringer, Experimental Resonance Raman Scattering, in D. A. Long, R. F. Barrow, and D. J. Millen (Eds.), *Chemical Society Specialist Periodical Reports*, No. 29, *Molecular Spectroscopy*, vol. 3, Chemical Society, London, 1975.
An exhaustive survey of experimental results.

W. Kiefer, Laser-excited Resonance Raman Spectra of Small Molecules and Ions, *Applied Spectroscopy*, **28**, 115, 1974.

T. G. Spiro, Raman Spectra of Biological Materials, in C. B. Moore (Ed.), *Chemical and Biological Applications of Lasers*, vol. 1, Academic Press, New York, 1974.
Includes resonance Raman studies.

V. Fawcett and D. A. Long, Biological Applications of Raman Spectroscopy, in D. A. Long, R. F. Barrow, and J. Sheridan (Eds.), *Chemical Society Specialist Periodical Reports*, No. 29, *Molecular Spectroscopy*, vol. 4, Chemical Society, London, 1976.
Includes resonance Raman studies.

R. J. H. Clark, Resonance Raman Spectra of Inorganic Molecules and Ions, in R. J. H. Clark and R. E. Hester (Eds.), *Advances in Infrared and Raman Spectroscopy*, vol. 1, Heyden, London, 1976.

T. G. Spiro and T. M. Loehr, Resonance Raman Spectra of Haem Proteins and other Biological Systems, in R. J. H. Clark and R. E. Hester (Eds.), *Advances in Infrared and Raman Spectroscopy*, vol. 1, Heyden, London, 1976.

7.9 *Macromolecules including those of biological interest*

M. J. Gall, P. J. Hendra, C. J. Peacock, and D. A. Watson, Recording Raman Spectra of Polymers, *Applied Spectroscopy*, **25**, 423, 1971.

J. L. Koenig, Raman Scattering of Synthetic Polymers, in E. G. Brame (Ed.), *Applied Spectroscopy Reviews*, vol. 4, Marcel Dekker, New York, 1971.

J. L. Koenig, Raman Scattering of Synthetic Polymers, *Journal of Polymer Science, Part D, Macromolecular Reviews*, **6**, 59, 1972.

W. L. Peticolas, Inelastic Light Scattering and the Raman Effect, *Annual Review of Physical Chemistry*, **23**, 93, 1972.

V. Fawcett and D. A. Long, Vibrational Spectroscopy of Macromolecules, in D. A. Long, R. F. Barrow, and D. J. Millen (Eds.), *Chemical Society Specialist Periodical Reports*, No. 29, *Molecular Spectroscopy*, vol. 1, Chemical Society, London, 1973.

F. S. Parker, Biochemical Applications of Infrared and Raman Spectroscopy, *Applied Spectroscopy*, **29**, 129, 1975.

V. Fawcett and D. A. Long, Biological Applications of Raman Spectroscopy, in D. A. Long, R. F. Barrow, and J. Sheridan (Eds.), *Chemical Society Specialist Periodical Reports*, No. 29, *Molecular Spectroscopy*, vol. 4, Chemical Society, London, 1976.

J. L. Koenig and B. G. Frushour, Raman Spectroscopy of Proteins, in R. J. H. Clark and R. E. Hester (Eds.), *Advances in Infrared and Raman Spectroscopy*, vol. 1, Heyden, London, 1976.

T. G. Spiro and T. M. Loehr, Resonance Raman Spectra of Haem Proteins and other Biological Systems, in R. J. H. Clark and R. E. Hester (Eds.), *Advances in Infrared and Raman Spectroscopy*, vol. 1, Heyden, London, 1976.

7.10 *Qualitative and quantitative analysis*

D. A. Long, Raman Spectroscopy, in L. A. K. Staveley (Ed.), *The Characterization of Chemical Purity, Organic Compounds*, Butterworths, London, 1971.

D. E. Irish and H. Chen, The Application of Raman Spectroscopy to Chemical Analysis, *Applied Spectroscopy*, **25**, 1, 1971.

M. Lapp and C. M. Penney (Eds.), *Laser Raman Gas Diagnostics*, Plenum Press, New York, 1974.

7.11 *Rapid-scan Raman spectroscopy*

M. Delhaye, Rapid Scanning Raman Spectroscopy, *Applied Optics*, **7**, 2195, 1968.

M. Bridoux, A. Chapput, M. Crunelle, and M. Delhaye, New Trends in Rapid Laser Raman Spectroscopy, in J. P. Mathieu (Ed.), *Advances in Raman Spectroscopy*, vol. 1, Heyden, London, 1973.

7.12 *Electronic transitions*

J. A. Koningstein and O. S. Mortensen, Electronic Raman Transitions, in A. Anderson (Ed.), *The Raman Effect*, vol. 2, Marcel Dekker, New York, 1973.

M. V. Klein, Electronic Raman Scattering, in M. Cardona (Ed.), *Light Scattering in Solids, Topics in Applied Physics*, vol. 8, Springer-Verlag, Berlin, 1975.

J. A. Koningstein, Electronic Raman Effect, in D. A. Long, R. F. Barrow, and J. Sheridan (Eds.), *Chemical Society Specialist Periodical Reports*, No. 29, *Molecular Spectroscopy*, vol. 4, Chemical Society, London, 1976.

7.13 *Bandwidths in liquids and molecular motions*

R. T. Bailey, Infrared and Raman Studies of Molecular Motion, in D. A. Long, R. F. Barrow, and D. J. Millen (Eds.), *Chemical Society Specialist Periodical Reports*, No. 29, *Molecular Spectroscopy*, vol. 2, Chemical Society, London, 1974.

7.14 *Bond parameters*

L. A. Woodward, General Introduction, in H. A. Szymanski (Ed.), *Raman Spectroscopy, Theory and Practice*, vol. 1, Plenum Press, New York, 1967.

R. E. Hester, Raman Intensities and the Nature of the Chemical Bond, in H. A. Szymanski (Ed.), *Raman Spectroscopy, Theory and Practice*, vol. 1, Plenum Press, New York, 1967.

G. W. Chantry, Polarizability Theory of the Raman Effect, in A. Anderson (Ed.), *The Raman Effect*, vol. 1, Marcel Dekker, New York, 1971.

R. E. Hester, Raman Intensities, in D. A. Long, R. F. Barrow, and D. J. Millen (Eds.), *Chemical Society Specialist Periodical Reports*, No. 29, *Molecular Spectroscopy*, vol. 2, Chemical Society, London, 1974.

7.15 and 7.16 *Raman and magnetic Raman optical activity*

L. D. Barron and A. D. Buckingham, Rayleigh and Raman Optical Activity, *Annual Reviews, Physical Chemistry*, **26**, 381, 1975.

L. D. Barron, Rayleigh and Raman Scattering of Polarized Light, in D. A. Long, R. F. Barrow, and J. Sheridan (Eds.), *Chemical Society Specialist Periodical Reports*, No. 29, *Molecular Spectroscopy*, vol. 4, Chemical Society, London, 1976.

7.17 *The solid state*

BOOKS AND CONFERENCE PROCEEDINGS

M. Born and K. Huang, *Dynamical Theory of Crystal Lattices*, Clarendon Press, Oxford, 1954.

S. Nudelman and S. S. Mitra, *Optical Properties of Solids*, Plenum Press, New York, 1969.

G. B. Wright (Ed.), *Light Scattering Spectra of Solids*, Springer-Verlag, New York, 1969.

H. Poulet and J. P. Mathieu, *Spectres de Vibration et Symétrie des Cristaux*, Gordon and Breach, Paris, 1970.

D. M. Adams and D. C. Newton, *Tables for Factor Group and Point Group Analysis*, Beckman-RIIC Ltd., Croydon, 1970.

J. A. Salthouse and M. J. Ware, *Point Group Character Tables and Related Data*, Cambridge University Press, Cambridge, 1972.
E. M. Balanski (Ed.), *Proc. 2nd International Conference on Light Scattering of Solids*, Flammarion Sciences, Paris, 1971.
P. M. A. Sherwood, *Vibrational Spectroscopy of Solids*, Cambridge University Press, Cambridge, 1972.
G. Turrell, *Infrared and Raman Spectra of Crystals*, Academic Press, New York, 1972.
M. Cardona (Ed.), *Light Scattering in Solids, Topics in Applied Physics*, vol. 8, Springer-Verlag, Berlin, 1975.

REVIEW ARTICLES

A. C. Menzies, Raman Effect in Solids, *Rep. Prog. Phys.*, **16**, 83, 1953.
R. Loudon, Theory of the First-order Raman Effect in Crystals, *Proc. Roy. Soc., A*, **275**, 218, 1963.
R. Loudon, The Raman Effect in Crystals, *Advances in Physics*, **13**, 423, 1964.
I. R. Beattie and T. R. Gilson, Single Crystal Laser Raman Spectroscopy, *Proc. Roy. Soc., A*, **307**, 407, 1968.
G. R. Wilkinson, A Study of Far Infrared and Raman Spectroscopy of Intermolecular Vibrations in Solids and Liquids, in P. Hepple (Ed.), *Molecular Spectroscopy*, Institute of Petroleum, London, 1968.
W. G. Fateley, N. T. McDevitt, and F. F. Bentley, Infrared and Raman Selection Rules for Lattice Vibrations: The Correlation Method, *Applied Spectroscopy*, **25**, 155, 1970.
R. A. Cowley, The Theory of Raman Scattering from Crystals, in A. Anderson (Ed.), *The Raman Effect*, vol. 1, Marcel Dekker, New York, 1971.
R. S. Krishnan, Raman Spectra of the Alkali Halides, in D. A. Long, A. J. Downs, and L. A. K. Staveley (Eds.), *Essays in Structural Chemistry*, Macmillan, London, 1971.
I. R. Beattie, Single Crystal and High Temperature Gas-phase Raman Spectroscopy, in D. A. Long, A. J. Downs, and L. A. K. Staveley (Eds.), *Essays in Structural Chemistry*, Macmillan, London, 1971.
A. S. Barker and R. Loudon, Response Functions in the Theory of Raman Scattering by Vibrational and Polariton Modes in Dielectric Crystals, *Revs. Mod. Phys.*, **44**, 18, 1972.
D. M. Adams and S. J. Payne, Vibrational Spectroscopy of Solids at High Pressures, *Annual Reports (London)*, **69A**, 3, 1972.
G. R. Wilkinson, Raman Spectra of Ionic, Covalent and Metallic Crystals, in A. Anderson (Ed.), *The Raman Effect*, vol. 2, Marcel Dekker, New York, 1973.
R. Savoie, Raman Spectra of Molecular Crystals, in A. Anderson (Ed.), *The Raman Effect*, vol. 2, Marcel Dekker, New York, 1973.
J. K. Burdett and M. Poliakoff, Spin-flip Lasers, *Chemical Society Reviews*, **3**, 293, 1974.
G. R. Wilkinson, Raman Spectra of Solids, in D. A. Long, R. F. Barrow, and D. J. Millen (Eds.), *Chemical Society Specialist Periodical Reports*, No. 29, *Molecular Spectroscopy*, vol. 3, Chemical Society, London, 1975.
R. M. Martin and L. M. Falicov, Resonant Raman Scattering, in M. Cardona (Ed.), *Light Scattering in Solids, Topics in Applied Physics*, vol. 8, Springer-Verlag, Berlin, 1975.
J. F. Scott, Soft Mode Spectroscopy, *Revs. Mod. Phys.*, **12**, 1, 1975.
M. Cardona (Ed.), *Light Scattering in Solids, Topics in Applied Physics*, vol. 8, Springer-Verlag, Berlin, 1975.
A Pinczuk and E. Burstein, Fundamentals of Inelastic Light Scattering in Semiconductors and Insulators, in M. Cardona (Ed.), *Light Scattering in Solids, Topics in Applied Physics*, vol. 8, Springer-Verlag, Berlin, 1975.
M. H. Brodsky, Raman Scattering in Amorphous Semiconductors, in M. Cardona (Ed.), *Light Scattering in Solids, Topics in Applied Physics*, vol. 8, Springer-Verlag, Berlin, 1975.

Chapter 8

The following books, listed at the beginning of this appendix, include substantial accounts of aspects of non-linear scattering: Schuler, Koningstein, and Sushchinskii. The theory is also treated in

A. Yariv, *Quantum Electronics*, Wiley, New York, 1967.
D. Marcuse, *Engineering Quantum Electrodynamics*, Harcourt, Brace and World, New York, 1970.

Useful review articles include

N. Bloembergen, The Stimulated Raman Effect, *American Journal of Physics*, **35**, 989, 1967.
D. A. Long, Spectroscopy in a New Light, *Chemistry in Britain*, **7**, 108, 1971.
P. Lallemand, The Stimulated Raman Effect, in A. Anderson (Ed.), *The Raman Effect*, vol. 1, Marcel Dekker, New York, 1971.
D. A. Long, The Hyper Raman Effect, in D. A. Long, A. J. Downs, and L. A. K. Staveley (Eds.), *Essays in Structural Chemistry*, Macmillan, London, 1971.
D. A. Long, Some Non-linear Spectroscopic Phenomena, in J. P. Mathieu (Ed.), *Advances in Raman Spectroscopy*, vol. 1, Heyden, London, 1973.
R. R. Alfano and S. L. Shapiro, Ultra-fast Phenomena in Liquids and Solids, *Scientific American*, **228**, 6, 1973.
Y. R. Shen, Stimulated Raman Scattering, in M. Cardona (Ed.), *Light Scattering in Solids, Topics in Applied Physics*, vol. 8, Springer-Verlag, Berlin, 1975.
R. F. Begley, A, B. Harvey, R. L. Byer, and B. S. Hudson, Coherent Anti-Stokes Raman Scattering, *International Laboratory*, **1**, 11, 1975.
M. J. French and D. A. Long, Non-linear Raman Effects, Part I, in D. A. Long, R. F. Barrow, and J. Sheridan (Eds.), *Chemical Society Specialist Periodical Reports*, No. 29, *Molecular Spectroscopy*, vol. 4, Chemical Society, London, 1976.
J. P. E. Taran, Coherent anti-Stokes Raman Spectroscopy, in E. D. Schmid (Ed.), *Proceedings of the Fifth International Conference on Raman Spectroscopy*, Schulz Verlag, Freiburg, 1976.
J. W. Nibler, J. R. McDonald and A. B. Harvey, Coherent anti-Stokes Raman Spectroscopy of Gases, in E. D. Schmid (Ed.), *Proceedings of the Fifth International Conference on Raman Spectroscopy*, Schulz Verlag, Freiburg, 1976.

Appendix II

Symmetry classes for x, y, z; R_x, R_y, R_z and components of $\boldsymbol{\alpha}$ and $\boldsymbol{\beta}$

The following tables give, for all the common point groups, the symmetry classes to which belong the Cartesian displacements x, y, z, the components of the tensor $\boldsymbol{\alpha}$, the components of the tensor $\boldsymbol{\beta}$, and the rotations R_x, R_y, and R_z. In general, $\boldsymbol{\alpha}$ and $\boldsymbol{\beta}$ are non-symmetric, but since they can become symmetric in a number of situations, the symmetric and non-symmetric parts are listed separately: the symmetric tensor components are denoted by α_{xy} and β_1 etc., and the non-symmetric tensor components by $\bar{\alpha}_{xy}$, $\bar{\beta}_1$ etc. To save space in the tables, numerical subscripts are used to denote the components of $\boldsymbol{\beta}$ and various linear combinations of them. The code is defined at the end of the tables.

The groups C_s, C_i and C_n ($n = 2, 3, 4, 5, 6$)

C_s	A'	x, y	$\alpha_{xx}, \alpha_{yy}, \alpha_{zz}, \alpha_{xy}$	$\bar{\alpha}_{xy}$	$\beta_1, \beta_2, \beta_4, \beta_6, \beta_8, \beta_9$	$\bar{\beta}_1, \bar{\beta}_3, \bar{\beta}_5, \bar{\beta}_6$	R_z
	A''	z	α_{yz}, α_{xz}	$\bar{\alpha}_{yz}, \bar{\alpha}_{xz}$	$\beta_3, \beta_5, \beta_7, \beta_{10}$	$\bar{\beta}_2, \bar{\beta}_4, \bar{\beta}_7, \bar{\beta}_8$	R_x, R_y
C_i	A_g		$\alpha_{xx}, \alpha_{yy}, \alpha_{zz}$ $\alpha_{xy}, \alpha_{yz}, \alpha_{zx}$	$\bar{\alpha}_{xy}, \bar{\alpha}_{yz}, \bar{\alpha}_{zx}$			R_x, R_y, R_z
	A_u	x, y, z			$\beta_1, \beta_2, \beta_3, \beta_4, \beta_5$ $\beta_6, \beta_7, \beta_8, \beta_9, \beta_{10}$	$\bar{\beta}_1, \bar{\beta}_2, \bar{\beta}_3, \bar{\beta}_4$ $\bar{\beta}_5, \bar{\beta}_6, \bar{\beta}_7, \bar{\beta}_8$	
C_2	A	z	$\alpha_{xx}, \alpha_{yy}, \alpha_{zz}, \alpha_{xy}$	$\bar{\alpha}_{xy}$	$\beta_3, \beta_5, \beta_7, \beta_{10}$	$\bar{\beta}_2, \bar{\beta}_4, \bar{\beta}_7, \bar{\beta}_8$	R_z
	B	x, y	α_{yz}, α_{zx}	$\bar{\alpha}_{yz}, \bar{\alpha}_{zx}$	$\beta_1, \beta_2, \beta_4, \beta_6, \beta_8, \beta_9$	$\bar{\beta}_1, \bar{\beta}_3, \bar{\beta}_5, \bar{\beta}_6$	R_x, R_y
C_3	A	z	$\alpha_{xx}+\alpha_{yy}, \alpha_{zz}$	$\bar{\alpha}_{xy}$	$\beta_3, \beta_{13}, \beta_{19}, \beta_{20}$	$\bar{\beta}_{11}, \bar{\beta}_{16}$	R_z
	E	(x, y)	$(\alpha_{xx}-\alpha_{yy}, \alpha_{xy})$, $(\alpha_{yz}, \alpha_{zx})$	$(\bar{\alpha}_{yz}, \bar{\alpha}_{zx})$	$(\beta_8, \beta_9), (\beta_{10}, \beta_{16})$, (β_{17}, β_{18})	$(\bar{\beta}_1, \bar{\beta}_3), (\bar{\beta}_5, \bar{\beta}_6)$, $(\bar{\beta}_{14}, \bar{\beta}_{15})$	(R_x, R_y)
C_4	A	z	$\alpha_{xx}+\alpha_{yy}, \alpha_{zz}$	$\bar{\alpha}_{xy}$	β_3, β_{13}	$\bar{\beta}_{11}, \bar{\beta}_{16}$	R_z
	B		$\alpha_{xx}-\alpha_{yy}, \alpha_{xy}$		β_{10}, β_{16}	$\bar{\beta}_{14}, \bar{\beta}_{15}$	
	E	(x, y)	$(\alpha_{yz}, \alpha_{zx})$	$(\bar{\alpha}_{yz}, \bar{\alpha}_{zx})$	$(\beta_1, \beta_2), (\beta_4, \beta_6), (\beta_8, \beta_9)$	$(\bar{\beta}_1, \bar{\beta}_3), (\bar{\beta}_5, \bar{\beta}_6)$	(R_x, R_y)
C_5	A	z	$\alpha_{xx}+\alpha_{yy}, \alpha_{zz}$	$\bar{\alpha}_{xy}$	β_3, β_{13}	$\bar{\beta}_{11}, \bar{\beta}_{16}$	R_z
	E_1	(x, y)	$(\alpha_{yz}, \alpha_{zx})$	$(\bar{\alpha}_{yz}, \bar{\alpha}_{zx})$	$(\beta_8, \beta_9), (\beta_{17}, \beta_{18})$	$(\bar{\beta}_1, \bar{\beta}_3), (\bar{\beta}_5, \bar{\beta}_6)$	(R_x, R_y)
	E_2		$(\alpha_{xx}-\alpha_{yy}, \alpha_{xy})$		$(\beta_{10}, \beta_{16}), (\beta_{19}, \beta_{20})$	$(\bar{\beta}_{14}, \bar{\beta}_{15})$	
C_6	A	z	$\alpha_{xx}+\alpha_{yy}, \alpha_{zz}$	$\bar{\alpha}_{xy}$	β_3, β_{13}	$\bar{\beta}_{11}, \bar{\beta}_{16}$	R_z
	B				β_{19}, β_{20}		
	E_1	(x, y)	$(\alpha_{yz}, \alpha_{zx})$	$(\bar{\alpha}_{yz}, \bar{\alpha}_{zx})$	$(\beta_8, \beta_9), (\beta_{17}, \beta_{18})$	$(\bar{\beta}_1, \bar{\beta}_3), (\bar{\beta}_5, \bar{\beta}_6)$	(R_x, R_y)
	E_2		$(\alpha_{xx}-\alpha_{yy}, \alpha_{xy})$		(β_{10}, β_{16})	$(\bar{\beta}_{14}, \bar{\beta}_{15})$	

The groups D_n ($n = 2, 3, 4, 5, 6$)

D_2	A		$\alpha_{xx}, \alpha_{yy}, \alpha_{zz}$		β_{10}	$\bar\beta_7, \bar\beta_8$	
	B_1	z	α_{xy}	$\bar\alpha_{xy}$	$\beta_3, \beta_5, \beta_7$	$\bar\beta_2, \bar\beta_4$	R_z
	B_2	y	α_{zx}	$\bar\alpha_{zx}$	$\beta_2, \beta_4, \beta_9$	$\bar\beta_1, \bar\beta_6$	R_y
	B_3	x	α_{yz}	$\bar\alpha_{yz}$	$\beta_1, \beta_6, \beta_8$	$\bar\beta_3, \bar\beta_5$	R_x
D_3	A_1[a]		$\alpha_{xx}+\alpha_{yy}, \alpha_{zz}$		β_{19}	$\bar\beta_{16}$	
	A_2[a]	z		$\bar\alpha_{xy}$	$\beta_3, \beta_{13}, \beta_{20}$	$\bar\beta_{11}$	R_z
	E	(x, y)	$(\alpha_{xx}-\alpha_{yy}, \alpha_{xy})$, $(\alpha_{yz}, \alpha_{zx})$	$(\bar\alpha_{yz}, \bar\alpha_{zx})$	$(\beta_8, \beta_9), (\beta_{10}, \beta_{16})$, (β_{17}, β_{18})	$(\bar\beta_1, \bar\beta_3), (\bar\beta_5, \bar\beta_6)$, $(\bar\beta_{14}, \bar\beta_{15})$	(R_x, R_y)
D_4	A_1		$\alpha_{xx}+\alpha_{yy}, \alpha_{zz}$			$\bar\beta_{16}$	
	A_2	z		$\bar\alpha_{xy}$	β_3, β_{13}	$\bar\beta_{11}$	R_z
	B_1		$\alpha_{xx}-\alpha_{yy}$		β_{10}	$\bar\beta_{15}$	
	B_2		α_{xy}		β_{16}	$\bar\beta_{14}$	
	E	(x, y)	$(\alpha_{yz}, \alpha_{zx})$	$(\bar\alpha_{yz}, \bar\alpha_{zx})$	$(\beta_1, \beta_2), (\beta_4, \beta_6), (\beta_8, \beta_9)$	$(\bar\beta_1, \bar\beta_3), (\bar\beta_5, \bar\beta_6)$	(R_x, R_y)
D_5	A_1		$\alpha_{xx}+\alpha_{yy}, \alpha_{zz}$			$\bar\beta_{16}$	
	A_2	z		$\bar\alpha_{xy}$	β_3, β_{13}	$\bar\beta_{11}$	R_z
	E_1	(x, y)	$(\alpha_{yz}, \alpha_{zx})$	$(\bar\alpha_{yz}, \bar\alpha_{zx})$	$(\beta_8, \beta_9), (\beta_{17}, \beta_{18})$	$(\bar\beta_1, \bar\beta_3), (\bar\beta_5, \bar\beta_6)$	(R_x, R_y)
	E_2		$(\alpha_{xx}-\alpha_{yy}, \alpha_{xy})$		$(\beta_{10}, \beta_{16}), (\beta_{19}, \beta_{20})$	$(\bar\beta_{14}, \bar\beta_{15})$	
D_6	A_1		$\alpha_{xx}+\alpha_{yy}, \alpha_{zz}$			$\bar\beta_{16}$	
	A_2	z		$\bar\alpha_{xy}$	β_3, β_{13}	$\bar\beta_{11}$	R_z
	B_1[b]				β_{19}		
	B_2[b]				β_{20}		
	E_1	(x, y)	$(\alpha_{yz}, \alpha_{zx})$	$(\bar\alpha_{yz}, \bar\alpha_{zx})$	$(\beta_8, \beta_9), (\beta_{17}, \beta_{18})$	$(\bar\beta_1, \bar\beta_3), (\bar\beta_5, \bar\beta_6)$	(R_x, R_y)
	E_2		$(\alpha_{xx}-\alpha_{yy}, \alpha_{xy})$		(β_{10}, β_{16})	$(\bar\beta_{14}, \bar\beta_{15})$	

The groups C_{nv} ($n = 2, 3, 4, 5, 6$)

C_{2v}	A_1	z	$\alpha_{xx}, \alpha_{yy}, \alpha_{zz}$		$\beta_3, \beta_5, \beta_7$	$\bar\beta_2, \bar\beta_4$	
	A_2		α_{xy}	$\bar\alpha_{xy}$	β_{10}	$\bar\beta_7, \bar\beta_8$	R_z
	B_1	x	α_{zx}	$\bar\alpha_{zx}$	$\beta_1, \beta_6, \beta_8$	$\bar\beta_3, \bar\beta_5$	R_y
	B_2	y	α_{yz}	$\bar\alpha_{yz}$	$\beta_2, \beta_4, \beta_9$	$\bar\beta_1, \bar\beta_6$	R_x
C_{3v}	A_1[c]	z	$\alpha_{xx}+\alpha_{yy}, \alpha_{zz}$		$\beta_3, \beta_{13}, \beta_{19}$	$\bar\beta_{11}$	
	A_2[c]			$\bar\alpha_{xy}$	β_{20}	$\bar\beta_{16}$	R_z
	E	(x, y)	$(\alpha_{xx}-\alpha_{yy}, \alpha_{xy})$, $(\alpha_{yz}, \alpha_{zx})$	$(\bar\alpha_{yz}, \bar\alpha_{zx})$	$(\beta_8, \beta_9), (\beta_{10}, \beta_{16})$, (β_{17}, β_{18})	$(\bar\beta_1, \bar\beta_3), (\bar\beta_5, \bar\beta_6)$, $(\bar\beta_{14}, \bar\beta_{15})$	(R_x, R_y)
C_{4v}	A_1	z	$\alpha_{xx}+\alpha_{yy}, \alpha_{zz}$		β_3, β_{13}	$\bar\beta_{11}$	
	A_2			$\bar\alpha_{xy}$		$\bar\beta_{16}$	R_z
	B_1		$\alpha_{xx}-\alpha_{yy}$		β_{16}	$\bar\beta_{14}$	
	B_2		α_{xy}		β_{10}	$\bar\beta_{15}$	
	E	(x, y)	$(\alpha_{yz}, \alpha_{zx})$	$(\bar\alpha_{yz}, \bar\alpha_{zx})$	$(\beta_1, \beta_2), (\beta_4, \beta_6), (\beta_8, \beta_9)$	$(\bar\beta_1, \bar\beta_3), (\bar\beta_5, \bar\beta_6)$	(R_x, R_y)
C_{5v}	A_1	z	$\alpha_{xx}+\alpha_{yy}, \alpha_{zz}$		β_3, β_{13}	$\bar\beta_{11}$	
	A_2			$\bar\alpha_{xy}$		$\bar\beta_{16}$	R_z
	E_1	(x, y)	$(\alpha_{yz}, \alpha_{zx})$	$(\bar\alpha_{yz}, \bar\alpha_{zx})$	$(\beta_8, \beta_9), (\beta_{17}, \beta_{18})$	$(\bar\beta_1, \bar\beta_3, (\bar\beta_5, \bar\beta_6)$	(R_x, R_y)
	E_2		$(\alpha_{xx}-\alpha_{yy}, \alpha_{xy})$		$(\beta_{10}, \beta_{16}), (\beta_{19}, \beta_{20})$	$(\bar\beta_{14}, \bar\beta_{15})$	
C_{6v}	A_1	z	$\alpha_{xx}+\alpha_{yy}, \alpha_{zz}$		β_3, β_{13}	$\bar\beta_{11}$	
	A_2			$\bar\alpha_{xy}$		$\bar\beta_{16}$	R_z
	B_1[c]				β_{19}		
	B_2[c]				β_{20}		
	E_1	(x, y)	$(\alpha_{yz}, \alpha_{zx})$	$(\bar\alpha_{yz}, \bar\alpha_{zx})$	$(\beta_8, \beta_9), (\beta_{17}, \beta_{18})$	$(\bar\beta_1, \bar\beta_3), (\bar\beta_5, \bar\beta_6)$	(R_x, R_y)
	E_2		$(\alpha_{xx}-\alpha_{yy}, \alpha_{xy})$		(β_{10}, β_{16})	$(\bar\beta_{14}, \bar\beta_{15})$	

The groups C_{nh} ($n = 2, 3, 4, 5, 6$)

C_{2h}	A_g		$\alpha_{xx}, \alpha_{yy}, \alpha_{zz}, \alpha_{xy}$	$\bar{\alpha}_{xy}$			R_z
	B_g		α_{yz}, α_{zx}	$\bar{\alpha}_{yz}, \bar{\alpha}_{zx}$			R_x, R_y
	A_u	z			$\beta_3, \beta_5, \beta_7, \beta_{10}$	$\bar{\beta}_2, \bar{\beta}_4, \bar{\beta}_7, \bar{\beta}_8$	
	B_u	x, y			$\beta_1, \beta_2, \beta_4, \beta_6, \beta_8, \beta_9$	$\bar{\beta}_1, \bar{\beta}_3, \bar{\beta}_5, \bar{\beta}_6$	
C_{3h}	A'		$\alpha_{xx} + \alpha_{yy}, \alpha_{zz}$	$\bar{\alpha}_{xy}$	β_{19}, β_{20}		R_z
	E'	(x, y)	$(\alpha_{xx} - \alpha_{yy}, \alpha_{xy})$		$(\beta_8, \beta_9), (\beta_{17}, \beta_{18})$	$(\bar{\beta}_1, \bar{\beta}_3), (\bar{\beta}_5, \bar{\beta}_6)$	
	A''	z			β_3, β_{13}	$\bar{\beta}_{11}, \bar{\beta}_{16}$	
	E''		$(\alpha_{yz}, \alpha_{zx})$	$(\bar{\alpha}_{yz}, \bar{\alpha}_{zx})$	(β_{10}, β_{16})	$(\bar{\beta}_{14}, \bar{\beta}_{15})$	(R_x, R_y)
C_{4h}	A_g		$\alpha_{xx} + \alpha_{yy}, \alpha_{zz}$	$\bar{\alpha}_{xy}$			R_z
	B_g		$\alpha_{xx} - \alpha_{yy}, \alpha_{xy}$				
	E_g		$(\alpha_{yz}, \alpha_{zx})$	$(\bar{\alpha}_{yz}, \bar{\alpha}_{zx})$			(R_x, R_y)
	A_u	z			β_3, β_{13}	$\bar{\beta}_{11}, \bar{\beta}_{16}$	
	B_u				β_{10}, β_{16}	$\bar{\beta}_{14}, \bar{\beta}_{15}$	
	E_u	(x, y)			$(\beta_1, \beta_2), (\beta_4, \beta_6), (\beta_8, \beta_9)$	$(\bar{\beta}_1, \bar{\beta}_3), (\bar{\beta}_5, \bar{\beta}_6)$	
C_{5h}	A'		$\alpha_{xx} + \alpha_{yy}, \alpha_{zz}$	$\bar{\alpha}_{xy}$			R_z
	E_1'	(x, y)			$(\beta_8, \beta_9), (\beta_{17}, \beta_{18})$	$(\bar{\beta}_1, \bar{\beta}_3), (\bar{\beta}_5, \bar{\beta}_6)$	
	E_2'		$(\alpha_{xx} - \alpha_{yy}, \alpha_{xy})$		(β_{19}, β_{20})		
	A''	z			β_3, β_{13}	$\bar{\beta}_{11}, \bar{\beta}_{16}$	
	E_1''		$(\alpha_{yz}, \alpha_{zx})$	$(\bar{\alpha}_{yz}, \bar{\alpha}_{zx})$			(R_x, R_y)
	E_2''				(β_{10}, β_{16})	$(\bar{\beta}_{14}, \bar{\beta}_{15})$	
C_{6h}	A_g		$\alpha_{xx} + \alpha_{yy}, \alpha_{zz}$	$\bar{\alpha}_{xy}$			R_z
	B_g						
	E_{1g}		$(\alpha_{yz}, \alpha_{zx})$	$(\bar{\alpha}_{yz}, \bar{\alpha}_{zx})$			(R_x, R_y)
	E_{2g}		$(\alpha_{xx} - \alpha_{yy}, \alpha_{xy})$				
	A_u	z			β_3, β_{13}	$\bar{\beta}_{11}, \bar{\beta}_{16}$	
	B_u				β_{19}, β_{20}		
	E_{1u}	(x, y)			$(\beta_8, \beta_9), (\beta_{17}, \beta_{18})$	$(\bar{\beta}_1, \bar{\beta}_3), (\bar{\beta}_5, \bar{\beta}_6)$	
	E_{2u}				(β_{10}, β_{16})	(β_{14}, β_{15})	

The groups D_{nh} $(n = 2, 3, 4, 5, 6)$

D_{2h}	A_g		$\alpha_{xx}, \alpha_{yy}, \alpha_{zz}$				
	B_{1g}		α_{xy}	$\bar{\alpha}_{xy}$			R_z
	B_{2g}		α_{zx}	$\bar{\alpha}_{zx}$			R_y
	B_{3g}		α_{yz}	$\bar{\alpha}_{yz}$			R_x
	A_u				β_{10}	$\bar{\beta}_7, \bar{\beta}_8$	
	B_{1u}	z			$\beta_3, \beta_5, \beta_7$	$\bar{\beta}_2, \bar{\beta}_4$	
	B_{2u}	y			$\beta_2, \beta_4, \beta_9$	$\bar{\beta}_1, \bar{\beta}_6$	
	B_{3u}	x			$\beta_1, \beta_6, \beta_8$	$\bar{\beta}_3, \bar{\beta}_5$	
D_{3h}	A_1'[a]		$\alpha_{xx} + \alpha_{yy}, \alpha_{zz}$		β_{19}		
	A_2'[a]			$\bar{\alpha}_{xy}$	β_{20}		R_z
	E'	(x, y)	$(\alpha_{xx} - \alpha_{yy}, \alpha_{xy})$		$(\beta_8, \beta_9), (\beta_{17}, \beta_{18})$	$(\bar{\beta}_1, \bar{\beta}_3), (\bar{\beta}_5, \bar{\beta}_6)$	
	A_1''					$\bar{\beta}_{16}$	
	A_2''	z			β_3, β_{13}	$\bar{\beta}_{11}$	
	E''		$(\alpha_{yz}, \alpha_{zx})$	$(\bar{\alpha}_{yz}, \bar{\alpha}_{zx})$	(β_{10}, β_{16})	$(\bar{\beta}_{14}, \bar{\beta}_{15})$	(R_x, R_y)
D_{4h}	A_{1g}		$\alpha_{xx} + \alpha_{yy}, \alpha_{zz}$				
	A_{2g}			$\bar{\alpha}_{xy}$			R_z
	B_{1g}		$\alpha_{xx} - \alpha_{yy}$				
	B_{2g}		α_{xy}				
	E_g		$(\alpha_{yz}, \alpha_{zx})$	$(\bar{\alpha}_{yz}, \bar{\alpha}_{zx})$			(R_x, R_y)
	A_{1u}					$\bar{\beta}_{16}$	
	A_{2u}	z			β_3, β_{13}	$\bar{\beta}_{11}$	
	B_{1u}				$\beta_{10},$	$\bar{\beta}_{15}$	
	B_{2u}				$\beta_{16},$	$\bar{\beta}_{14}$	
	E_u	(x, y)			$(\beta_1, \beta_2), (\beta_4, \beta_6), (\beta_8, \beta_9)$	$(\bar{\beta}_1, \bar{\beta}_3), (\bar{\beta}_5, \bar{\beta}_6)$	
D_{5h}	A_1'		$\alpha_{xx} + \alpha_{yy}, \alpha_{zz}$				
	A_2'			$\bar{\alpha}_{xy}$			R_z
	E_1'	(x, y)			$(\beta_8, \beta_9), (\beta_{17}, \beta_{18})$	$(\bar{\beta}_1, \bar{\beta}_3), (\bar{\beta}_5, \bar{\beta}_6)$	
	E_2'		$(\alpha_{xx} - \alpha_{yy}, \alpha_{xy})$		(β_{19}, β_{20})		
	A_1''					$\bar{\beta}_{16}$	
	A_2''	z			β_3, β_{13}	$\bar{\beta}_{11}$	
	E_1''		$(\alpha_{yz}, \alpha_{zx})$	$(\bar{\alpha}_{yz}, \bar{\alpha}_{zx})$			(R_x, R_y)
	E_2''				(β_{10}, β_{16})	$(\bar{\beta}_{14}, \bar{\beta}_{15})$	
D_{6h}	A_{1g}		$\alpha_{xx} + \alpha_{yy}, \alpha_{zz}$				
	A_{2g}			$\bar{\alpha}_{xy}$			R_z
	B_{1g}						
	B_{2g}						
	E_{1g}		$(\alpha_{yz}, \alpha_{zx})$	$(\bar{\alpha}_{yz}, \bar{\alpha}_{xz})$			(R_x, R_y)
	E_{2g}		$(\alpha_{xx} - \alpha_{yy}, \alpha_{xy})$				
	A_{1u}					$\bar{\beta}_{16}$	
	A_{2u}	z			β_3, β_{13}	$\bar{\beta}_{11}$	
	B_{1u}[b]				β_{19}		
	B_{2u}[b]				β_{20}		
	E_{1u}	(x, y)			$(\beta_8, \beta_9), (\beta_{17}, \beta_{18})$	$(\bar{\beta}_1, \bar{\beta}_3), (\bar{\beta}_5, \bar{\beta}_6)$	
	E_{2u}				(β_{10}, β_{16})	$(\bar{\beta}_{14}, \bar{\beta}_{15})$	

The groups D_{nd} ($n = 2, 3, 4, 5, 6$)

D_{2d}	A_1		$\alpha_{xx}+\alpha_{yy}, \alpha_{zz}$		β_{10}	$\bar{\beta}_{15}$	
	A_2			$\bar{\alpha}_{xy}$	β_{16}	$\bar{\beta}_{14}$	R_z
	B_1		$\alpha_{xx}-\alpha_{yy}$			$\bar{\beta}_{16}$	
	B_2	z	α_{xy}		β_3, β_{13}	$\bar{\beta}_{11}$	
	E	(x, y)	$(\alpha_{yz}, \alpha_{zx})$	$(\bar{\alpha}_{yz}, \bar{\alpha}_{zx})$	$(\beta_1, \beta_2), (\beta_4, \beta_6), (\beta_8, \beta_9)$	$(\bar{\beta}_1, \bar{\beta}_3), (\bar{\beta}_5, \bar{\beta}_6)$	(R_x, R_y)
D_{3d}	A_{1g}		$\alpha_{xx}+\alpha_{yy}, \alpha_{zz}$				
	A_{2g}			$\bar{\alpha}_{xy}$			R_z
	E_g		$(\alpha_{xx}-\alpha_{yy}, \alpha_{xy})$, $(\alpha_{yz}, \alpha_{zx})$	$(\bar{\alpha}_{yz}, \bar{\alpha}_{zx})$			(R_x, R_y)
	$A_{1u}{}^a$				β_{19}	$\bar{\beta}_{16}$	
	$A_{2u}{}^a$	z			$\beta_3, \beta_{13}, \beta_{20}$	$\bar{\beta}_{11}$	
	E_u	(x, y)			$(\beta_8, \beta_9), (\beta_{10}, \beta_{16})$, (β_{17}, β_{18})	$(\bar{\beta}_1, \bar{\beta}_3), (\bar{\beta}_5, \bar{\beta}_6)$ $(\bar{\beta}_{14}, \bar{\beta}_{15})$	
D_{4d}	A_1		$\alpha_{xx}+\alpha_{yy}, \alpha_{zz}$				
	A_2			$\bar{\alpha}_{xy}$			R_z
	B_1					$\bar{\beta}_{16}$	
	B_2	z			β_3, β_{13}	$\bar{\beta}_{11}$	
	E_1	(x, y)			$(\beta_8, \beta_9), (\beta_{17}, \beta_{18})$	$(\bar{\beta}_1, \bar{\beta}_3), (\bar{\beta}_5, \bar{\beta}_6)$	
	E_2		$(\alpha_{xx}-\alpha_{yy}, \alpha_{xy})$		(β_{10}, β_{16})	$(\bar{\beta}_{14}, \bar{\beta}_{15})$	
	E_3		$(\alpha_{yz}, \alpha_{zx})$	$(\bar{\alpha}_{yz}, \bar{\alpha}_{zx})$	(β_{19}, β_{20})		(R_x, R_y)
D_{5d}	A_{1g}		$\alpha_{xx}+\alpha_{yy}, \alpha_{zz}$				
	A_{2g}			$\bar{\alpha}_{xy}$			R_z
	E_{1g}		$(\alpha_{yz}, \alpha_{zx})$	$(\bar{\alpha}_{yz}, \bar{\alpha}_{zx})$			(R_x, R_y)
	E_{2g}		$(\alpha_{xx}-\alpha_{yy}, \alpha_{xy})$				
	A_{1u}					$\bar{\beta}_{16}$	
	A_{2u}	z			β_3, β_{13}	$\bar{\beta}_{11}$	
	E_{1u}	(x, y)			$(\beta_8, \beta_9), (\beta_{17}, \beta_{18})$	$(\bar{\beta}_1, \bar{\beta}_3), (\bar{\beta}_5, \bar{\beta}_6)$	
	E_{2u}				$(\beta_{10}, \beta_{16}), (\beta_{19}, \beta_{20})$,	$(\bar{\beta}_{14}, \bar{\beta}_{15})$	
D_{6d}	A_1		$\alpha_{xx}+\alpha_{yy}, \alpha_{zz}$				
	A_2			$\bar{\alpha}_{xy}$			R_z
	B_1					$\bar{\beta}_{16}$	
	B_2	z			β_3, β_{13}	$\bar{\beta}_{11}$	
	E_1	(x, y)			$(\beta_8, \beta_9), (\beta_{17}, \beta_{18})$	$(\bar{\beta}_1, \bar{\beta}_3), (\bar{\beta}_5, \bar{\beta}_6)$	
	E_2		$(\alpha_{xx}-\alpha_{yy}, \alpha_{xy})$				
	E_3				(β_{19}, β_{20})		
	E_4				(β_{10}, β_{16})	$(\bar{\beta}_{14}, \bar{\beta}_{15})$	
	E_5		$(\alpha_{yz}, \alpha_{zx})$	$(\bar{\alpha}_{yz}, \bar{\alpha}_{zx})$			(R_x, R_y)

The groups S_n ($n = 4, 6, 8$)

S_4	A		$\alpha_{xx}+\alpha_{yy}, \alpha_{zz}$	$\bar{\alpha}_{xy}$	β_{10}, β_{16}	$\bar{\beta}_{14}, \bar{\beta}_{15}$	R_z
	B	z	$\alpha_{xx}-\alpha_{yy}, \alpha_{xy}$		β_3, β_{13}	$\bar{\beta}_{11}, \bar{\beta}_{16}$	
	E	(x, y)	$(\alpha_{yz}, \alpha_{zx})$	$(\bar{\alpha}_{yz}, \bar{\alpha}_{zx})$	$(\beta_1, \beta_2), (\beta_4, \beta_6), (\beta_8, \beta_9)$	$(\bar{\beta}_1, \bar{\beta}_3), (\bar{\beta}_5, \bar{\beta}_6)$	(R_x, R_y)
S_6	A_g		$\alpha_{xx}+\alpha_{yy}, \alpha_{zz}$	$\bar{\alpha}_{xy}$			R_z
	E_g		$(\alpha_{xx}-\alpha_{yy}, \alpha_{xy})$, $(\alpha_{yz}, \alpha_{zx})$	$(\bar{\alpha}_{yz}, \bar{\alpha}_{zx})$			(R_x, R_y)
	A_u	z			$\beta_3, \beta_{13}, \beta_{19}, \beta_{20}$	$\bar{\beta}_{11}, \bar{\beta}_{16}$	
	E_u	(x, y)			$(\beta_8, \beta_9), (\beta_{10}, \beta_{16})$, (β_{17}, β_{18})	$(\bar{\beta}_1, \bar{\beta}_3), (\bar{\beta}_5, \bar{\beta}_6)$ $(\bar{\beta}_{14}, \bar{\beta}_{15})$	
S_8	A		$\alpha_{xx}+\alpha_{yy}, \alpha_{zz}$	$\bar{\alpha}_{xy}$			R_z
	B	z			β_3, β_{13}	$\bar{\beta}_{11}, \bar{\beta}_{16}$	
	E_1	(x, y)			$(\beta_8, \beta_9), (\beta_{17}, \beta_{18})$	$(\bar{\beta}_1, \bar{\beta}_3), (\bar{\beta}_5, \bar{\beta}_6)$	
	E_2		$(\alpha_{xx}-\alpha_{yy}, \alpha_{xy})$		(β_{10}, β_{16})	$(\bar{\beta}_{14}, \bar{\beta}_{15})$	
	E_3		$(\alpha_{yz}, \alpha_{zx})$	$(\bar{\alpha}_{yz}, \bar{\alpha}_{zx})$	(β_{19}, β_{20})		(R_x, R_y)

The groups T, T_h, T_d, O, O_h

T	A		$\alpha_{xx}+\alpha_{yy}+\alpha_{zz}$		β_{10}		
	E		$(\alpha_{xx}+\alpha_{yy}-2\alpha_{zz},$ $\alpha_{xx}-\alpha_{yy})$			$(\bar{\beta}_{15}, \bar{\beta}_{16})$	
	F	(x, y, z)	$(\alpha_{xy}, \alpha_{yz}, \alpha_{zx})$	$(\bar{\alpha}_{xy}, \bar{\alpha}_{yz}, \bar{\alpha}_{zx})$	$(\beta_1, \beta_2, \beta_3), (\beta_5, \beta_6, \beta_9),$ $(\beta_4, \beta_7, \beta_8)$	$(\bar{\beta}_2, \bar{\beta}_3, \bar{\beta}_6),$ $(\bar{\beta}_1, \bar{\beta}_4, \bar{\beta}_5)$	(R_x, R_y, R_z)
T_h	A_g		$\alpha_{xx}+\alpha_{yy}+\alpha_{zz}$				
	E_g		$(\alpha_{xx}+\alpha_{yy}-2\alpha_{zz},$ $\alpha_{xx}-\alpha_{yy})$				
	F_g		$(\alpha_{xy}, \alpha_{yz}, \alpha_{zx})$	$(\bar{\alpha}_{xy}, \bar{\alpha}_{yz}, \bar{\alpha}_{zx})$			(R_x, R_y, R_z)
	A_u				β_{10}		
	E_u					$(\bar{\beta}_{15}, \bar{\beta}_{16})$	
	F_u	(x, y, z)			$(\beta_1, \beta_2, \beta_3), (\beta_5, \beta_6, \beta_9),$ $(\beta_4, \beta_7, \beta_8)$	$(\bar{\beta}_2, \bar{\beta}_3, \bar{\beta}_6),$ $(\bar{\beta}_1, \bar{\beta}_4, \bar{\beta}_5)$	
T_d	A_1		$\alpha_{xx}+\alpha_{yy}+\alpha_{zz}$		β_{10}		
	A_2						
	E		$(\alpha_{xx}+\alpha_{yy}-2\alpha_{zz},$ $\alpha_{xx}-\alpha_{yy})$			$(\bar{\beta}_{15}, \bar{\beta}_{16})$	
	F_1			$(\bar{\alpha}_{xy}, \bar{\alpha}_{yz}, \bar{\alpha}_{zx})$	$(\beta_{14}, \beta_{15}, \beta_{16})$	$(\bar{\beta}_{12}, \bar{\beta}_{13}, \bar{\beta}_{14})$	(R_x, R_y, R_z)
	F_2	(x, y, z)	$(\alpha_{xy}, \alpha_{yz}, \alpha_{zx})$		$(\beta_1, \beta_2, \beta_3), (\beta_{11}, \beta_{12}, \beta_{13})$	$(\bar{\beta}_9, \bar{\beta}_{10}, \bar{\beta}_{11})$	
O	A_1		$\alpha_{xx}+\alpha_{yy}+\alpha_{zz}$				
	A_2				β_{10}		
	E		$(\alpha_{xx}+\alpha_{yy}-2\alpha_{zz},$ $\alpha_{xx}-\alpha_{yy})$			$(\bar{\beta}_{15}, \bar{\beta}_{16})$	
	F_1	(x, y, z)		$(\bar{\alpha}_{xy}, \bar{\alpha}_{yz}, \bar{\alpha}_{zx})$	$(\beta_1, \beta_2, \beta_3), (\beta_{11}, \beta_{12}, \beta_{13})$	$(\bar{\beta}_9, \bar{\beta}_{10}, \bar{\beta}_{11})$	(R_x, R_y, R_z)
	F_2		$(\alpha_{xy}, \alpha_{yz}, \alpha_{zx})$		$(\beta_{14}, \beta_{15}, \beta_{16})$	$(\bar{\beta}_{12}, \bar{\beta}_{13}, \bar{\beta}_{14})$	
O_h	A_{1g}		$\alpha_{xx}+\alpha_{yy}+\alpha_{zz}$				
	A_{2g}						
	E_g		$\alpha_{xx}+\alpha_{yy}-2\alpha_{zz},$ $(\alpha_{xx}-\alpha_{yy})$				
	F_{1g}			$(\bar{\alpha}_{xy}, \bar{\alpha}_{yz}, \bar{\alpha}_{zx})$			(R_x, R_y, R_z)
	F_{2g}		$(\alpha_{xy}, \alpha_{yz}, \alpha_{zx})$				
	A_{1u}						
	A_{2u}				β_{10}		
	E_u					$(\bar{\beta}_{15}, \bar{\beta}_{16})$	
	F_{1u}	(x, y, z)			$(\beta_1, \beta_2, \beta_3), (\beta_{11}, \beta_{12}, \beta_{13})$	$(\bar{\beta}_9, \bar{\beta}_{10}, \bar{\beta}_{11})$	
	F_{2u}				$(\beta_{14}, \beta_{15}, \beta_{16})$	$(\bar{\beta}_{12}, \bar{\beta}_{13}, \bar{\beta}_{14})$	

The groups $C_{\infty v}$ and $D_{\infty h}$

$C_{\infty v}$	Σ^+	z	$\alpha_{xx}+\alpha_{yy}, \alpha_{zz}$		β_3, β_{13}	$\bar{\beta}_{11}$	
	Σ^-			$\bar{\alpha}_{xy}$		$\bar{\beta}_{16}$	R_z
	Π	(x, y)	$(\alpha_{yz}, \alpha_{zx})$	$(\bar{\alpha}_{yz}, \bar{\alpha}_{zx})$	$(\beta_8, \beta_9), (\beta_{17}, \beta_{18})$	$(\bar{\beta}_1, \bar{\beta}_3), (\bar{\beta}_5, \bar{\beta}_6)$	(R_x, R_y)
	Δ		$(\alpha_{xx}-\alpha_{yy}, \alpha_{xy})$		(β_{10}, β_{16})	$(\bar{\beta}_{14}, \bar{\beta}_{15})$	
	Φ				(β_{19}, β_{20})		
$D_{\infty h}$	Σ_g^+		$\alpha_{xx}+\alpha_{yy}, \alpha_{zz}$				
	Σ_g^-			$\bar{\alpha}_{xy}$			R_z
	Π_g		$(\alpha_{yz}, \alpha_{zx})$	$(\bar{\alpha}_{yz}, \bar{\alpha}_{xz})$			(R_x, R_y)
	Δ_g		$(\alpha_{xx}-\alpha_{yy}, \alpha_{xy})$				
	Φ_g						
	Σ_u^+	z			β_3, β_{13}	$\bar{\beta}_{11}$	
	Σ_u^-					$\bar{\beta}_{16}$	
	Π_u	(x, y)			$(\beta_8, \beta_9), (\beta_{17}, \beta_{18})$	$(\bar{\beta}_1, \bar{\beta}_3), (\bar{\beta}_5, \bar{\beta}_6)$	
	Δ_u				(β_{10}, β_{16})	$(\bar{\beta}_{14}, \bar{\beta}_{15})$	
	Φ_u				(β_{19}, β_{20})		

$a = x$-axis along C_2 $\quad$ $b = x$-axis along C_2' $\quad$ $c = x$-axis in σ_v

Code for numerical subscripts on β and $\bar{\beta}$

$\beta_1 = \beta_{xxx}$	$\bar{\beta}_1 = \bar{\beta}_{xxy}$
$\beta_2 = \beta_{yyy}$	$\bar{\beta}_2 = \bar{\beta}_{xxz}$
$\beta_3 = \beta_{zzz}$	$\bar{\beta}_3 = \bar{\beta}_{yyx}$
$\beta_4 = \beta_{xxy}$	$\bar{\beta}_4 = \bar{\beta}_{yyz}$
$\beta_5 = \beta_{zxx}$	$\bar{\beta}_5 = \bar{\beta}_{zzx}$
$\beta_6 = \beta_{xyy}$	$\bar{\beta}_6 = \bar{\beta}_{zzy}$
$\beta_7 = \beta_{yyz}$	$\bar{\beta}_7 = \bar{\beta}_{xyz}$
$\beta_8 = \beta_{zzx}$	$\bar{\beta}_8 = \bar{\beta}_{yzx}$
$\beta_9 = \beta_{yzz}$	$\bar{\beta}_9 = \bar{\beta}_{yyx} + \bar{\beta}_{zzx}$
$\beta_{10} = \beta_{xyz}$	$\bar{\beta}_{10} = \bar{\beta}_{xxy} + \bar{\beta}_{zzy}$
$\beta_{11} = \beta_{xyy} + \beta_{zzx}$	$\bar{\beta}_{11} = \bar{\beta}_{xxz} + \bar{\beta}_{yyz}$
$\beta_{12} = \beta_{xxy} + \beta_{yzz}$	$\bar{\beta}_{12} = \bar{\beta}_{yyx} - \bar{\beta}_{zzx}$
$\beta_{13} = \beta_{zxx} + \beta_{yyz}$	$\bar{\beta}_{13} = \bar{\beta}_{xxy} - \bar{\beta}_{zzy}$
$\beta_{14} = \beta_{xyy} - \beta_{zzx}$	$\bar{\beta}_{14} = \bar{\beta}_{xxz} - \bar{\beta}_{yyz}$
$\beta_{15} = \beta_{xxy} - \beta_{yzz}$	$\bar{\beta}_{15} = \bar{\beta}_{xyz} + \bar{\beta}_{yzx}$
$\beta_{16} = \beta_{zxx} - \beta_{yyz}$	$\bar{\beta}_{16} = \bar{\beta}_{xyz} - \bar{\beta}_{yzx}$
$\beta_{17} = \beta_{xxx} + \beta_{xyy}$	
$\beta_{18} = \beta_{yyy} + \beta_{xxy}$	
$\beta_{19} = \beta_{xxx} - 3\beta_{xyy}$	
$\beta_{20} = \beta_{yyy} - 3\beta_{xxy}$	

Note: In the tables given by Cyvin, Decius and Rauch[1] there is a misprint in the components of $\boldsymbol{\beta}$ for class B_{1u} in group D_{2h} which has been carried over into the tables given by Long[2] and by Salthouse and Ware[3]; the group D_{2h} is not listed by Christie and Lockwood.[4]

1. S. J. Cyvin, J. E. Rauch, and J. C. Decius, *J. Chem. Phys.*, **43**, 4083, 1965.

2. D. A. Long, The Hyper Raman Effect, in D. A. Long, A. J. Downs and L. A. K. Staveley (Eds.), *Essays in Structural Chemistry*, Macmillan, London, 1971.

3. J. A. Salthouse and M. J. Ware, *Point Group Character Tables and Related Data*, Cambridge University Press, Cambridge, 1972.

4. J. H. Christie and D. J. Lockwood, *J. Chem. Phys.*, **54**, 1141, 1971.

Appendix III

Worked example of a Raman spectroscopic study of an oriented single crystal

To illustrate in more detail how the symmetry of vibrations may be determined unambiguously from the Raman spectroscopic studies of oriented single crystals, we consider the case of the $Co(CN)_6{}^{3-}$ ion in $Cs_2LiCo(CN)_6$. A crystal of this compound has a face-centred cubic structure, space group $Fm3m$ (i.e., O_h^5) with $z = 4$; and there is one formula unit per primitive unit cell. The Co and Li atoms occupy O_h sites and the Cs atoms occupy T_d sites (see Fig. A.III.1).

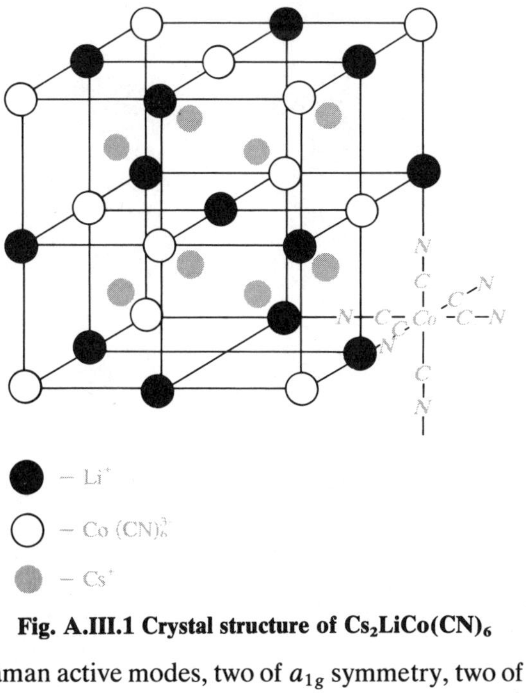

Fig. A.III.1 Crystal structure of $Cs_2LiCo(CN)_6$

There are six Raman active modes, two of a_{1g} symmetry, two of e_g symmetry, and two of f_{2g} symmetry; and one Raman active lattice mode of f_{2g} symmetry. The number and the symmetry of the internal modes is the same as for the 'free' complex ion in solution; the Raman active lattice mode is associated with movement of the Cs^+ ions. In solution, measurements of $\rho_\perp(\pi/2)$ enable the a_{1g} modes ($\rho_\perp(\pi/2)\approx 0$) to be distinguished from the e_g and f_{2g} modes (both of which have $\rho_\perp(\pi/2)=\frac{3}{4}$). Oriented single crystal studies can, however, distinguish between the e_g and f_{2g} modes as we shall now demonstrate.

Following our usual practice, we define the illumination–observation geometry with respect to a space-fixed Cartesian axis system, x, y, z, where z is the direction of propagation of the incident radiation and x the direction of observation. We note that the cubic crystallographic axes x', y', z' are coincident with the Co–C–N directions, and we consider two orientations of the crystal axes with respect to the space-fixed axes x, y, z.

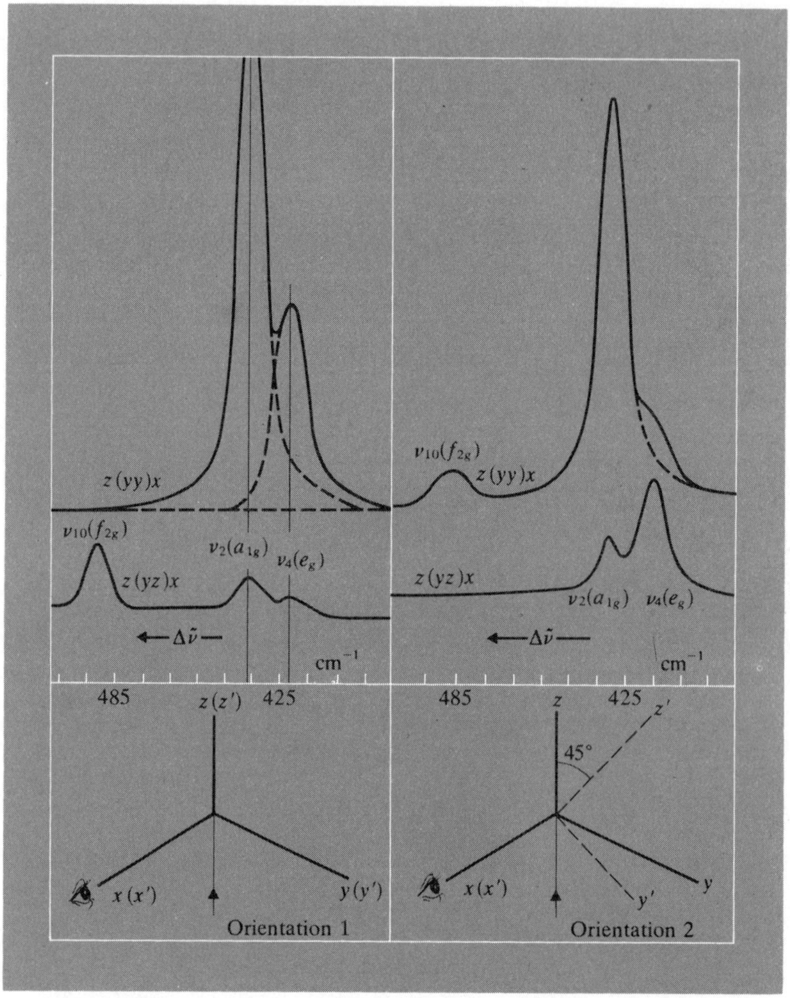

Fig. A.III.2 ***z(yy)x*** **and** ***z(yz)x*** **Stokes Raman spectra ($\Delta\tilde{\nu}$=385–505 cm^{-1}) for $Cs_2LiCo(CN)_6$ crystal in two orientations; 6328 Å (632·8 nm) excitation**

Orientation 1 The crystal axes x', y', z' are coincident with the space-fixed axes x, y, z (see Fig. A.III.2). The characteristics of the observed Raman scattering are determined by the components of the scattering tensor referred to the space-fixed axes, and in this case correspond to the tensor components referred to the crystal axes, which are also the C_4 axes of the complex ion itself. Thus the required tensor components are obtained directly from the activity tables. After normalization the matrices $[\alpha]$ of the tensor components have the following forms:

$$a_{1g}: \quad \begin{bmatrix} a & 0 & 0 \\ 0 & a & 0 \\ 0 & 0 & a \end{bmatrix}$$

$$e_g: \quad \begin{bmatrix} -b & 0 & 0 \\ 0 & -b & 0 \\ 0 & 0 & 2b \end{bmatrix} \quad \text{and} \quad \begin{bmatrix} \sqrt{3}b & 0 & 0 \\ 0 & -\sqrt{3}b & 0 \\ 0 & 0 & 0 \end{bmatrix}$$

$$f_{2g}: \quad \begin{bmatrix} 0 & c & 0 \\ c & 0 & 0 \\ 0 & 0 & 0 \end{bmatrix}, \quad \begin{bmatrix} 0 & 0 & c \\ 0 & 0 & 0 \\ c & 0 & 0 \end{bmatrix}, \text{ and } \begin{bmatrix} 0 & 0 & 0 \\ 0 & 0 & c \\ 0 & c & 0 \end{bmatrix}$$

We note immediately that for modes of f_{2g} symmetry there are no diagonal entries in the scattering tensors, whereas for modes of a_{1_g} and e_g symmetry the tensors contain diagonal entries. Thus f_{2_g} modes can be distinguished from a_{1_g} and e_g modes by comparison of scattering involving diagonal and off-diagonal tensor components. For example, comparison of the $z(yy)x$ and $z(yz)x$ Raman spectra (using Porto notation see p. 64) will yield the desired information. In the $z(yy)x$ spectra, the a_{1g} modes have intensity proportional to a^2, the e_g modes have intensity proportional to $4b^2$ and the f_{2g} modes have zero intensity; in the $z(yz)x$ spectra, the a_{1g} and e_g modes have zero intensity and the f_{2g} modes have intensity proportional to c^2. However these measurements do not distinguish a_{1g} and e_g modes.

Orientation 2 The crystal axis x' is coincident with the space-fixed axis x, but the crystal axes y' and z' make angles of 45° with the space-fixed axes y and z respectively (see Fig. A.III.2). The two axis systems are related as follows:

$$\begin{bmatrix} x \\ y \\ z \end{bmatrix} = \begin{bmatrix} 1 & 0 & 0 \\ 0 & \frac{1}{\sqrt{2}} & \frac{1}{\sqrt{2}} \\ 0 & -\frac{1}{\sqrt{2}} & \frac{1}{\sqrt{2}} \end{bmatrix} \begin{bmatrix} x' \\ y' \\ z' \end{bmatrix}$$

If we denote the transformation matrix by T, then it can be shown that the matrix $[\alpha]$ with components α_{xy}, referred to the space-fixed axis system x, y, z, is related to the matrix $[\alpha']$ with components $\alpha_{x'y'}$, referred to the crystal (or ion) axes by the transformation

$$[\alpha]=T[\alpha']T^{-1}$$

Applying this transformation and noting that, since T is an orthogonal matrix, its inverse is given by its transpose, we obtain the following matrices of the tensor components referred to x, y, z, for orientation 2.

$$a_{1g}:\quad \begin{bmatrix} a & 0 & 0 \\ 0 & a & 0 \\ 0 & 0 & a \end{bmatrix}$$

$$e_g:\quad \begin{bmatrix} -b & 0 & 0 \\ 0 & \frac{b}{2} & \frac{3b}{2} \\ 0 & \frac{3b}{2} & \frac{b}{2} \end{bmatrix} \quad \text{and} \quad \begin{bmatrix} \sqrt{3}b & 0 & 0 \\ 0 & -\frac{\sqrt{3}b}{2} & \frac{\sqrt{3}b}{2} \\ 0 & \frac{\sqrt{3}b}{2} & -\frac{\sqrt{3}b}{2} \end{bmatrix}$$

$$f_{2g}:\quad \begin{bmatrix} 0 & \frac{c}{\sqrt{2}} & -\frac{c}{\sqrt{2}} \\ \frac{c}{\sqrt{2}} & 0 & 0 \\ -\frac{c}{\sqrt{2}} & 0 & 0 \end{bmatrix}, \quad \begin{bmatrix} 0 & \frac{c}{\sqrt{2}} & \frac{c}{\sqrt{2}} \\ \frac{c}{\sqrt{2}} & 0 & 0 \\ \frac{c}{\sqrt{2}} & 0 & 0 \end{bmatrix}, \text{ and } \begin{bmatrix} 0 & 0 & 0 \\ 0 & c & 0 \\ 0 & 0 & -c \end{bmatrix}$$

We see from these matrices that in this crystal orientation the a_{1g} and e_g modes may be distinguished by comparing the $z(yy)x$ and $z(yz)x$ Raman spectra. The e_g modes appear in the $z(yy)x$ Raman spectrum but with reduced intensity proportional to b^2 and in the $z(yz)x$ Raman spectrum with intensity proportional to $3b^2$; the f_{2g} modes now have zero intensity in the $z(yz)x$ Raman spectrum but appear in the $z(yy)x$ Raman spectrum with intensity proportional to c^2; the behaviour of the a_{1g} modes is unchanged and they again appear only in the $z(yy)x$ Raman spectrum with intensity proportional to a^2.

The behaviour of the a_{1g}, e_g, and f_{2g} in the two orientations is summarized in the Table on p. 265.

The Stokes Raman spectra (wavenumber shift region 385–505 cm^{-1}) in Fig. A.III.2 clearly exhibit this pattern of intensity behaviour and enable $\nu_4(e_g)$, $\nu_2(a_{1g})$ and $\nu_{10}(f_{2g})$ to be identified at 417, 432, and 485 cm^{-1}, respectively. Similarly, measurements in other wavenumber shift regions enable the remaining Raman active modes to be identified as follows: $\nu_1(a_{1g})$, 2161 cm^{-1}; $\nu_3(e_g)$, 2150 cm^{-1}; $\nu_{11}(f_{2g})$, 189 cm^{-1}; $\nu_{\text{lattice}}(f_{2g})$, 40 cm^{-1}.

Mode	Raman spectrum	*Orientation 1* Intensity proportional to	*Orientation 2* Intensity proportional to
a_{1g}	$z(yy)x$	a^2	a^2
	$z(yz)x$	0	0
e_g	$z(yy)x$	$4b^2$	b^2
	$z(yz)x$	0	$3b^2$
f_{2g}	$z(yy)x$	0	c^2
	$z(yz)x$	c^2	0

The arguments above have been presented using the axis definitions generally adopted in this book, *viz.* x, y, z space-fixed and x', y', z' molecule-fixed. In crystal studies, x', y', z' are often used to define laboratory or space-fixed axes and x, y, z crystal-fixed axes. In this convention, orientation 2 could be regarded as a rotation of the space-fixed axes relative to the crystal-fixed axes, and the scattering is then determined by components of a tensor $\boldsymbol{\alpha}'$ obtained by transformation of the tensor $\boldsymbol{\alpha}$ which relates to the crystal axes. Of course, the conclusions regarding observed intensities are the same whichever convention is used.

This example is based on unpublished work of Dr. B. M. Chadwick and Dr. J. R. Armstrong.

Index

This index relates to the main text and does not cover the Appendices and Central reference section. The numbers of pages on which spectra are reproduced are shown in italics. Readers will find the very detailed list of contents at the front of the book valuable as a supplement to this index.